T0201204

3G, 4G AND BEYOND–BRINGING NETWORKS, DEVICES AND THE WEB TOGETHER

3G, 4G AND BEYOND–BRINGING NETWORKS, DEVICES AND THE WEB TOGETHER

Second Edition

Martin Sauter

WirelessMoves, Germany

A John Wiley & Sons, Ltd., Publication

Library of Congress Cataloging-in-Publication Data

Sauter, Martin.
  3g, 4g and beyond : bringing networks, devices, and the web together / Martin Sauter. — 2nd ed.
    p. cm.
  Includes bibliographical references and index.
  ISBN 978-1-118-34148-3 (cloth)
1. Wireless Internet. 2. Wireless communication systems. 3. Mobile communication systems. 4. Smartphones. 5. Mobile computing. 6. Long-Term Evolution (Telecommunications) I. Title. II. Title: Three g, four g and beyond.
  TK5103.4885.S38 2013
  384.5–dc23

                                                                                    2012032302

Hardback ISBN: 9781118341483

A catalogue record for this book is available from the British Library.

Set in 10/12pt Times by Laserwords Private Limited, Chennai, India
Printed and bound in Singapore by Markono Print Media Pte Ltd

Contents

Preface   xi

1   **Evolution from 2G over 3G to 4G**   **1**
1.1   First Half of the 1990s — Voice-Centric Communication   1
1.2   Between 1995 and 2000: The Rise of Mobility and the Internet   1
1.3   Between 2000 and 2005: Dot Com Burst, Web 2.0, Mobile Internet   2
1.4   Between 2005 and 2010: Global Coverage, Fixed Line VoIP,
        and Mobile Broadband   4
1.5   2010 and Beyond   5
1.6   All over IP in Mobile — The Biggest Challenge   6
1.7   Summary   6

2   **Beyond 3G Network Architectures**   **9**
2.1   Overview   9
2.2   UMTS, HSPA, and HSPA+   10
        2.2.1   *Introduction*   10
        2.2.2   *Network Architecture*   10
        2.2.3   *Air Interface and Radio Network*   19
        2.2.4   *HSPA (HSDPA and HSUPA)*   28
        2.2.5   *HSPA+ and other Improvements: Competition for LTE*   34
        2.2.6   *Competition for LTE in 5 MHz*   43
2.3   LTE   43
        2.3.1   *Introduction*   43
        2.3.2   *Network Architecture*   44
        2.3.3   *Air Interface and Radio Network*   49
        2.3.4   *Basic Procedures*   64
        2.3.5   *Summary and Comparison with HSPA*   67
        2.3.6   *LTE-Advanced*   68
2.4   802.11 Wi-Fi   74
        2.4.1   *Introduction*   74
        2.4.2   *Network Architecture*   76
        2.4.3   *The Air Interface — From 802.11b to 802.11n*   78
        2.4.4   *Air Interface and Resource Management*   83

2.4.5	Basic Procedures	86
2.4.6	Wi-Fi Security	87
2.4.7	Quality of Service: 802.11e	89
2.4.8	Gigabit Speeds with 802.11ac and 802.11ad	90
2.4.9	Summary	91

3 **Network Capacity and Usage Scenarios** — **95**
3.1	Usage in Developed Markets and Emerging Economies	95
3.2	How to Control Mobile Usage	96
3.2.1	Per Minute Charging	97
3.2.2	Volume Charging	97
3.2.3	Split Charging	97
3.2.4	Small Screen Flat Rates	97
3.2.5	Strategies to Inform Users when their Subscribed Data Volume is Used Up	98
3.2.6	Mobile Internet Access and Prepaid	98
3.3	Measuring Mobile Usage from a Financial Point of View	99
3.4	Cell Capacity in Downlink	100
3.5	Current and Future Frequency Bands for Cellular Wireless	105
3.6	Cell Capacity in Uplink	106
3.7	Per-User Throughput in Downlink	109
3.8	Per-User Throughput in Uplink	114
3.9	Traffic Estimation Per User	116
3.10	Overall Wireless Network Capacity	117
3.11	Network Capacity for Train Routes, Highways, and Remote Areas	124
3.12	When will GSM be Switched Off?	125
3.13	Cellular Network VoIP Capacity	127
3.14	Wi-Fi VoIP Capacity	130
3.15	Wi-Fi and Interference	132
3.16	Wi-Fi Capacity in Combination with DSL, Cable, and Fiber	134
3.17	Backhaul for Wireless Networks	138
3.18	A Hybrid Cellular/Wi-Fi Network Today and in the Future	143

4 **Voice over Wireless** — **149**
4.1	Circuit-Switched Mobile Voice Telephony	150
4.1.1	Circuit Switching	150
4.1.2	A Voice-Optimized Radio Network	151
4.1.3	The Pros of Circuit Switching	151
4.1.4	The Bearer Independent Core Network Architecture	151
4.2	Packet-Switched Voice Telephony	153
4.2.1	Network and Applications are Separate in Packet-Switched Networks	153
4.2.2	Wireless Network Architecture for Transporting IP Packets	154
4.2.3	Benefits of Migrating Voice Telephony to IP	155
4.2.4	Voice Telephony Evolution and Service Integration	155
4.2.5	Voice Telephony over IP: The End of the Operator Monopoly	156

4.3   SIP Telephony over Fixed and Wireless Networks   157
       4.3.1   SIP Registration   157
       4.3.2   Establishing a SIP Call between Two SIP Subscribers   160
       4.3.3   Session Description   162
       4.3.4   The Real-Time Transfer Protocol   164
       4.3.5   Establishing a SIP Call between a SIP and a PSTN Subscriber   165
       4.3.6   Proprietary Components of a SIP System   167
       4.3.7   Network Address Translation and SIP   168
4.4   Voice and Related Applications over IMS   169
       4.4.1   IMS Basic Architecture   173
       4.4.2   The P-CSCF   173
       4.4.3   The S-CSCF and Application Servers   175
       4.4.4   The I-CSCF and the HSS   177
       4.4.5   Media Resource Functions   180
       4.4.6   User Identities, Subscription Profiles, and Filter Criteria   181
       4.4.7   IMS Registration Process   183
       4.4.8   IMS Session Establishment   187
       4.4.9   Voice Telephony Interworking with Circuit-Switched Networks   192
       4.4.10 Push-to-Talk, Presence, and Instant Messaging   197
       4.4.11 Voice Call Continuity, Dual Radio, and Single Radio
               Approaches   200
       4.4.12 IMS with Wireless LAN Hotspots and Private Wi-Fi Networks   203
       4.4.13 IMS and TISPAN   207
       4.4.14 IMS on the Mobile Device   211
       4.4.15 Rich Communication Service (RCS-e)   213
       4.4.16 Voice over LTE (VoLTE)   215
       4.4.17 Challenges for IMS Rollouts   217
       4.4.18 Opportunities for IMS Rollouts   221
4.5   Voice over DSL and Cable with Femtocells   223
       4.5.1   Femtocells from the Network Operator's Point of View   225
       4.5.2   Femtocells from the User's Point of View   226
       4.5.3   Conclusion   227
4.6   Unlicensed Mobile Access and Generic Access Network   228
       4.6.1   Technical Background   228
       4.6.2   Advantages, Disadvantages, and Pricing Strategies   230
4.7   Network Operator Deployed Voice over IP Alternatives   231
       4.7.1   CS Fallback   232
       4.7.2   Voice over LTE via GAN   235
       4.7.3   Dual-Radio Devices   236
4.8   Over-the-Top (OTT) Voice over IP Alternatives   236
4.9   Which Voice Technology will Reign in the Future?   237

5   Evolution of Mobile Devices and Operating Systems   241
5.1   Introduction   241
       5.1.1   The ARM Architecture   243
       5.1.2   The x86 Architecture for Mobile Devices   244

	5.1.3	Changing Worlds: Android on x86, Windows on ARM	245
	5.1.4	From Hardware to Software	246
5.2	The System Architecture for Voice-Optimized Devices		246
5.3	The System Architecture for Multimedia Devices		248
5.4	Mobile Graphics Acceleration		253
	5.4.1	2D Graphics	253
	5.4.2	3D Graphics	254
5.5	Hardware Evolution		256
	5.5.1	Chipset	257
	5.5.2	Process Shrinking	259
	5.5.3	Displays	260
	5.5.4	Batteries	261
	5.5.5	Camera and Optics	261
	5.5.6	Global Positioning, Compass, 3D Orientation	263
	5.5.7	Wi-Fi	265
	5.5.8	Bluetooth	267
	5.5.9	NFC, RFID, and Mobile Payment	268
	5.5.10	Physical Keyboards	271
	5.5.11	TV Receivers	272
	5.5.12	TV-Out, Mobile Projectors, and DLNA	272
5.6	Multimode, Multifrequency Terminals		273
5.7	Wireless Notebook Connectivity		276
5.8	Impact of Hardware Evolution on Future Data Traffic		277
5.9	Power Consumption and User Interface as the Dividing Line in Mobile Device Evolution		279
5.10	Feature Phone Operating Systems		280
	5.10.1	Java Platform Micro Edition	281
	5.10.2	BREW	281
5.11	Smartphone Operating Systems		282
	5.11.1	Apple iOS	282
	5.11.2	Google Android	283
	5.11.3	Android, Open Source, and its Positive Influence on Innovation	285
	5.11.4	Other Smartphone Operating Systems	285
	5.11.5	Fracturization	287
5.12	Operating System Tasks		288
	5.12.1	Multitasking	288
	5.12.2	Memory Management	288
	5.12.3	File Systems and Storage	290
	5.12.4	Input and Output	290
	5.12.5	Network Support	291
	5.12.6	Security	291
6	**Mobile Web 2.0, Apps, and Owners**		**297**
6.1	Overview		297
6.2	(Mobile) Web 1.0—How Everything Started		298

6.3	Web 2.0 — Empowering the User	299
6.4	Web 2.0 from the User's Point of View	299
	6.4.1 Blogs	300
	6.4.2 Media Sharing	300
	6.4.3 Podcasting	300
	6.4.4 Advanced Search	301
	6.4.5 User Recommendation	302
	6.4.6 Wikis — Collective Writing	302
	6.4.7 Social Networking Sites	303
	6.4.8 Web Applications	304
	6.4.9 Mashups	304
	6.4.10 Virtual Worlds	305
	6.4.11 Long-Tail Economics	305
6.5	The Ideas behind Web 2.0	306
	6.5.1 The Web as a Platform	306
	6.5.2 Harnessing Collective Intelligence	306
	6.5.3 Data is the next Intel Inside	307
	6.5.4 End of the Software Release Cycle	308
	6.5.5 Lightweight Programing Models	308
	6.5.6 Software above the Level of a Single Device	309
	6.5.7 Rich User Experience	309
6.6	Discovering the Fabrics of Web 2.0	310
	6.6.1 HTML	310
	6.6.2 AJAX	311
	6.6.3 Aggregation	314
	6.6.4 Tagging and Folksonomy	316
	6.6.5 Open Application Programing Interfaces	318
	6.6.6 Open Source	320
6.7	Mobile Web 2.0 — Evolution and Revolution of Web 2.0	321
	6.7.1 The Seven Principles of Web 2.0 in the Mobile World	322
	6.7.2 Advantages of Connected Mobile Devices	325
	6.7.3 Access to Local Resources for Web Apps	328
	6.7.4 2D Barcodes and Near Field Communication (NFC)	329
	6.7.5 Web Page Adaptation for Mobile Devices	330
6.8	(Mobile) Web 2.0 and Privacy and Security Considerations	334
	6.8.1 On-Page Cookies	334
	6.8.2 Inter-Site Cookies	336
	6.8.3 Flash Shared Objects	336
	6.8.4 Session Tracking	337
	6.8.5 HTML5 Security and Privacy Considerations	338
	6.8.6 Private Information and Personal Data in the Cloud	338
6.9	Mobile Apps	340
	6.9.1 App Stores and Ecosystem Approaches	341
6.10	Android App Programing Introduction	342
	6.10.1 The Eclipse Programing Environment	342
	6.10.2 Android and Object Oriented Programing	342

6.10.3   *A Basic Android Program*   344
6.11   Impact of Mobile Apps on Networks and Power Consumption   349
6.12   Mobile Apps Security and Privacy Considerations   351
6.12.1   *Wi-Fi Eavesdropping*   352
6.12.2   *Access to Private Data by Apps*   352
6.12.3   *User Tracking by Apps and the Operating System*   353
6.12.4   *Third-Party Information Leakage*   354
6.13   Summary   354

7   **Conclusion**   **357**

Index   **361**

Preface

In recent years, cellular voice networks have transformed into powerful packet-switched access networks for both voice communication and Internet access. Evolving Universal Mobile Telecommunication System (UMTS) networks and first Long Term Evolution (LTE) installations now deliver bandwidths of several megabits per second to individual users, and mobile access to the Internet from handheld devices and notebooks is no longer perceived as slower than a Digital Subscriber Line (DSL) or cable connection. Bandwidth and capacity demands, however, keep rising because of the increasing number of people using the networks and because of bandwidth-intensive applications such as video streaming. Thus, network manufacturers and network operators need to find ways to continuously increase the capacity and performance of their cellular networks while reducing the cost.

In the past, network evolution mainly involved designing access networks with more bandwidth and capacity. As we go beyond 3G network architectures, there is now also an accelerated evolution of core networks and, most importantly, user devices and applications. This evolution follows the trends that are already in full swing in the "fixed-line" Internet world today. Circuit-switched voice telephony is being replaced by voice over IP technologies, and Web 2.0 has empowered consumers to become creators, to communicate with their friends and to share their own information with a worldwide audience. With connected smartphones having become a mainstream phenomenon in recent years, they will have a major impact on this trend, as they are an ideal tool for creating and consuming content. The majority of mobile phones today have advanced camera and video capabilities, and together with fast wireless access technologies, it has become possible to share information with others instantly.

While all these trends are already occurring, few resources are available that describe them from a technical perspective. This book therefore aims to introduce the technology behind this evolution. Chapter 1 gives an overview of how mobile networks have evolved in the past and what trends are emerging today. Chapter 2 then takes a look at radio access technologies such as High-Speed Packet Access (HSPA+), LTE, and the evolution of the Wi-Fi standard. Despite the many enhancements next-generation radio systems will bring, bandwidth on the air interface is still the limiting factor. Chapter 3 takes a look at the performance of next-generation systems in comparison to today's networks, shows where the limits are, and discusses how Wi-Fi can help to ensure future networks can meet the rising demand for bandwidth and integrated home networking. Voice over IP is already widely used in fixed line networks today, and "Beyond 3G" networks have enough capacity and performance to bring about this change in the wireless world as well. Chapter 4 thus focuses on Voice over IP architectures, such as the IP Multimedia Subsystem (IMS) and the Session Initiation Protocol (SIP) and discusses the impacts of these systems on future voice and multimedia communication. Just as important as wireless networks are the mobile devices using them, and Chapter 5 gives an overview of current mobile device architectures and their evolution. Finally, mobile devices are only as useful as the applications running on them. So Chapter 6 discusses how "mobile Web 2.0" applications and native apps are changing the way we communicate today and in the future.

Since the publication of the first edition of this book, many predictions have become a reality and new challenges and opportunities have arisen. While LTE was only on the distant horizon when the first edition was published, it is a reality today, and HSPA networks have undergone significant evolution as well. New spectrum bands have been assigned and auctioned in the meantime and many network operators around the globe have since made use of them to increase the coverage and capacity of their networks. Perhaps the biggest evolution over the past five years has been on the mobile device side. Mobile operating systems dominating the market only a few years ago have almost vanished and new entrants such as Android and iOS have taken the mobile world by storm. And finally on the web and application programming side, significant advances triggered an update of this chapter as well. As a consequence, about half the content of the previous edition of this book was updated or entirely rewritten to reflect the current state of the art and to give an outlook of what is to come in the next five years.

No book is written in isolation and many of the ideas that have gone into this manuscript are the result of countless conversations over the years with people from across the industry. Specifically, I would like to thank Debby Maxwell, Prashant John, Kevin Wriston, Peter van den Broek, and John Edwards for the many insights they have provided to me over the years in their areas of expertise and for their generous help with reviewing the manuscript. A special thank you goes to Berenike for her love, her passion for life, and for inspiring me to always go one step further. And last but not least I would like to thank Mark Hammond, Susan Barclay, and Sandra Grayson of John Wiley & Sons for the invaluable advice they gave me throughout this and previous projects.

1

Evolution from 2G over 3G to 4G

In the past 20 years, fixed line and wireless telecommunication as well as the Internet have developed both very quickly and very slowly depending on how one looks at the domain. To set current and future developments into perspective, the first chapter of this book gives a short overview of major events that have shaped these three sectors in the previous two decades. While the majority of the developments described below took place in most high-tech countries, local factors, and national regulation delayed or accelerated events. Therefore, the time frame is split up into a number of periods and specific dates are only given for country-specific examples.

1.1 First Half of the 1990s — Voice-Centric Communication

Twenty years ago, in 1993, Internet access was not widespread and most users were either studying or working at universities or in a few select companies in the IT industry. At this time, whole universities were connected to the Internet with a data rate of 9.6 kbit/s. Users had computers at home but dial-up to the university network was not yet widely used. Distributed bulletin board networks such as the Fidonet [1] were in widespread use by the few people who were online then.

It can therefore be said that telecommunication 20 years ago was mainly voice-centric from a mass market point of view. An online telecom news magazine [2] gives a number of interesting figures on pricing around that time, when the telecom monopolies were still in place in most European countries. A 10-min "long-distance" call in Germany during office hours, for example, cost €3.25.

On the wireless side, first-generation analog networks had been in place for a number of years, but their use was even more expensive and mobile devices were bulky and unaffordable except for business users. In 1992, GSM networks had been launched in a number of European countries, but only few people noticed the launch of these networks.

1.2 Between 1995 and 2000: The Rise of Mobility and the Internet

Around 1998, telecom monopolies came to an end in many countries in Europe. At the time, many alternative operators were preparing themselves for the end of the monopoly and prices went down significantly in the first week and months after the new regulation

3G, 4G and Beyond–Bringing Networks, Devices and the Web Together, Second Edition. Martin Sauter.
© 2013 John Wiley & Sons, Ltd. Published 2013 by John Wiley & Sons, Ltd.

came into effect. As a result, the cost of the 10 min long-distance call quickly fell to only a fraction of the former price. This trend has continued to this day and the current price is in the range of a few cents. Also, European and even intercontinental phone calls to many countries, like the USA and other industrialized countries, can be made at a similar cost.

At around the same time, another important milestone was reached. About five years after the start of GSM mobile networks, tariffs for mobile phone calls and mobile phone prices had reached a level that stimulated mass market adoption. Although the use of a mobile phone was perceived as a luxury and mainly for business purposes in the first years of GSM, adoption quickly accelerated at the end of the decade and the mobile phone was quickly transformed from a high-price business device to an indispensable communication tool for most people.

Fixed line modem technology had also evolved somewhat during that time, and modems with speeds of 30–56 kbit/s were slowly being adopted by students and other computer users for Internet access either via the university or via private Internet dial-up service providers. Around this time, text-based communication also started to evolve and Web browsers appeared that could show Web pages with graphical content. Also, e-mail leapt beyond its educational origin. Content on the Internet at the time was mostly published by big news and IT organizations and was very much a top-down distribution model, with the user mainly being a consumer of information. Today, this model is known as Web 1.0.

While voice calls over mobile networks quickly became a success, mobile Internet access was still in its infancy. At the time, GSM networks allowed data rates of 9.6 and 14.4 kbit/s. over circuit-switched connections. Few people at the time made use of mobile data, however, mainly due to high costs and missing applications and devices. Nevertheless, the end of the decade saw the first mobile data applications such as Web browsers and mobile e-mail on devices such as Personal Digital Assistants (PDAs), which could communicate with mobile phones via an infrared port.

1.3 Between 2000 and 2005: Dot Com Burst, Web 2.0, Mobile Internet

Developments continued and even accelerated in all three sectors despite the dot com burst in 2001, which sent both the telecoms and the Internet industry into a downward spiral for several years. Despite this downturn, a number of new important developments took place during this period.

One of the major breakthroughs during this period was the rise of Internet access via Digital Subscriber Lines (DSL) and TV cable modems. These quickly replaced dial-up connections as they became affordable and offered speeds of 1 Mbit/s and higher. Compared with the 56 kbit/s analog modem connections, the download times for web pages with graphical content and larger files improved significantly. At the end of this period, the majority of people in many countries had access to broadband Internet that allowed them to view more and more complex Web pages. Also, new forms of communication like Blogs and Wikis appeared, which quickly revolutionized the creator–consumer imbalance.

Suddenly, users were no longer only consumers of content, but could also be creators for a worldwide audience. This is one of the main properties of what is popularly called Web 2.0 and will be further discussed later on in this book.

In the fixed line telephony world, prices for national and international calls continued to decline. Toward the end of this period, initial attempts were also made to use the Internet for transporting voice calls. Early adopters discovered the use of Internet telephony to make phone calls over the Internet via their DSL lines. Proprietary programs like Skype suddenly allowed users to call any Skype subscriber in the world for free, in many cases with superior voice quality. "Free" in this regard is a relative term, however, as both parties in the call have to pay for access to the Internet, telecom operators still benefit from such calls because of the monthly charge for DSL or cable connections. Additionally, many startup companies started to offer analog telephone to Internet Protocol (IP) telephone converters, which used the standardized SIP (Session Initiation Protocol) protocol to transport phone calls over the Internet. Gateways ensured that such subscribers could be reached via an ordinary fixed line telephone number and could call any legacy analog phone in the world. Alternative long-distance carriers also made active use of the Internet to tunnel phone calls between countries and thus offered cheaper rates.

Starting in 2001, the General Packet Radio Service (GPRS) was introduced in public GSM networks for the first time. When the first GPRS-capable mobile phones quickly followed, mobile Internet access became practically feasible for a wider audience. Until then, mobile Internet access had only been possible via circuit-switched data calls. However, the data rate, call establishment times and the necessity of maintaining the channel even during times of inactivity were not suitable for most Internet applications. These problems, along with the small and monochrome displays in mobile phones and mobile software being in its infancy, meant that the first wireless Internet services (WAP 1.0) never became popular. Toward 2005, devices matured, high-resolution color displays made it into the mid-range mobile phone segment and WAP 2.0 mobile Web browsers and easy-to-use mobile e-mail clients in combination with GPRS as a packet-switched transport layer finally allowed mobile Internet access to cross the threshold between niche and mass market. Despite these advances, pricing levels and the struggle between open and closed Internet gardens slowed down progress considerably.

At this point, it should be noted that throughout this book the terms "mobile access to the Internet" and "mobile Internet access" are used rather than "mobile Internet." This is done on purpose since the latter term implies that there might be a fracture between a "fixed line" and a "mobile" Internet. While it is true that some services are specifically tailored for use on mobile devices and even benefit and make use of the user's mobility, there is a clear trend for the same applications, services, and content to be offered and useful on both small mobile devices and bigger nomadic or stationary devices. This will be discussed further in Chapter 6.

Another important milestone for wireless Internet access during this time frame was 3G networks going online in many countries in 2004 and 2005. While GPRS came close to analog modem speeds, Universal Mobile Telecommunication System (UMTS) brought data rates of up to 384 kbit/s in practice, and the experience became similar to DSL.

1.4   Between 2005 and 2010: Global Coverage, Fixed Line VoIP, and Mobile Broadband

From 2005 to 2010, the percentage of people in industrialized countries accessing the Internet via broadband DSL or cable connections continued to rise. Additionally, many network operators started to roll out ADSL2+, and new modems enabled download speeds beyond 15 Mbit/s for users living close to a central exchange. VDSL and fiber to the curb/fiber to the home deployments offered even higher data rates. Another trend that accelerated since 2005 is Voice over Internet Protocol (VoIP) via a telephone port in the DSL or cable modem router. This effectively circumvents the traditional analog telephone network and traditional network fixed line telephony operators saw a steady decline in their customer base.

In this period, the number of mobile phone users had reached 3 billion. This means that almost every second person on Earth owned a mobile phone, a trend which only a few people foresaw only five years earlier. In 2007, network operators registered 1000 new users per minute [3]. Most of this growth has been driven by the rollout of second-generation GSM/GPRS networks in emerging markets. Owing to global competition between network vendors, network components reached a price that made it feasible to operate wireless networks in countries with very low revenue per user per month. Another important factor for this rapid growth was ultra-low-cost GSM mobile phones, which became available for less than $50. In only a few years, mobile networks have changed working patterns and access to information for small entrepreneurs like taxi drivers and tradesmen in emerging markets [4] as GSM networks were by that time available in most parts of the world. In industrialized countries, third-generation networks continued to evolve and 2006 saw the first upgrades of UMTS networks to High Speed Data Packet Access (HSDPA). In a first step, this allowed user data speeds between 1 and 3 Mbit/s which was particularly useful in combination with notebooks to give users broadband Internet almost anywhere.

While 3G networks had been available for some time, take-up was sluggish until around 2006/2007, when mobile network operators finally introduced attractive price plans. Prices fell below €40–50 for wireless broadband Internet access and monthly transfer volumes of around 5 Gbytes. This was more than enough for everything but file sharing and substantial video streaming. Operators also started to offer smaller packages in the range of €6–15 a month for occasional Internet access with notebooks. Packages in a similar price range were also offered for unlimited Web browsing and e-mail on mobile phones. Pricing and availability varied in different countries. In 2006, mobile data revenue in the USA alone reached a $15.7 billion, of which 50–60% was non-SMS (Short Message Service) revenue [5]. In some countries, mobile data revenues accounted for between 20% and 30% of the total operator revenue.

While wireless data roaming was still in its infancy, wireless Internet access via prepaid Subscriber Identity Module (SIM) cards was already offered in many countries at similar prices to those for customers with a monthly bill. This is another important step, as it opened the door to anytime and anywhere Internet access for creative people such as students, who favor prepaid SIMs to monthly bills. In addition, it made life much easier for travelers, who until then had no access to the Internet while traveling, except for wireless hotspots at airports and hotels. An updated list of such offers is maintained by the Web community on the prepaid wireless Internet access Wiki [6].

1.5   2010 and Beyond

At the time of publication, the state of the mobile industry has once again changed significantly compared to the 2005–2010 area. In 2013, the number of mobile subscriptions is estimated to surpass 7 billion [7], up from 3 billion subscriptions only five years earlier. This means that the vast majority of the global population now owns a mobile device.

After Apple successfully entered the mobile domain with their iPhone in 2007 and Google following with their Android mobile operating system about a year later, innovation on mobile devices has decoupled from the telecoms industry and has moved toward IT-based software companies. It only took little time for this effect to become mainstream. In 2012, half the mobile phones sold in Germany, for example, were smartphones [8]. While smartphones for many years only covered the high end of mobile device sales, this has significantly changed in the meantime as well, with sophisticated Android devices now available for little more than €100.

While the growing number of voice minutes was previously the main driver for increasing network capacity, it is now the growing use of smartphones and Internet-based services on them that require constant investment on the network side to increase the capacity and reach. In the meantime, capabilities of UMTS networks increase further and data rates of up to 42 MBit/s can now be reached under ideal conditions. To further increase the capacity, Long Term Evolution (LTE) networks were launched in several countries early in this decade.

While the amount of data transferred in wireless networks is increasing year over year, this trend seems to be somewhat slowing. The German regulatory authority reported, for example, that year-over-year data growth in mobile networks has slowed to a factor of 0.5 in 2011, down from having doubled the year earlier and a three times increase in 2009 [9]. A 50% increase of the amount of data sent through mobile networks in one year is still impressive and requires network operators to spend a sizable amount of their revenue to increase the capacity of their networks.

Like with UMTS less than a decade earlier, early LTE networks were first used with data sticks. Mobile devices were only introduced in 2011 and like UMTS phones a decade earlier suffered from increased size, weight, heat generation, and higher power consumption, often leading to an inferior experience compared to high end High-speed Packet Access (HSPA+) devices. Unlike early UMTS devices, however, one significant difference is the absence of circuit-switched voice capabilities of LTE networks, which require LTE devices to use 2G or 3G networks for voice calls. Several VoIP over LTE solutions have been in the making for many years but no network operator has yet deployed a solution for the mass market. This is one of the biggest technical challenges ahead for the mobile industry in the years to come.

Another trend is the emergence of new mobile device types. Led by the success of netbooks and tablets that bridge the gap between smartphones with small screens and limited input functionalities on the one hand and notebooks with large screens and full keyboards on the other leaves the industry experimenting of how to best unify the user experience across the different screen sizes and input capabilities. Already now it can be observed that websites previously designed for PCs and notebooks are modified for touch input, slightly smaller screen sizes, and less processing power. Graphical user interfaces for PCs are also undergoing change to adapt to non-keyboard based devices. Two examples are Windows 8 with its "Metro" tiles approach and Ubuntu with its "Unity" desktop environment.

1.6   All over IP in Mobile — The Biggest Challenge

A process now almost completed in fixed line networks is still in its infancy in mobile networks, moving the traditional telephony service to an IP-based infrastructure. This will be the most challenging task of mobile network operators in many years to come as they are on the one hand facing competition from Internet-based voice service providers and significant implementation challenges on the mobile network side on the other. Mobile network operators, however, have little choice but to face this challenge in one way or another, as 4G LTE radio and core networks are fully based on IP technology and do not support traditional mobile voice telephony. As this topic is of central importance and significant issues have to be overcome, a complete chapter in this book will describe this challenge in-depth.

1.7   Summary

This chapter presented how fixed and wireless networks evolved in the past 20 years from circuit-switched voice-centric systems to packet-switched Internet access systems. Due to the additional complexity of wireless systems, enhancements are usually introduced in fixed-line systems first and only some years later in wireless systems as well. To date, fixed-line networks offer data rates to the customer premises of tens to hundreds of megabits per second, in some cases already going beyond this. Wireless 3.5G networks are capable of data rates in the order of several megabits per second and 4G networks promise even higher data rates and much higher overall capacity. As current wireless systems will continue to play a major role in the evolution of mobile networks, this book therefore not only concentrates on 4G systems, but also discusses the evolution of 3G systems.

References

1. Background on Fidonet (2008) http://www.fidonet.org (accessed 2012).
2. Neuhetzki, T. (2005) German Long Distance Tariffs in the 1990s, December 2005, http://www.teltarif.de /arch/2005/kw52/s19950.html (accessed 2012).
3. Sauter, M. (2006) 1000 New Mobile Phone Users a Minute, August 2006, http://mobilesociety.typepad.com /mobile_life/2006/08/1000_new_mobile.html (accessed 2012).
4. Andersen, T. (2007) Mobile Phone Lifeline for World's Poor, February 19, 2007, http://news.bbc.co.uk /1/hi/business/6339671.stm (accessed 2012).
5. Sharma, C. (2007) Global Wireless Data Market, September 2007, http://www.chetansharma.com /globalmarketupdate1H07.htm (accessed 2012).
6. The Prepaid Wireless Internet Access Wiki (2008) http://prepaid-wireless-internet-access.wetpaint.com (accessed 2012).

7. Ahonen, T. (2012) Massive Milestones in Mobile – Will These Numbers Change Your Mobile Strategy? June 2012, http://communities-dominate.blogs.com/brands/2012/06/massive-milestones-in-mobile -will-these-numbers-change-your-mobile-strategy.html (accessed 2012).
8. Sauter, M. (2012) Half The Phones Sold in Germany In 2012 Will Be Smartphones, March 2012, http://mobilesociety.typepad.com/mobile_life/2012/03/half-the-phones-sold-in-germany-in-2012-will -be-smartphones.html (accessed 2012).
9. Sauter, M. (2012) Slowing Mobile Data Volume Growth, May 2012, http://mobilesociety.typepad.com /mobile_life/2012/05/slowing-mobile-data-volume-growth.html (accessed 2012).

2

Beyond 3G Network Architectures

2.1 Overview

As discussed in Chapter 1, the general trend in telecommunications is to move all applications to a common transmission protocol, the Internet Protocol. The tremendous advantage of this approach is that applications no longer require a specific network technology but can be used over different kinds of networks. This is important since, depending on the situation, an application might be used best over a cellular network while at other times it is more convenient and cheaper to use a wireless home or office networking technology such as Wi-Fi. Today, there are few, if any, smartphones that do not support this trend. Today and even more so in the future, a number of wireless technologies are deployed in parallel. This is necessary as the deployment of a new network requires a considerable amount of time and there are usually only a small number of devices supporting a new network technology at first. It is therefore important that different network technologies are deployed not only in parallel but also at the same location. As well as the introduction of new technologies, existing network technologies continue to evolve to offer improved performance while the new technology is not yet deployed or is just in the process of being rolled out. For these reasons, this chapter looks at a number of different Beyond 3G network. In this context, the term "Beyond 3G networks" is used for cellular networks that offer higher speeds than the original Universal Mobile Telecommunication System (UMTS) networks with their maximum data rate of 384 kbit/s per user.

In the cellular world, the UMTS with its High-speed Packet Access (HSPA) evolution is currently the Beyond 3G system with the broadest deployment. This system, together with its further evolution, HSPA+, is therefore discussed first.

Next, the chapter focuses on the successor technology of HSPA and HSPA+, which is commonly known as Long Term Evolution (LTE). In the standards, LTE is referred to as the Evolved Packet System (EPS), which is divided into the Evolved Packet Core (EPC) and the Enhanced-UMTS Terrestrial Radio Access Network (E-UTRAN).

As will be shown throughout this book, 802.11 Wi-Fi networks play an important role in overall wireless network architectures today and in the future as well. Consequently, this chapter also introduces Wi-Fi and the latest enhancements built around the original standard, such as an evolved air interface with speeds of several hundred megabits per second, security enhancements for home and enterprise use and quality of service (QoS) extensions.

3G, 4G and Beyond–Bringing Networks, Devices and the Web Together, Second Edition. Martin Sauter.
© 2013 John Wiley & Sons, Ltd. Published 2013 by John Wiley & Sons, Ltd.

To give an initial idea about the performance of each system, some general observations for each system in terms of bandwidth, speed, and latency are discussed. Since these parameters are of great importance, and often grossly exaggerated by marketing departments, Chapter 3 will then look at this topic in much more detail.

2.2   UMTS, HSPA, and HSPA+

2.2.1   Introduction

Initial drafts of UMTS standards documents appeared in working groups of the Third Generation Partnership Project (3GPP) at the end of 1999, but work on feasibility studies for the system began much earlier. A few UMTS networks were opened to the public in 2003, but it was not until the end of 2004, when adequate UMTS mobile phones became available and networks were rolled out to more than just a few cities, that even early adopters could afford and actually use UMTS. A time frame of five years from a first set of specifications to first deployments is not uncommon due to the complexity involved. The development and deployment of LTE has seen a similar time frame.

2.2.2   Network Architecture

Figure 2.1 shows an overview of the network architecture of a UMTS network. The upper-left side of the figure shows the radio access part of the network, referred to in the 3GPP standards as the UTRAN.

2.2.2.1   The Base Stations

The UTRAN consists of two components. At the edge of the network, base stations, referred to in the standards as the NodeB, communicate with mobile devices over the air. In cities, a base station usually covers an area with a radius of about 500 m to 1 km, sometimes less, depending on the population density and bandwidth requirements. To increase the amount of data and the number of simultaneous voice calls per base station, the coverage area is usually split into three sectors. Each sector has its own directional antenna and transceiver equipment. In the standards, a sector is sometimes also referred to as a cell. A NodeB with three sectors therefore consists of three individual cells. If a user walked around such a base station during an ongoing voice call or while data was exchanged, he would be consecutively served by each of the cells. During that time, the radio network would hand the connection over from one cell to the next once radio conditions deteriorated. From a technical point of view, there is thus little difference between a handover between cells of the same base station and between cells of different base stations. These and other mobility management scenarios will be discussed in more detail in Section 2.2.3. The radio link between mobile devices and the base station is also referred to as the "air interface" and this term will also be used throughout this book. A device using a UMTS network is referred to in the standard as User Equipment (UE). In this book, however, the somewhat less technical terms "mobile," "mobile device," and "connected mobile device" are used instead.

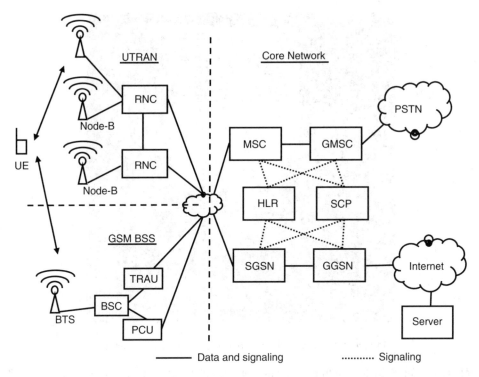

Figure 2.1 Common GSM/UMTS network. (Reproduced from *Communication Systems for the Mobile Information Society*, Martin Sauter, 2006, John Wiley & Sons, Ltd, Ref. [1].)

Today, base stations are connected to the network in one of two ways. The traditional approach is to connect base stations via one or more 2 Mbit/s links, referred to as E-1 connections in Europe and T-1 connections in the USA (with a slightly lower transmission speed). Each E-1 or T-1 link is carried over a pair of copper cables. An alternative to copper cables is a microwave connection, which can carry several logical E-1 links over a single microwave connection. This is preferred by many operators as they do not have to pay monthly line rental fees to the owner of the copper cable infrastructure. To make full use of the air interface capacity of a multisector base station, several E-1 links are required. The protocol used over these links is ATM (Asynchronous Transfer Mode), a robust transmission technology widely used in many fixed and wireless telecommunication networks around the world today. Figure 2.2 shows a typical base station cabinet located at street level. In practice, base stations are also frequently installed on flat rooftops close to the antennas, as there is often no space at ground level and as this significantly reduces the length and thus the cost for the cabling between the base station cabinet and the antennas.

In recent years, the use of E-1 links over copper cables has become very uneconomical and limiting because of the rising transmission speeds over the air interface. As the number of copper cables leading to a base station is limited and, more significantly, the line rental costs per month are high, network operators are in a transition phase or have

Figure 2.2   A typical GSM or UMTS base station cabinet.

already transitioned to a number of alternative technologies to connect base stations to the network:

- High bandwidth microwave links — the latest equipment is capable of speeds of several hundred megabits per second [2].
- Fiber links — especially in dense urban areas, many fixed line carriers are currently deploying additional fiber cables for providing very high data rate Internet access to businesses and homes. This infrastructure is also ideal for connecting base stations to the rest of the infrastructure of a wireless network. In practice, however, only a fraction of deployed base stations already have a fiber laid up to the cabinet.
- Asymmetric Digital Subscriber Line (ADSL)/Very-high-bit-rate Digital Subscriber Line (VDSL) — a viable alternative to directly using fiber is to connect base stations via a high-speed Digital Subscriber Line (DSL) link to an optical transmission network. T-Mobile in Germany is one operator that has chosen this solution [3].

One characteristic all types of modern access technologies have in common is that data transmission is no longer based on E1 connections and the corresponding legacy protocols. Instead, Ethernet and IP have become the protocols of choice. For this purpose, the Ethernet protocol was extended somewhat to cater for the specific needs of metro networks. In general, however, the technology works in the same way as Ethernet at home and in company networks today for interconnecting PCs to company servers and the Internet. This is reflected in the current designs for UMTS/HSPA base stations, which can now not only be equipped with E-1 ATM-based interfaces but also with IP over Ethernet cards. The Ethernet interface is either based on the standard 100 Mbit/s or 1 Gbit/s twisted pair copper cable interface commonly used with other IT equipment such as PCs and notebooks or via an optical port. In the case of copper cabling, additional equipment is usually required to transport the Ethernet frames over longer distances, as the length of the twisted pair copper cable is limited to 100 m.

Over the next few years, it is likely that most E-1-based links will be replaced with IP-based technologies because of the rising bandwidth capabilities of base stations and much lower cost of such links. Another reason is that GSM, UMTS, and LTE base stations are mostly co-located and it is desirable to have a single high-capacity backhaul link for the data exchanged via all three radio technologies. As the technology used for backhauling data from base stations has a significant impact on the bandwidth and cost of a network, this topic will be discussed in more detail in Section 3.17.

2.2.2.2 The Radio Network Controllers

The second component of the radio access network is the Radio Network Controller (RNC). It is responsible for the following management and control tasks:

- The establishment of a radio connection, also referred to as bearer establishment.
- The selection of bearer properties such as the maximum bandwidth, based on current available radio capacity, type of required bearer (voice or data), QoS requirements and subscription options of the user.
- Mobility management while a radio bearer is established, that is, handover control between different cells and different base stations of a network.
- Overload control in the network and on the radio interface. In situations when more users want to communicate than there are resources available, the RNC can block new connection establishment requests to prevent other connections from breaking up. Another option is to reduce the bandwidth of established bearers. A new data connection might, for example, be blocked by the network if the load in a cell is already at the limit, while for a new voice call, the bandwidth of an ongoing data connection might be reduced to allow the voice call to be established. In practice, blocking the establishment of a radio bearer for data transmission is very rare, as most network operators monitor the use of their networks and remove bottlenecks, for example, by installing additional transceivers in a base station, by increasing backhaul capacity between the base station and the RNC or by installing additional base stations to reduce the coverage area and thus the number of users per base station. Capacity management will be discussed in more detail in Chapter 3.

2.2.2.3 The Mobile Switching Center

Moving further to the right in Figure 2.1, it can be seen that the RNCs of the network are connected to gateway nodes between the radio access network and the core network. In UMTS, there are two independent core network entities. The upper right of the figure shows the Message Service Center (MSC), which is the central unit of the circuit-switched core network. It handles voice and video calls and forwards Short Message Service (SMS) messages via the radio network to subscribers. As discussed in Chapter 1, circuit switching means that a dedicated connection is established for a call between two parties via the

MSC that remains in place while the call is ongoing. Large mobile networks usually have several MSCs, each responsible for a different geographical area. All RNCs located in this area are then connected to the MSC. Each MSC in the network is responsible for the management of all users of the network in its region and for the establishment of circuit-switched channels for incoming and outgoing calls. When a mobile device requests the establishment of a voice call, the RNC forwards the request to the MSC. The MSC then checks if the user is allowed to make an outgoing call and instructs the RNC to establish a suitable radio bearer. At the same time, it informs the called party of the call establishment request or, if the called party is located in a different area or different network, establishes a circuit-switched connection to another MSC. If the subscriber is in the same network it might be possible to contact the MSC responsible for the called party directly. In many cases, however, the called party is not in the same network or not a mobile subscriber at all. In this case, a circuit-switched connection is established to a Gateway Message Service Center (GMSC), shown in Figure 2.1 on the top right. Based on the telephone number of the called party, the GMSC then forwards the call to an external fixed or mobile telephone network. In practice, a MSC usually serves mobile subscribers and also acts as a GMSC.

To allow the MSC to manage subscribers and to alert them about incoming calls, mobile devices need to register with the MSC when they are switched on. At the beginning of the registration process, the mobile device sends its International Mobile Subscriber Identity (IMSI), which is stored on a SIM card (Subscriber Identity Module), to the MSC. If the IMSI is not known to the MSC's Visitor Location Register (VLR) database from a previous registration request, the network's main user database, the Home Location Register, is queried for the user's subscription record and authentication information. The authentication information is used to verify the validity of the request and to establish an encrypted connection for the exchange of signaling messages. The authentication information is also used later on during the establishment of a voice or video call to encrypt the speech path of the connection. Note that the exchange of these messages is not based on the IP protocol but on an out-of-band signaling protocol stack called Signaling System Number 7 (SS7). Out-of-band means that messages are exchanged in dedicated signaling connections, which are not used for transporting circuit-switched voice and video.

For completeness, it should be mentioned at this point that many network operators today have exchanged their circuit-switched MSCs for an IP-based combination of MSC Servers and media gateways as described in the previous chapter. This is fully transparent to the mobile device, however, and the same procedures inside the network still apply from a high level point of view, even if signaling and the voice connections are transported over IP links inside the core network.

2.2.2.4 The SIM Card

An important component of UMTS networks, even though it is very small, is the SIM card. It allows the network subscription to be separate from the mobile device. A user can thus buy the SIM card and the mobile device separately. It is therefore possible to use the SIM card with several devices or to use several SIM cards with a single device. This encourages competition between network operators, as users can change from one network to another quickly if prices are no longer competitive. When traveling abroad, it is also possible to buy and use a local prepaid SIM card to avoid prohibitive roaming charges. Separating network subscriptions from mobile devices has the additional benefit that mobile devices can not only be bought from a network operator but also from independent shops, for example, electronic stores and mobile phone shops that sell subscriptions for several network operators. This stimulates competitive pricing for mobile devices, which would not happen if a device could only be bought from a single source. A further discussion of this topic can be found in [4].

2.2.2.5 The SMSC

A data service that became very popular long before the rise of current high-speed wireless Internet access technologies is the SMS, used to send text messages between users. As the service dates back to the mid 1990s, it is part of the circuit-switched core network. SMS messages are transported in a store and forward fashion. When a subscriber sends a message, it is sent via the signaling channel, the main purpose of which is to transport messages for call establishment and mobility management purposes, to the Short Message Service Center (SMSC). The SMSC stores the message and queries the Home Location Register database to find the MSC which is currently responsible for the destination subscriber. Afterwards, it forwards the message, again in an SS-7 signaling link, to the MSC. When receiving the text message, the MSC locates the subscriber by sending a paging message. This is necessary, as in most cases the subscriber is not active when a text message arrives and therefore the user's current serving cell is not known to the MSC. On the air interface, the paging message is sent on a broadcast channel (BCH) that is observed by all devices attached to the network. The mobile device can thus receive the paging message and send an answer to the network despite not having being in active communication with the network. The network then authenticates the subscriber, activates encryption, and delivers the text message. In case the subscriber is not reachable, the delivery attempt fails and the SMSC stores the message until the subscriber is reachable again.

2.2.2.6 Service Control Points

Optional, but very important, components in circuit-switched core networks are integrated databases and control logic on Service Control Points (SCPs). An SCP is required, for example, to offer prepaid voice services that allow users to top-up an account with a voucher and then use the credit to make phone calls and send SMS messages. For each call or SMS, the MSC requests permission from the prepaid service logic on an SCP. The SCP then checks and modifies the balance on the user's account and allows or denies the request. MSCs communicate with SCPs via SS-7 connections. When a prepaid user roams to another country, foreign MSCs also need to communicate with the SCP in the home network of the user. As there are many MSC vendors, the interaction model and protocol between MSCs and SCPs have been specified in the CAMEL (Customized Applications for Mobile Enhanced Logic) standard [5].

For providing the actual service (e.g., prepaid), only signaling connections between SCPs and MSCs are required. Some services, such as prepaid, however, also require an interface to allow a user to check his balance and to top up their account. In practice there are several possibilities. Most operators use some form of scratch card and an automated voice system for this purpose. Therefore, there are usually also voice circuits required between SCP-controlled interactive voice gateways and the MSCs. In addition, most prepaid services also let users top up or check their current balance via short codes (e.g., *100#), which do not require the establishment of a voice call. Instead, such short codes are sent to the SCP via an SS-7 signaling link.

2.2.2.7 Billing

In addition to the billing of prepaid users, which is performed in real time on SCPs, further equipment is required in the core network to collect billing information from the MSCs for subscribers who receive a monthly invoice. This is the task of billing servers, which are not shown in Figure 2.2. In essence, the billing server collects Call Detail Records (CDRs) from the MSCs and SMSCs in the network and assembles a monthly invoice for each user based on the selected tariff. CDRs contain information such as the identity of the calling party, the identity of the called party, date and duration of the call, and the identity of the cell from which the call was originated. Location information is required as calls placed from foreign networks while the user is roaming are charged differently from calls originated in the home network. Some network operators also use location information for zone-based billing, that is, they offer cheaper calls to users while they are at home or in the office. Another popular billing approach is to offer cheaper rates at certain times. Most operators combine many different options into a single tariff and continuously change their billing options. This requires a flexible rule-based billing service.

2.2.2.8 The Packet-Switched Core Network

The core network components discussed so far have been designed for circuit-switched communication. For communicating with services on the Internet, which is based on packet switching, a different approach is required. This is why a packet-switched core network was added to the circuit-switched core network infrastructure. As can be seen in Figure 2.2, the RNC connects to both the circuit-switched core network and the packet-switched core network. UMTS devices are even capable of having circuit-switched and packet-switched connections established at the same time. A user can therefore establish a voice call while at the same time using his device as a modem for a PC, or for downloading content such as a podcast to the mobile device without interrupting the connection to the Internet while the voice call is ongoing. Another example of the benefits of being connected to both the packet-switched and circuit-switched networks is that an ongoing instant messaging session is not interrupted during a voice call.

Before a mobile device can exchange data with an external packet-switched network such as the Internet, it has to perform two tasks. First, the mobile device needs to attach to the packet-switched core network and perform an authentication procedure. This is usually done after the device is switched on and once it has registered with the circuit-switched core network. In a second step, the mobile device can then immediately, or at any time later on, request an IP address from the packet-switched side of the network. This process is referred to as establishing a data call or as establishing a PDP (Packet Data Protocol) context. The expression "establishing a data call" is interesting because it suggests that establishing a connection to the Internet or another external packet network is similar to setting up a voice call. From a signaling point of view, the two actions are indeed similar. The connections that are established as a result of the two requests, however, are very different. While voice calls require a connection with a constant bandwidth and delay that remain in place while the call is ongoing, data calls only require a physical connection while packets are transmitted. During times in which no data is transferred, the channel for the connection is either modified or completely released. Nevertheless, the logical connection of the data call remains in place so the data transfer can be resumed at any time. Furthermore, the IP address remains in place even though no resources are assigned on the air interface. In UMTS, separating the attachment to the packet-switched core network from the establishment of a data connection makes sense when looked at from a historical and practical perspective. The majority of mobile devices today are still mostly used for voice communication for which no Internet connection is required. Therefore, the mobile device can perform its main duty without establishing a data call. This is changing quickly, however, with smartphones becoming more popular. At the time LTE was initially conceived, this trend was foreseen and taken into account. As a consequence, attaching to the network and requesting an IP address is part of the same procedure in this system and the notion of a "data call" is no longer part of the system design.

2.2.2.9   The Serving GPRS Support Node

The packet-switched UMTS core network was, like the circuit-switched core network, adapted from GSM with only a few modifications. This is the reason why the gateway node to the radio access network is still referred to as the Serving GPRS Support Node (SGSN). GPRS stands for General Packet Radio Service and is the original name of the packet-switched service introduced in GSM networks. Like the MSC, the SGSN is responsible for subscriber and mobility management. To enable users to move between the coverage areas of different RNCs, the SGSN keeps track of the location of users and changes the route of IP packets arriving from the core network when they change their locations. SGSNs can be connected to RNCs in several ways. The traditional way is to use ATM-based links. Since RNCs often control several hundred cells and are physically distant from an SGSN, optical connections are used. In practice, a single RNC is connected to an SGSN via one or more Optical Carrier-3 (OC-3) optical links with a speed of 155 Mbit/s each or an Optical Carrier-12 (OC-12) optical connection with a speed of 622 Mbit/s [6]. Above the ATM layer, IP is used as a transport protocol. In recent years, many network operators have also made the change to Ethernet and IP-based technologies on this interface, which no longer require the ATM protocol below the IP layer.

The SGSN also has a signaling connection to the Home Location Register of the network that in addition to the data required for the circuit-switched network also contains subscription information for the packet-switched network. This includes, for example, if the user is allowed to use packet-switched services, their QoS settings such as the maximum transmission speed that is granted by the network and which access points to the Internet they are allowed to use. The record also shows whether a user has a prepaid contract, in which case the SGSN has to contact a prepaid SCP before a connection request is granted.

2.2.2.10   The Gateway GPRS Support Node

At the edge of the wireless core network, Gateway GPRS Support Nodes (GGSNs) connect the wireless network to the Internet. Their prime purpose is to hide the mobility of the users from routers on the Internet. This is required as IP routers forward packets based on the destination IP address and a routing table. For each incoming packet, each router on the Internet consults its routing table and forwards the packet to the next router via the output port specified in the routing table. Eventually, the packets for wireless subscribers end up at the GGSN. Here, the routing mechanism is different. As RNCs and SGSNs can change at any time due to the mobility of the subscriber, a static database containing the next hop based on the IP address is not suitable. Instead, the GGSN has a database table which lists the IP address of the SGSN currently responsible for a subscriber with a certain IP address. The IP packet for the subscriber is then enclosed in an IP packet with the destination address of the SGSN. This principle is called tunneling because each IP packet to the subscriber is encapsulated in an IP packet to the responsible SGSN. On the SGSN the original IP packet is restored and once again tunneled to the RNC.

Figure 2.3 shows how tunneling an IP packet in the core network works in practice. The protocol stack shown uses the Ethernet protocol on layer 2 which is followed by IP on layer 3. The source and destination address of the IP packet belong to the SGSN and GGSN of the network. Afterwards, the GPRS Tunneling Protocol (GTP) encapsulates the original IP packet. Here, the source and destination address represent the subscriber and a host on the Internet such as a Web server.

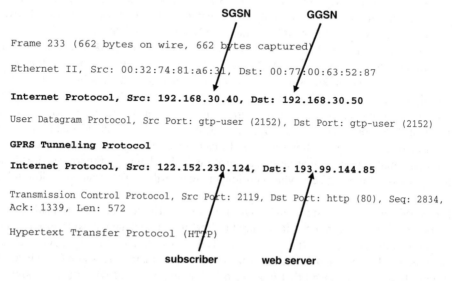

Figure 2.3   Encapsulated IP packet in a packet-switched core network.

In addition to tunneling, the GGSN is also responsible for assigning IP addresses to subscribers. During the connection establishment, the SGSN verifies the request of the subscriber with the information from the Home Location Register (HLR) and then requests the establishment of a tunnel and an assignment of an IP address from the GGSN. The GGSN then assigns an IP address from a pool and establishes the tunnel. In practice, there are two types of IP addresses. Many network operators use private IP addresses that are not valid outside the network. This is similar to the use of private IP addresses in home networks where the DSL or cable router assigns private IP addresses to all devices in the home network. To the outside world, the network is represented with only a single IP address and the DSL or cable router has to translate internal addresses in combination with Transmission Control Protocol (TCP) and User Datagram Protocol (UDP) port numbers into the external IP address with new TCP and UDP port numbers. This process is referred to as Network Address Translation (NAT). If private IP addresses are used in wireless networks, the same process has to be performed by the GGSN. The advantage for the operator is that fewer public IP addresses are required, of which there is a shortage today. Other operators use public IP addresses for their subscribers by default. For subscribers this has the advantage that they are directly reachable from the Internet, which is required for applications such as hosting a Web server or for remote desktop applications. It should be noted, however, that a public IP address also has disadvantages, especially for mobile devices, as unsolicited packets arriving from the Internet can drain batteries of mobile devices [7]. In practice, a subscriber can have several connection profiles, referred to as Access Point Names (APNs). The operator can therefore choose which subscribers and applications to use private or public IP addresses for.

2.2.2.11   Interworking with GSM

The lower left side of Figure 2.2 shows a GSM radio network, which is also connected to the circuit-switched and packet-switched core network. Depending on the network vendor, the GSM radio network can either be connected to the same MSCs and SGSNs or to separate ones. This is possible since the functionality of these components is very similar for GSM and UMTS. The interfaces to the 2G and 3G radio networks, however, are different. For the connections to the 3G radio network, ATM or IP over Ethernet is used while the traditional type of connectivity between the SGSN and the 2G radio network was based on the Frame Relay protocol. In a later version of the standard, an IP-based interface between SGSNs and the 2G radio network has been specified as well and is now used in practice as well. The MSC is connected to the GSM radio network via circuit-switched links or over Ethernet-based IP links as well. These have become more popular for this interface in recent years as well as in an attempt to focus on a single fixed line technology for all interfaces to streamline the network and reduce costs.

2.2.3   *Air Interface and Radio Network*

2.2.3.1   The CDMA Principle

2G radio systems such as GSM are based on timeslots and channels on different carrier frequencies so a single base station can serve many users simultaneously. While such an approach is well suited for the transmission of circuit-switched voice calls, it only offers limited flexibility for packet-switched data transmission, as the GSM channel bandwidth is only 200 kHz and thus only a limited amount of bandwidth can be assigned to a single user at a time. For 3G systems such as UMTS, it was therefore decided to use a different transmission scheme for the air interface. An alternative transmission technology that overcomes the limitations of a narrow band and was already well understood at the end of the 1990s, when work on UMTS started, was Code Division Multiple Access (CDMA). Instead of a narrow channel bandwidth of 200 kHz as in GSM, the Wideband Code Division Multiple Access (W-CDMA) channel used in UMTS has a bandwidth of 5 MHz. Furthermore, instead of assigning timeslots, all users communicate with the base station at the same time but using a different code. These codes are also referred to as spreading codes, as a single bit is represented on the air interface by a codeword. Each binary unit of the codeword is referred to as a chip to clearly distinguish it from a bit. For a transmission speed of 384 kbit/s, a single bit is encoded in eight chips. The length of the spreading code thus equals 8. A voice call with a typical data rate of 12.2 kbit/s uses a code with a spreading factor of 128. The codes used by different users for transmitting simultaneously are mathematically orthogonal to each other. The base station can separate several simultaneous transmissions from mobile devices by applying the reverse algorithm to the one used by the mobile devices, as shown in Figure 2.4, and by knowing the codes used by each mobile device.

To accommodate the different bandwidth requirements of different users, the system can use several different spreading code lengths simultaneously. For voice calls, which require a transmission rate of only 12.2 kbit/s, a spreading code length of 128 is used. This means that each bit is represented on the air interface by 128 chips and that up to 128 users can send their data streams simultaneously under ideal conditions. In practice, the number of

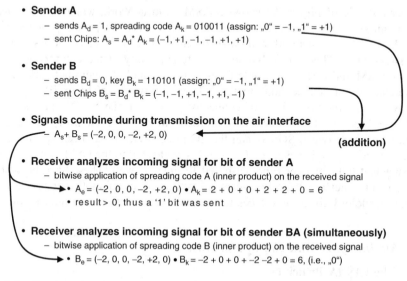

Figure 2.4 Simultaneous data streams of two users to a single base station. (Reproduced from *Communication Systems for the Mobile Information Society*, Martin Sauter, 2006, John Wiley & Sons, Ltd, Ref. [1].)

voice calls per cell is smaller, due to interference from neighboring cells and transmission errors due to nonideal signal conditions. This requires a higher signal-to-noise ratio than could be achieved if there were 128 simultaneous transmissions. Furthermore, some of the codes are reserved for broadcasting system information channels to all devices in the cell, as will be discussed in more detail later. In practice, it is estimated in [8] that each UMTS cell can host up to 60 simultaneous voice calls, excluding subscribers using the cell for packet-switched Internet access.

To increase the data rate of UMTS beyond 384 kbit/s, HSPA and HSPA+ introduce, among other enhancements, the simultaneous use of more than one spreading code per mobile device. This is discussed in more detail below.

2.2.3.2   UMTS Channel Structure

In wired Ethernet networks, which are commonly used in home and business environments today, detecting the network and sending packets is straightforward for devices. As soon as the network cable is plugged in, the network card senses the transmission of packets over the cable and can start transmitting as well, as soon as it detects that no other device is sending a packet. In wireless networks, however, things are more complicated. First of all, there are usually several networks visible at the same time so the device needs to get information about which network belongs to which operator. Also, devices need to detect neighboring cells to be able to quickly react to changing signal levels when the user moves. As keeping the receiver switched on at all times is not very power efficient, it is also necessary to have a mechanism in place that allows the device to power down the radio in situations where little or no data is transmitted and only wake up periodically to check for new data. Especially for voice calls, it is furthermore important to ensure a certain QoS in order to prevent the network from breaking down in overload situations. For these reasons, the radio channel is split up into a number of individual channels. Access to these channels is controlled by the network and can be denied during overload situations.

In UMTS networks, a physical channel is represented by a certain spreading code. The following list gives an overview of the most important channels in the downlink direction (network to mobile device) and the uplink direction (mobile device to the network). All of these channels are transmitted simultaneously:

- **The Primary Common Control Physical Channel (P-CCPCH)** — this channel carries the logical broadcast control channel (BCCH) which is monitored by all mobile devices while they do not have an active connection established to the network. Information distributed via this channel includes the identity of the cell, how the network can be accessed, which spreading codes are used for other channels in the cell, which codes are used by neighboring cells, timers for network access, and so on.
- **The Secondary Common Control Physical Channel (S-CCPCH)** — this physical channel serves several purposes and thus carries a number of different logical channels. First, it carries paging messages that are used by the network to search for mobile devices for incoming calls, SMS messages, or when data packets arrive after a long period of inactivity. The channel is also used to deliver IP packets and control messages to devices which only exchange small quantities of data at a particular time and thus do not need to be put into a more active state.

- **Physical Random Access Channel (PRACH)** — the random access channel only exists in the uplink direction and is the only channel that a mobile device is allowed to transmit without prior permission of the network. Its purpose is to allow the mobile to request the establishment of a connection to set up a voice call, to establish a data call, to react to a paging message, or to send an SMS message. If a data call is already established and the mobile only sends or receives a small amount of data, the channel can also be used to send IP packets in combination with the S-CCPCH in the downlink direction. However, the data rate in both directions is very low and round trip delay times are high (in the order of 200 ms). The network therefore usually assigns different channels that are dedicated to packet-switched data transfers as soon as it detects renewed network activity from a device.
- **Dedicated Physical Data Channel and Dedicated Physical Control Channel (DPDCH, DPCCH)** — once a mobile has contacted the network via the random access channel, the network usually decides to establish a full connection to the mobile. In case of a voice call, the network assigns a dedicated connection. On the air interface, the connection uses a dedicated spreading code that is assigned by the network. In the early days, UMTS networks also used dedicated connections for packet-switched data transmissions. In practice, however, the speed of such dedicated connections was limited to 384 kbit/s. Only a few users could get such a connection, also referred to as a bearer, in a cell simultaneously. Another downside of using a dedicated bearer for packet-switched connections is that the network has to frequently reassign spreading codes to different users. This is necessary, as most packet-switched data transmissions are bursty in nature and therefore only require a high bandwidth bearer for a limited amount of time. In order to reuse the spreading code for other users, the network needs to change the spreading code length of a connection as soon as it is not fully used any more or even put the connection on a common control channel in order to have the resources available for other users. In practice, this creates a high signaling overhead and a mediocre user experience. It was thus decided that a new concept was needed that offers higher data rates and more flexibility for bursty data transmissions. The outcome of this process is known today as High-speed Packet Access, which many people also refer to as 3.5G. HSPA will be discussed in Section 2.2.4.

2.2.3.3 Radio Resource Control States

To save power and to only assign resources to mobile devices when necessary, a connection to the network can be in one of the following Radio Resource Control (RRC) states:

- **Idle state** — devices not actively communicating with the network are in this state. Here, they periodically listen to the paging channel for incoming voice or video calls and SMS messages.
- **Cell-FACH (Forward Access Channel) state** — if a mobile device wants to contact the network, it moves to the Cell-FACH state. In this state the mobile sends its control messages via the random access channel and the network replies on the FACH, which is sent over the S-CCPCH described earlier. Power requirements also increase as the mobile device now needs to monitor the downlink for incoming control messages.

- **Cell-DCH (Dedicated Channel) state** — once the network decides to establish a voice or data connection, the mobile is instructed to use a dedicated channel and is therefore moved to the Cell-DCH State. If the device is HSPA-capable, the network selects a shared channel instead, as will be discussed in more detail below. From a RRC point of view, however, the mobile is still treated as being in Cell-DCH state. Round trip delay times in Cell-DCH state range from 160 ms with a dedicated bearer to 60 ms for an HSPA+ connection. In the case of a packet-switched data connection, the network can decide to move a device back to Cell-FACH state if its activity, that is, the amount of data transferred, decreases. Once activity increases again, the network moves the connection back to Cell-DCH state. For even longer periods of inactivity during a data session, the network can even decide to put the mobile device back into idle state to reduce power consumption of the mobile device and to reduce the management processing load for the network. Despite being in Idle state, the mobile device can resume sending IP packets at any time via the random access channel. However, there is a noticeable delay when moving to a more active state. Depending on the radio network vendor, initial round trip delay times between 2 and 4 seconds can be observed [1].
- **Cell-PCH (Physical Channel) and URA-PCH (UTRAN Registration Area-Physical Channel) states** — even while in Cell-FACH state, the mobile device requires a considerable amount of power to listen to the FACH despite its minimal activity. Because of the noticeable delay when resuming data transmission from the Idle state, two further states have been specified which are between Idle and Cell-FACH state. In Cell-PCH and URA-PCH state, the mobile only needs to periodically listen to the paging channel while the logical connection between the radio network and the device remains in place. If there is renewed activity, the connection can be quickly resumed. The difference between the two states is that in Cell-PCH state the mobile has to report cell changes to the network because the paging message is only sent into one cell, whereas in URA-PCH state the mobile can roam between several cells that belong to the same routing area without reporting a cell change to the network since the paging message is sent into all cells belonging to the routing area. While in the early years of UMTS, little use was made of Cell-PCH and URA-PCH, these states have become more popular recently. Not only is the number of signaling messages reduced in the network to switch the mobile device back into Cell-DCH state when necessary but with around 700 ms delay, the procedure takes far less time than moving to the fully active state from the idle state.

2.2.3.4 Mobility Management in the Radio Network for Dedicated Connections

An important task in cellular networks is to handle the mobility of the user. In Cell-DCH state, the base station actively monitors the air interface connection to and from a mobile device and adjusts the transmission power of the base station and the transmission power of the mobile device 1500 times per second. In CDMA systems, such very quick power adjustments are necessary, as the transmissions of all mobile devices should be received by the base station with the same power level whether they are very close or very far away. In the direction from the network to a mobile device, as little transmission power as possible should be used as the transmission power of the base station is limited. In practice, data transfers to and from mobile devices close to a base station only require a small amount of transmission power while devices far away or inside buildings require significantly more transmission power. Adjusting the transmission power so frequently also means that, along with the user data, the mobile device has to continuously report to the network how well the user data is received. The network then processes this information and instructs the mobile device, at the same rate, whether it should increase, hold, or decrease its current transmission power. For this purpose, a dedicated control channel (DCCH) is always established alongside a dedicated traffic channel (DTCH). While the content of the DTCH is given to higher layers of the protocol stack, which eventually extract a voice packet or an IP packet, the information transported in the control channel remains in the radio software stack and is not visible to higher-layer applications.

When users move, they eventually leave the coverage area of a base station. While a DCH is established, the RNC takes care that the connection to the mobile device is transferred to a more suitable cell. This decision is based on reception quality of the current cell and the neighboring cells, which is measured by the mobile device and sent to the network. In CDMA-based networks such as UMTS, there are two different kinds of handovers. The first type is referred to as a hard handover. When the network detects that there is a more suitable cell for a mobile device, it prepares the new cell for the subscriber and afterwards instructs the mobile to change to the new cell. The mobile device then interrupts the connection to the current cell and uses the handover parameters sent by the network, which include information on the frequency and spreading codes of the new cell, for establishing a connection. A hard handover is therefore a break-before-make handover, that is, the old connection is cut before the new one is established.

The second type of handover is a soft handover, which is the most commonly used type of handover in CDMA networks. Soft handovers make use of the fact that neighboring cells transmit on the same frequency as the current cell. It is thus possible to perform a make-before-break handover, that is, the mobile device communicates with more than a single cell at a time. A mobile enters the soft handover state when the network sends control information to instruct it to listen to more than a single spreading code in the downlink direction. Each spreading code represents a transmission of a different cell and the mobile combines the signals it receives from the different cells. In the uplink direction, the mobile device continues using only a single spreading code for a dedicated channel. All cells taking part in the soft handover are instructed by the RNC to decode the data sent with this code and forward it to the RNC. During soft handover, the RNC thus receives several copies of the user's uplink data stream. This increases the likelihood that a packet is received correctly and does not have to be retransmitted. The cells which are involved in a soft handover for a dedicated channel are referred to as the active set. Up to six cells can be part of the active set. In practice, however, most networks limit the number of cells in the active set to two or three. Otherwise, the benefit of multiple transmissions that can be combined in both the network and the mobile station is outweighed by the additional capacity required in both the radio network and on the air interface. To improve the capacity of the network, around 30–40% of all dedicated connections in a network are in soft handover state, even if the users are not moving.

Figure 2.5 shows how hard and soft handovers are performed for a case in which not all cells are connected to the same RNC. In this example, the uplink and downlink

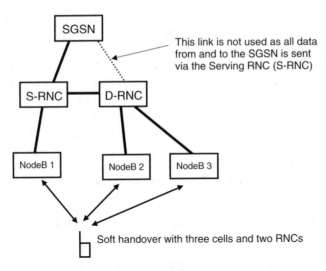

Figure 2.5 Soft handover with several RNCs. (Reproduced from *Communication Systems for the Mobile Information Society*, Martin Sauter, 2006, John Wiley & Sons, Ltd, Ref. [1].)

data are collected and distributed by two RNCs. However, only one of the two RNCs, referred to as the Serving-Radio Network Controller (S-RNC), communicates with the core network. The other RNC, referred to as the Drift-Radio Network Controller (D-RNC), communicates with the S-RNC.

If the user moves further into the area covered by NodeB 2 and NodeB 3, the mobile will at some point lose contact with NodeB 1. At this point, the involvement of the RNC controlling NodeB 1 is no longer required. The current S-RNC then requests the SGSN to promote the current D-RNC as the new S-RNC.

If the optional interface between the RNCs is not present, a soft handover as shown in Figure 2.5 is not possible. In this case, a hard handover is required when the mobile device moves out of the coverage area of the first NodeB and into the coverage area of the other two NodeB. Furthermore, the change of S-RNC has to be performed at the same time, which further complicates the handover procedure and potentially increases the period during which no data can be exchanged between the mobile device and the network. When the user moves between the areas controlled by two different SGSNs, it is additionally necessary to change the serving SGSN. To make matters even more complicated, handovers can also take place while both the circuit-switched and a packet-switched connections are active. In such a case, not only the RNCs need to be changed, but also the SGSN and MSC. This requires close coordination between the packet-switched and circuit-switched core network. While in practice typical handover scenarios only involve cells controlled by the same RNC, all other types of handovers have to be implemented as well.

Since 3G/3.5G networks did not have the same coverage area as the already existing 2G GSM networks, and often still lack the same coverage area and indoor penetration today, it is very important to be able to handover active connections to GSM networks. This was and still is especially important for voice calls. Mobile devices, therefore, do not only search for neighboring 3G/3.5G cells but can also be instructed by the network to search for GSM cells. While a dedicated channel is established this is very difficult since in the standard transmission mode the mobile device sends and receives continuously. For cells at the edge of the coverage area, the RNC can activate a compressed transmission mode. The compressed mode opens up predefined transmission gaps that allow the mobile device's receiver to retune to the frequencies of neighboring GSM cells, receive their signal, synchronize to them and return to the UMTS cell to resume communication and to report the result of the neighboring cell signal measurements to the network. Based on these measurements, the network can then instruct the mobile to perform an Inter-Radio Access Technology (Inter-RAT) hard handover to GSM. During an ongoing voice call, such a hard handover is usually not noticed by the user. A data connection handed over to a GSM/GPRS/EDGE network, however, gets noticeably slower.

2.2.3.5   Radio Network Mobility Management in Idle, Cell-FACH, and Cell/URA-PCH State

When the user moves and the mobile device is in Idle or Cell/URA-PCH state, that is, no voice call is established and the physical connection of a data call has been removed after a phase of inactivity, the mobile device can decide on its own to move to a cell with a better signal.

Cells are grouped into location areas and mobile devices in the Idle state changing to a different cell only need to contact the network to report their new position when selecting a cell in a new location area. This means that paging messages have to be broadcasted in all cells belonging to a location area. In practice, a location area contains about 20–30 base stations, providing a satisfactory trade-off between the reduced number of location updates and the increased overhead of sending paging messages into all cells of a location area.

A mobile device in Cell-PCH state reports every cell change to the network with a cell update message. This way, the mobile device only needs to be paged in a single cell. A compromise between the need for paging in several cells and the number of cell update messages that have to be sent is the URA-PCH state. Here, a number of cells are grouped into a URA, which can be smaller than a location area. Mobile devices only have to send a cell update message if they cross such a boundary.

In Cell-FACH state, which is used if the mobile enters a phase of lower activity (i.e., it sends or receives only a few IP packets), the cell changes are also controlled by the mobile device and not the network. While the connection is still fully active and (slow) uplink and downlink transmissions can occur at any time, an efficient transition from one cell to another is deemed to be not important enough to justify the required processing capacity in the network and the higher power consumption in the mobile device. As data transmissions can occur any time, a cell update message is required when moving from one cell to another.

2.2.3.6   Mobility Management in the Packet-Switched Core Network

In the core network, the SGSN is responsible for keeping track of the location and reachability of a mobile device. In PMM (Packet Mobility Management) Detached state, the mobile is not registered in the network and consequently does not have an IP address.

In PMM Connected state the mobile has a signaling connection to the SGSN (e.g., during a location update to report its new position) and also an IP connection. The SGSN, however, only knows the current RNC responsible for forwarding the packet and not the cell identity. This knowledge is not necessary, as the SGSN just forwards the data packet to the RNC and the RNC is responsible for forwarding it via the current cell to the user. In addition the SGSN does not know if the radio connection to the mobile is in Cell-FACH, Cell-DCH, Cell-PCH, or URA-PCH state. This information is not necessary since it is the RNC's task to decide in which state to keep the mobile based on the QoS requirements of a connection and the current amount of transmitted data. Note that PMM Connected state is also entered for short periods of time during signaling exchanges such as a location update even if no IP connection is established.

If the RNC sets the radio connection to Idle state, it also informs the SGSN that the mobile is no longer directly reachable. The SGSN then modifies its control state to PMM Idle. For the SGSN this means that, if an IP packet arrives from the Internet later on, the mobile device has to be paged first. The IP packet is then buffered until a response is received and a signaling connection has been set up.

Figure 2.6 shows the different radio and core network states and how they are related to each other while a packet-switched data connection is established. When moving from left to right in the figure, it can be seen that, the deeper the node is in the network, the less information it has about the current mobility state and the radio network connection of the mobile device. Note that the state of a mobile device is also PMM Idle when the mobile has attached to the packet-switched core network after power on but has not yet established a data connection. The radio network mobility management states also apply when the mobile device communicates with the circuit-switched network. During a voice or video call via the MSC, the mobile device is in Cell-DCH state. The mobile is also set into Cell-DCH or Cell-FACH state during signaling message exchanges, for example, during a location update with the MSC and SGSN.

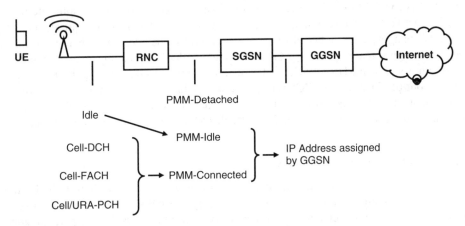

Figure 2.6 Radio network and core network mobility management states for an active packet-switched connection.

2.2.4   HSPA (HSDPA and HSUPA)

Soon after the first deployments of UMTS networks, it became apparent that the use of dedicated channels on the air interface for packet-switched data transmission was too inflexible in a number of ways. As per the specification, the highest data rate that could be achieved was 384 kbit/s with a spreading code length of 8. This limited the number of people who could use such a bearer simultaneously to eight in theory and to two or three in practice, as some spreading codes were required for the BCHs of the cell and for voice calls of other subscribers. Because of the bursty nature of many packet-switched applications, the bearer could seldom be fully used by a mobile device and therefore a lot of capacity remained unused. This was countered somewhat by first assigning longer code lengths to the user's connection and then only upgrading the connection when it was detected that the bearer was fully used for some time (e.g., for a file download). Also, short spreading codes were quickly replaced by longer ones once it became apparent to the RNC that the capacity was no longer fully used. Despite these mechanisms, efficiency remained rather low. As a consequence, 3GPP started to specify a package of enhancements which are now referred to as High-Speed Downlink Packet Access (HSDPA) [9]. The specification of HSDPA started as early as 2001 but it took until late 2006 for the first networks and devices to support HSDPA in practice. Once standardization of HSDPA was a good way on, a number of enhancements were specified for the uplink direction as well. These are referred to as High-Speed Uplink Packet Access (HSUPA). Together, HSDPA and HSUPA are known today as HSPA.

The following paragraphs are structured as follows: first, an overview of the downlink enhancements is given. Afterwards, the performance improvements for the uplink direction are discussed.

2.2.4.1   Shared Channels

To improve the utilization of the air interface, the concept of shared channels has been (re-) introduced. HSDPA introduces a High-Speed Downlink Physical Shared Channel (HS-DPSCH), which several devices can use quasi simultaneously as it is split into timeslots. The network can thus quickly react to the changing bandwidth requirements of a device by changing the number of timeslots assigned to a device. Thus, there is no longer a need to assign spreading codes or to modify the length of a spreading code to change the bandwidth assignment. HS-DPSCH timeslots are assigned to devices via the High-Speed Shared Control Channel (HS-SCCH). This channel has to be monitored by all mobile devices that the RNC has instructed to use a shared channel. The HS-DPSCH uses a fixed spreading code length of 16 and the HS-SCCH uses a spreading code length of 256.

2.2.4.2   Multiple Spreading Codes

To increase transmission speeds, mobile devices can listen to more than a single high-speed downlink shared channel. While older category 6 HSDPA devices can listen to up to five high-speed downlink channels simultaneously, current devices that support category 9 or higher can receive and process up to 15 channels simultaneously. Multiple high-speed downlink channels can also be used to transmit data to several devices simultaneously.

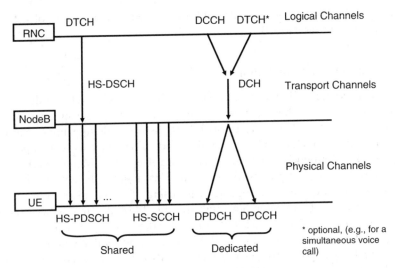

Figure 2.7 Simplified HSDPA channel overview in the downlink direction. (Reproduced from *Communication Systems for the Mobile Information Society*, Martin Sauter, 2006, John Wiley & Sons, Ltd, Ref. [1].)

For this purpose the network can configure up to four simultaneous shared control channels in the downlink direction and up to four mobile devices can therefore receive data on one or more shared channels simultaneously.

Figure 2.7 shows how user data is transferred through the radio network and which channels an HSDPA device has to decode simultaneously while it is in Cell-DCH state. At the RNC, the user's data packets arrive from the core network in a DTCH. From there, the RNC repackages the data into a single High-speed Dedicated Shared Channel (DSCH) and forwards it to the NodeB. In addition, the RNC sends control information to the device via a DCCH. This channel is not shared between subscribers.

It is also possible for the user to establish or receive a voice call during an ongoing HSDPA data transfer. In this case, the RNC additionally uses a DTCH to forward the circuit-switched voice channel to the user.

At the base station (NodeB), data arriving via the HS-DSCH channel is buffered for a short while and then transmitted over one or several High-speed Physical Downlink Shared Channels (HS-PDSCHs). In addition to those 5–15 shared channels, the device has to be able to monitor up to four high-speed downlink shared control channels to receive the timeslot assignments for the shared channels. Finally, the mobile has to decode control information arriving on the DPCCH and potentially also the voice data stream arriving on the DPDCH.

In the uplink direction, an HSDPA-capable device also has to send a number of different data streams, each with a different spreading code. The first stream is the HSDPA control channel, which is used to acknowledge the correct reception of downlink data packets. In addition, a dedicated channel is required for IP packets transmitted in uplink direction. A further channel is necessary to transmit or to reply to radio resource commands from the RNC (e.g., a cell change command) and for the uplink direction of an ongoing voice call.

Finally, the mobile needs to send control information (reception quality of the downlink dedicated channel) to the NodeB.

2.2.4.3 Higher Order Modulation

To further increase transmission speeds, HSDPA introduces a higher order modulation scheme to improve throughput for devices that receive the transmissions of the base station with a good signal quality. Basic UMTS uses Quadrature Phase Shift Keying (QPSK) modulation, which encodes two basic information elements (chips) per transmission step. HSDPA introduces 16QAM (Quadrature Amplitude Modulation) modulation, which encodes four chips per transmission step. This doubles the achievable bandwidth under good reception conditions. Together with using 15 simultaneous downlink shared channels, the theoretical maximum downlink speed is 14 Mbit/s. In practice, transmission speeds range between 1.5 and 2.5 Mbit/s for older devices that can bundle five shared channels (HSDPA category 6) and over 10 Mbit/s for devices that can bundle 15 shared channels (HSDPA category 10). Even higher speeds can be reached with a number of further enhancements as part of HSPA+. These are described in the following.

2.2.4.4 Scheduling, Modulation, and Coding, HARQ

The higher the transmission speed, the more important it is to detect transmission errors and react to them as quickly as possible. Otherwise, connection-oriented higher-layer protocols such as TCP misinterpret air interface packet errors as congestion and slow down the transmission. To react to transmission errors more quickly, it was decided not to implement the HSDPA scheduler in the RNC, as was done before for dedicated channels, but to assign this task to the base station. HSDPA uses the Hybrid Automated Retransmission Request (HARQ) scheme for this purpose. With HARQ, data frames with a fixed length of 2 ms have to be immediately acknowledged by the mobile device. If a negative acknowledgment (ACK) is sent, the packet can be repeated within 10 ms. The next data frame is only sent once the previous one has been positively confirmed. The mobile device must be able to handle up to eight simultaneous HARQ processes as the mobile device is allowed up to 5 ms to decode the packet before it has to send the ACK or the NACK (Negative Acknowledgment) to the network. During this time, two further data frames of other HARQ processes can arrive at the mobile station. In case data was not received correctly in one HARQ process, frames received via other HARQ processes have to be stored in the mobile device until all previous frames have been successfully received, so that the data can be forwarded in the correct order to higher protocol layers.

Another reason to implement the scheduler in the base station is to be able to quickly react to changing signal conditions. Based on the feedback of the mobile device, the base station's scheduler decides for each frame which modulation (QPSK or 16QAM) to use and how many error correction bits to insert. This process is referred to as Adaptive Modulation and Coding (AMC). Advanced algorithms in the base station use the knowledge about reception conditions of all mobile devices currently served on the high-speed shared channel. Devices with better signal conditions can be preferred by the scheduler, while devices that are in temporary deep fading situations receive fewer packets. This helps to reduce transmission errors and improves overall throughput of the cell as, on average, frames are transmitted with more efficient modulation and coding schemes. Studies in [10, 11] have shown that an efficient scheduler can increase overall cell capacity by up to 30% compared with a simple scheduler that assigns timeslots in a round robin fashion.

2.2.4.5 Cell Updates and Handovers

As has been shown earlier, a mobile device in HSDPA Cell-DCH state receives its packet-switched data via shared channels. Nevertheless, additional dedicated channels are required alongside the shared channels to transport control information in both directions. Furthermore, IP packets in the uplink direction are also transported in a dedicated channel. Finally, a parallel circuit-switched voice call is also transported in a dedicated channel. These dedicated connections can be in soft handover state to improve reception conditions and to better cope with the user's mobility. The shared channels, however, are only sent from one base station of a device's active set. Based on the feedback received from the mobile device, the RNC can at any time promote another base station to forward the IP packets in the downlink. This procedure is referred to as a change of the serving cell. Since the procedure is controlled by the network, the interruption of the data traffic in the downlink direction is only very short.

2.2.4.6 HSUPA

In early 3G networks, uplink speeds were limited to 64–128 kbit/s. With the introduction of HSDPA, some vendors also included a radio network software update to allow dedicated uplink radio bearers with speeds of up to 384 kbit/s, if permitted by the conditions on the radio interface to a user. In practice, it can be observed that in many situations the mobile's transmission power is sufficient to actually make use of such an uplink bearer. This is good for sending large files or e-mails with large file attachments, as well as for many Web 2.0 applications (cf. Chapter 6), where user-generated content is sent from a mobile device to a database in the network. As for dedicated downlink transmissions, however, assigning resources in the uplink in this way is not very flexible. The 3GPP standards body thus devised a number of improvements for the uplink direction, which are referred to as High-Speed Uplink Packet Access.

Unlike in the downlink, which was enhanced by using high-speed shared channels, the companies represented in 3GPP decided to keep the dedicated channel approach for uplink transmissions. There were several reasons for this decision. While in the downlink, the base station has an overview of how much data is in the buffers to be sent to all mobile devices, there is no knowledge in the base station about the buffer states of the mobile devices requesting to send data in the uplink. This makes assigning timeslots on a shared channel difficult as mobile devices continuously have to indicate if they have more data. The second reason to re-use the dedicated channel concept was that the soft handover concept is especially valuable for the uplink direction as the transmission power of mobile devices is limited. In the standards, an HSUPA dedicated uplink channel is referred to as an Enhanced-Dedicated Channel or E-DCH.

A standard uplink dedicated channel is controlled by the RNC and spreading codes can only be changed to reach maximum data transfer rates of 64, 128, or 384 kbit/s. As the

RNC is in control of the connection, the bearer parameters are only changed very slowly, for example, only every few seconds once the RNC detects that the pipe is too big or too small for the current traffic load or once it detects that signal conditions have significantly changed. With the E-DCH concept, the control of the radio interface channel has been moved from the RNC to the base station, in a similar way to the solution for HSDPA. The base station controls overall scheduling of all E-DCH uplink transmissions by assigning bandwidth grants to all active E-DCH devices. Bandwidth grants are transmitted in the downlink direction via a new shared control channel, the Enhanced Absolute Grant Channel (E-AGCH). Access grants are translated into the maximum amount of transmission power each device is allowed to use. This way, all devices can still transmit their data to the base station at the same time while their maximum transmission speeds can be quickly adjusted as necessary. Staying with the dedicated channel concept and hence allowing all mobile devices to transmit at the same time has the additional benefit of shorter packet delay times compared with an approach where devices have to wait for their timeslot to transmit. Optionally, a second control channel, the Enhanced Relative Grant Channel (E-RGCH) can be used to quickly increase or decrease uplink transmission power step by step. This enables each base station involved in an E-DCH soft handover to decrease the interference caused by a mobile device, if this creates too much interference for communication with other devices.

The HARQ concept first introduced with HSDPA is also used with HSUPA for uplink transmissions. This means that the base station immediately acknowledges each data frame it receives. Faulty frames can thus be retransmitted very quickly. For this purpose, an additional dedicated downlink channel has been created, the Enhanced HARQ information channel (E-HICH). Each HSUPA device gets its own E-HICH while it is in Cell-DCH mode. If an E-DCH connection is in soft handover state, each of the contributing base stations sends its own ACK for a frame to the mobile device. If only one ACK is positive, the transfer is seen as successful and the HARQ process moves on to the next frame.

While it was decided to introduce a new modulation scheme for the downlink to further increase transmission speeds, it was shown with simulations that there would be no such effect for uplink transmissions. Uplink transmissions are usually power limited, which means that a higher order modulation cannot be used as the error rate would become too high. Instead, it was decided to introduce multicode transmissions in a single dedicated channel to allow the mobile to split its data into several code channels which are then transported simultaneously. This concept is similar to the concept of using several (shared) channels in the downlink to increase the data rates.

The highest terminal category defined in 3GPP Release 6 when E-DCH was first specified could support data rates of up to 5.76 Mbit/s on the physical layer. In practice, data rates are lower in a similar way to those described earlier for HSDPA. Many networks still only support up to 2 Mbit/s in the uplink direction (E-DCH category 5), but it can be observed that some networks have now been upgraded to also support E-DCH category 6 with data rates in practice of around 4 Mbit/s.

Figures 2.8 and 2.9 show the channels used in uplink and downlink direction for a terminal that supports both HSDPA and HSUPA for the most complicated case when a voice call is ongoing in parallel. In practice, encoding and decoding so many channels at the same time is very demanding and has only been made possible by the ever increasing processing power of mobile device chipsets, as is further discussed in Chapter 5.

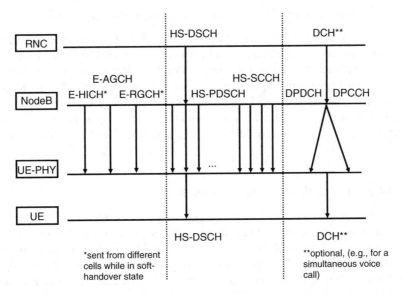

Figure 2.8 Channels for a combined HSDPA and HSUPA transmission in the downlink direction. (Reproduced from *Communication Systems for the Mobile Information Society*, Martin Sauter, 2006, John Wiley & Sons, Ltd, Ref. [1].)

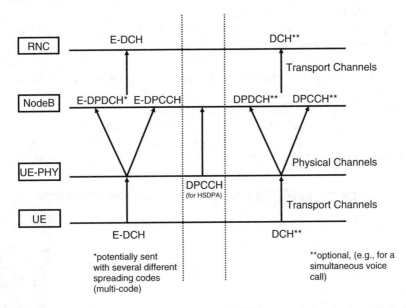

Figure 2.9 Channels for a combined HSDPA and HSUPA transmission in the uplink direction. (Reproduced from *Communication Systems for the Mobile Information Society*, Martin Sauter, 2006, John Wiley & Sons, Ltd, Ref. [1].)

2.2.5   HSPA+ and other Improvements: Competition for LTE

The quest to improve UMTS and make it even faster and more power-efficient and to allow more devices to use a cell simultaneously has not ended with HSPA. For instance, there are a number of initiatives in Release 7 and beyond of the 3GPP standard to further improve the system. Improvements of the air interface are referred to as HSPA+. In addition, the network architecture has also received an optional overhaul to be more efficient with a feature referred to as "one-tunnel." The following sections give an overview of these improvements. Some of them have already been introduced in life networks, which are mentioned accordingly.

2.2.5.1   64QAM Modulation in the Downlink

One of the key parameters of a wireless system that is often cited in articles is the maximum transmission speed. HSPA up to 3GPP Release 6 uses QPSK and 16QAM modulation to reach theoretical data rates of up to 14.4 Mbit/s in the downlink direction, which translates to 10–11 Mbit/s in practice today under very good radio conditions. To increase transmission rates further, 3GPP Release 7 introduces 64QAM modulation in the downlink, which transmits six chips per transmission step compared with four chips with 16QAM.

At its inception it was expected that introducing 64QAM will result in a 30% throughput gain for users close to the center of the cell [12] and a theoretical peak data rate of 21.1 Mbit/s. This has been confirmed and even exceeded in practice with achievable speeds of around 14–16 Mbit/s measured with an HSDPA category 14 device in a life-network-supporting 64QAM modulation and under very good signal conditions [13]. At these speeds, not every web server is capable of delivering data this fast. For speed testing, it is thus sometimes required to download several files from different servers simultaneously. Figure 2.10 shows the download speeds when receiving several files at once with an aggregate throughput of around 16 Mbit/s.

Most users, however, will not be able to use 64QAM modulation, as their signal-to-interference ratio will be too low. This will be discussed in more detail in Chapter 3. As the overall throughput in the cell increases, these users will nevertheless also benefit

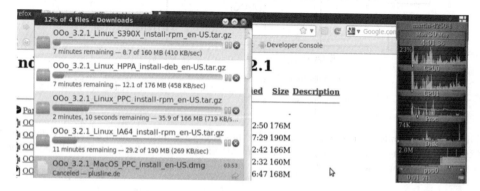

Figure 2.10   16 Mbit/s data throughput with a HSPA+ category 14 device in a life network.

from this measure, as the cell will have more time to transmit their data, because data for users closer to the cell center can be sent faster. Furthermore, 64QAM modulation is also beneficial for micro cell deployments in public places such as shopping malls, where users are close to small cells and therefore create little interference.

2.2.5.2 Dual Carrier Operation in the Downlink

Another method to further improve per-user data rates is to bundle two adjacent 5-MHz UMTS carriers. This is referred to as dual carrier operation. In combination with 64QAM modulation, the theoretical data rate that can be achieved is 42 Mbit/s with HSDPA category 20 devices. In practice, speeds of beyond 20 Mbit/s can be observed in loaded cells during daytime and peak data rates of beyond 32 Mbit/s have been observed in life networks in unloaded cells with little neighboring cell interference [14].

While such high data rates in practice are impressive, there are additional advantages of dual carrier operation, which are perhaps even more important. When network operators initially began to deploy a second carrier to increase the capacity in cities, one could observe in practice that many were using the first 5-MHz carrier for voice and the second carrier for HSPA. While HSPA traffic thus became independent of the voice traffic in the area, a lot of capacity of the first carrier was unused. With dual carrier operation, network operators are now using both carriers for HSPA, which increases the overall capacity and individual user data rates.

In cell edge scenarios when a user is at a coverage edge or in the middle between two cells, data rates are much lower compared to the peak data rates offered by the system as the signal level is low or because there is significant interference from neighboring cells. While dual carrier operation cannot counter those effects directly, user data rates in such scenarios double as two carriers are used to transmit the data to the user. So it is in these scenarios, where most users will notice a significant difference to single-carrier operation.

2.2.5.3 MIMO

Another emerging technique to increase throughput under good signal conditions is Multiple Input Multiple Output, or MIMO for short. In essence, MIMO transmission uses two or more antennas at both the transmitter and the receiver side to transmit two independent data streams simultaneously over the same frequency band. This linearly increases data rates with the number of antennas. Two transmitter antennas and two receiver antennas (2×2) as currently specified for HSPA+ can thus double the data rate of the system under ideal signal conditions. Further technical background on MIMO can be found in Section 2.3. Release 7 of the 3GPP standards has combined the use of MIMO and 64QAM modulation in the downlink and Release 9 combines MIMO, 64QAM modulation, and multicarrier operation for a theoretical peak data rate of 84 Mbit/s with a category 28 device.

Depending on the signal conditions, available antennas and device capabilities, the network can choose between a single data stream with 64QAM or two streams with 16QAM. Release 8 combines MIMO with 64QAM, which results in a peak data rate of 43.2 Mbit/s in the downlink. However, it should be once more noted at this point that such data rates can only be achieved by very few users of a cell. Details are given in Chapter 3.

As uplink transmissions are usually power limited, MIMO has only been considered for the downlink direction.

2.2.5.4   16QAM Modulation and Dual-Carrier Operation in the Uplink

In the uplink direction, no changes were made to the modulation scheme in 3GPP Release 6. Despite earlier predictions that uplink transmissions would not benefit from a higher modulation and coding scheme, 3GPP has reconsidered its position and Release 7 then introduced 16QAM. More advanced transceivers and a focus on microcell environments in places with a high-data throughput demand may be responsible for this change in opinion. E-DCH category 7 devices are specified for an uplink speed of up to 11.5 Mbit/s. Details can be found in 3GPP TS 25.306, figure 5.1g [15]. In 3GPP Release 8, uplink transmission capabilities were further enhanced by the introduction of dual-carrier operation, which bundles two adjacent 5-MHz carriers. As only QPSK was specified as modulation scheme in combination with carrier aggregation (CA), uplink speeds remained at 11.5 Mbit/s. 3GPP Release 9 then went one step further and combined dual-carrier uplink operation with 16QAM modulation. E-DCH category 9 devices are therefore specified for speeds of up to 23 Mbit/s in the uplink direction.

At the time of publication, these enhancements were not yet widely deployed. It can be expected, however, that over the next few years, these new features will be introduced as the benefits are likely to justify the additional expense in development and deployment of the functionality. Furthermore, uplink data rates in life LTE networks are significantly higher than those of current UMTS networks because of higher order modulation and larger channel bandwidth compared to a single 5-MHz UMTS channel.

2.2.5.5   Multicarrier and Multiband Operation

To further increase the downlink data rates, it is necessary to combine even more 5 MHz channels to a single mobile device. While some network operators have more than two consecutive channels available in a single band, many network operators are restricted to merely two. As a consequence, a number of enhancements were specified to combine nonadjacent channels in different frequency bands to form a single downlink channel.

3GPP Release 9 introduces dual-carrier HSDPA on different frequency bands. In Europe, for example, some network operators have spectrum in the 2100 MHz band for UMTS and have also started to use some of the 900 MHz spectrum, which is previously only used by 2G networks for UMTS. At the time of publication, a mobile device can either use the channel on the 900 MHz band or the channel on the 2100 MHz band but not both simultaneously. This enhancement defines on how the two channels can be used simultaneously in a way almost identical to the already existing dual-carrier operation of two adjacent channels in the same band. While this measure does not increase the theoretical maximum speed of the dual-carrier operation mode already used in life networks today, it offers the interesting possibility to use two channels in bands with very different propagation characteristics simultaneously. With intelligent schedulers in the base station, this can increase the average transmission speeds as fading is band dependent. Also, it gives the base station scheduler further options for users moving deeper indoors where a 900 MHz channel is less affected by walls and objects than a channel on the 2100 MHz band.

To further the increase data rates, 3GPP Release 10 introduces triple-channel and quad-channel operation in up to two different frequency bands [16]. In each band, all channels need to be adjacent. With 64QAM modulation peak data rates rise from the previous 42.2 to 63.3 and 84.4 Mbit/s, respectively. If 64QAM is used in combination with MIMO, data rates rise to 126.6 and 168.8 Mbit/s, respectively.

3GPP Release 11 then makes the next steps and defines methods to combine up to eight 5 MHz carriers in two different frequency bands for an aggregate peak data rate of 337.5 Mbit/s if 64QAM and MIMO are used on all channels. For further flexibility, the specification foresees that MIMO can be used selectively on some channels but not on others.

In summary, it should be noted that while the list of supporters of these features in 3GPP includes major network operators and manufacturers, it is not yet certain if network operators will actually attempt to use UMTS on more than two or three channels. Today, few, if any, network operators have the necessary spectrum on two frequency bands to combine more than three or four 5 MHz channels. New spectrum has become available in recent years in different bands, but it is very likely that most network operators will use this spectrum for LTE.

2.2.5.6 Continuous Packet Connectivity

Continuous Packet Connectivity (CPC) is a package of features introduced in the 3GPP standards to improve handling of mobile subscribers while they have a packet connection established, that is, while they have an IP address assigned. Taken together, they aim at reducing the number of state changes to minimize delay and signaling overhead by introducing enhancements to keep a device on the high-speed channels (in HSPA Cell-DCH state) for as long as possible, even while no data transfer is ongoing. For this, it is necessary to reduce power consumption while mobiles listen to the shared channels and, at the same time, reduce the bandwidth requirements for radio layer signaling to increase the number of mobile devices per cell that can be held in HSPA Cell-DCH state.

CPC does not introduce new revolutionary features. Instead, already existing features are modified to achieve the desired results. To understand how these enhancements work, it is necessary to dig a bit deeper into the standards. 3GPP TR 25.903 [17] gives an overview of the proposed changes and the following descriptions refer to the chapters in the document which have been selected for implementation.

Feature 1: A New Uplink Control Channel Slot Format (Section 4.1 of [17])
While a connection is established between the network and a mobile device, several channels are used simultaneously. This is necessary as there is not only user data sent over the connection but also control information to keep the link established, to control transmit power, and so on. Currently, the radio control channel in the uplink direction (the Uplink Dedicated Control Channel, UL DPCCH) is transmitted continuously, even during times of inactivity, in order not to lose synchronization. This way, the terminal can resume uplink transmissions without delay whenever required.

The control channel carries four parameters:

- Transmit Power Control (TPC);
- pilot (used for channel estimation of the receiver);
- TFCI (Transport Format Combination Identifier);
- FBI (Feedback Indicator).

The pilot bits are always the same and allow the receiver to get a channel estimate before decoding user data frames. While no user data frames are received, however, the pilot bits are of little importance. What remains important is the TPC. The idea behind the new slot format is to increase the number of bits to encode the TPC and decrease the number of pilot bits while the uplink channel is idle. This way, additional redundancy is added to the TPC field. As a consequence, the transmission power for the control channel can be lowered without risking corruption of the information contained in the TPC. Once user data transmission resumes, the standard slot format is used again and the transmission power used for the control channel is increased again.

Feature 2: CQI Reporting Reduction (Section 4.4 of [17]), Uplink Discontinuous Transmission (Section 4.2 of [11]) in Combination with Downlink Control Information Transmission Enhancements

- **CQI reporting reduction:** to make the best use of the current signal conditions in the downlink, the mobile has to report to the network how well its transmissions are received. The quality of the signal is reported to the network with the Channel Quality Indicator (CQI) alongside the user data in the uplink. To reduce the transmit power of the terminal while data is being transferred in the uplink but not in the downlink, this feature reduces the number of CQI reports.
- **UL HS-DPCCH gating (gating = switch off):** when no data is being transmitted in either the uplink or downlink, the uplink control channel (UL DPCCH) for HSDPA is switched off. Periodically, it is switched on for a short time to transmit bursts to the network in order to maintain synchronization. This improves battery life for applications such as Web browsing. This solution also lowers battery consumption for Voice over Internet Protocol (VoIP) and reduces the noise level in the network (i.e., allowing more simultaneous VoIP users). Figure 2.11 shows the benefits of this approach.
- **F-DPCH (Fractionated Dedicated Physical Control Channel) gating:** terminals in HSDPA active mode always receive a Dedicated Physical Channel in the downlink, in addition to high-speed shared channels, which carries power control information and Layer 3 radio resource (RRC) messages, for example, for handovers, channel modifications, and so on. The Fractional-DPCH feature puts the RRC messages on the HSDPA shared channels and the mobile thus only has to decode the power control information from the DPCH. At all other times, that is, when the terminal is not in HSDPA active mode, the DPCH is not used by the mobile (thus it is fractional). During these times, power control information is transmitted for other mobiles using the same spreading code. Consequently, several mobiles use the same spreading code for the dedicated physical channel but listen to it at different times. This means that fewer

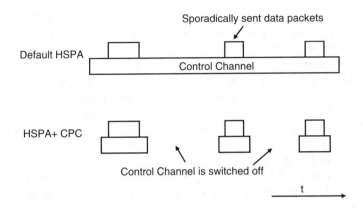

Figure 2.11 Control channel switch-off during times with little activity.

spreading codes are used by the system for this purpose, which in turn leaves more resources for the high-speed downlink channels or allows more users to be kept in HSPA Cell-DCH state simultaneously.

Feature 3: Discontinuous Reception (DRX) in the Downlink (Based on Section 4.5 of [17])

While a mobile is in HSPA mode, it has to monitor one or more HS-SCCHs to see when packets are delivered to it on the high-speed shared channels. This monitoring is continuous, that is, the receiver can never be switched off. For situations when no data is transmitted, or the average data transfer rate is much lower than that which could be delivered over the high-speed shared channels, the base station can instruct the mobile to only listen to selected slots of the shared control channel. The slots which the mobile does not have to observe are aligned as much as possible with the uplink control channel gating (switch-off) times. Therefore, there are times when the terminal can power down its receiver to conserve energy. Once more data arrives from the network than can be delivered with the selected Discontinuous Reception (DRX) cycle, the DRX mode is switched off and the network can once again schedule data in the downlink continuously.

Feature 4: HS-SCCH-Less Operation (Based on Sections 4.7 and 4.8 of [17])

This feature is not intended to improve battery performance but to increase the number of simultaneous real-time VoIP users in the network. VoIP service, for example, via the IP Multimedia Subsystem (IMS) (cf. Chapter 4), requires relatively little bandwidth per user and thus the number of simultaneous users can be high. On the radio link, however, each connection has a certain signaling overhead. Therefore, more users mean more signaling overhead which decreases overall available bandwidth for user data. In the case of HSPA, the main signaling resources are the HS-SCCHs. The more active users there are, the more they proportionally require of the available bandwidth.

HS-SCCH-less operation aims at reducing this overhead. For real-time users who require only limited bandwidth, the network can schedule data on high-speed downlink channels without prior announcements on a shared control channel. This is done as follows: the network instructs the mobile not only to listen to the HS-SCCH but in addition to all packets being transmitted on one of the high-speed downlink shared channels. The terminal then attempts to blindly decode all packets received on that shared channel. To make blind decoding easier, packets which are not announced on a shared control channel can only have one of four transmission formats (number of data bits) and are always modulated using QPSK. These restrictions are not an issue for performance, since HS-SCCH-less operation is only intended for low-bandwidth real-time services.

The checksum of a packet is additionally used to identify for which device the packet is intended. This is done by using the terminal's Medium Access Control (MAC) address as an input parameter for the checksum algorithm in addition to the data bits. If the device can decode a packet correctly and if it can reconstruct the checksum, it is the intended recipient. If the checksum does not match then either the packet is intended for a different terminal or a transmission error has occurred. In both cases the packet is discarded.

In case of a transmission error, the packet is automatically retransmitted since the mobile did not send an HARQ ACK. Retransmissions are announced on the shared control channel, which requires additional resources but should not happen frequently as most packets should be delivered properly on the first attempt.

2.2.5.7 Enhanced Cell-FACH, Cell/URA PCH States

The CPC features described earlier aim to reduce power consumption and signaling overhead in HSPA Cell-DCH state. The CPC measures therefore increase the number of mobile devices that can be in Cell-DCH state simultaneously and allow a mobile device to remain in this state for a longer period of time even if there is little or no data being transferred. Eventually, however, there is so little data transferred that it no longer makes sense to keep the mobile in Cell-DCH state, that is, it does not justify even the reduced signaling overhead and power consumption. In this case, the network puts the connection into Cell-FACH state as described earlier or even into Cell-PCH or URAPCH state to reduce energy consumption even further. The downside of this is that a state change back into Cell-DCH state takes a long time and that little or no data can be transferred during the state change. In Release 7 and 8, the 3GPP standards were thus extended to also use the high-speed downlink shared channels for these states, as described in [18, 19]. In practice this is done as follows:

- **Enhanced Cell-FACH** — in the standard Cell-FACH state the mobile device listens to the S-CCPCH in the downlink as described earlier for incoming RRC messages from the RNC and for user data (IP packets). With the Enhanced Cell-FACH feature, the network can instruct a mobile device to observe a high-speed downlink control channel or the shared data channel directly for incoming RRC messages from the RNC and for user data. The advantage of this approach is that, in the downlink direction, information can be sent much faster. This reduces latency and speeds up the Cell-FACH to Cell-DCH state change procedure. Unlike in Cell-DCH state, no other uplink or downlink control channels are used. In the uplink, the mobile still uses the random access channel to respond to RRC messages from the RNC and to send its own IP packets. This limits the use of AMC since the mobile cannot send frequent measurement reports to the base station to indicate the downlink reception quality. Furthermore, it is also not possible to acknowledge proper receipt of frames. Instead, the RNC informs the base station when it receives measurement information in radio resource messages from the mobile.

• **Enhanced Cell/URA-PCH states** — in these two states, the mobile device is in a deep sleep state and only observes the paging information channel to be alerted of an incoming paging message which is transmitted on the paging channel. To transfer data, the mobile device is then moved to Cell-FACH or Cell-DCH state. If the mobile device and the network support Enhanced Cell/URA-PCH states, the network can instruct the mobile device not to use the slow paging channel to receive paging information but to use a high-speed downlink shared channel instead. The high-speed downlink channel is then also used for subsequent RRC commands which are required to move the device back into a more active state. Like the measure above, this significantly decreases the wakeup time.

Figure 2.14 shows how this works in practice. While the message exchange to notify the mobile device of incoming data and to move it to another activity state remains the same, using the high-speed downlink shared channels for the purpose speeds up the procedure by several hundred milliseconds.

Which of the described enhancements will make it into networks in the future remains to be seen and will also depend on how quickly LTE and other competing network technologies are rolled out. While CPC and enhanced mobility management states increase

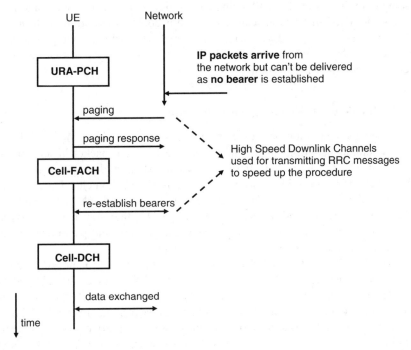

Figure 2.12 Message exchange to move a mobile device from URA-PCH state back to cell-DCH state when IP packets arrive from the network.

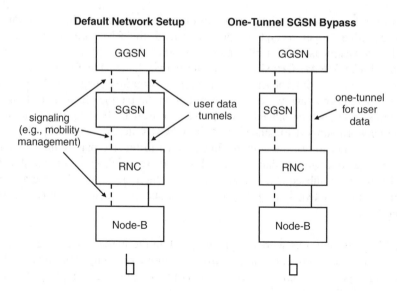

Figure 2.13   Current network architecture vs the one-tunnel enhancement.

the efficiency of the system, they also significantly increase the complexity of the air interface, as both old and new mobile devices have to be supported simultaneously. This rising complexity is especially challenging for the development of devices and networks, as it creates additional interaction scenarios which become more and more difficult to test and debug before. Already today, devices are tested with network equipment of several vendors and different software versions. Adding yet another layer of features will make this even more complex in the future.

2.2.5.8   Radio Network Enhancement: One-Tunnel

Figure 2.13 shows the default path of user data between a mobile device and the Internet through the cellular network. In the current architecture, the packet is sent through the GGSN, the SGSN, the RNC, and the base station. All user data packets are tunneled through the network as described earlier, since the user's location can change at any time. The current architecture uses a tunnel between the GGSN and the SGSN and a second tunnel between the SGSN and the RNC. All data packets therefore have to pass through the SGSN, which terminates one tunnel, extracts the packets, and puts them into another tunnel. This requires both time and processing power.

Since both the RNC and the GGSN are IP routers, this process is not required in most cases. The one-tunnel approach, now standardized in 3GPP (see [20, 21]), allows the SGSN to create a direct tunnel between the RNC and the GGSN. It thus removes itself from the transmission chain. Mobility management, however, remains on the SGSN which means, for example, that it continues to be responsible for mobility management and tunnel modifications in case the mobile device is moved to an area served by another RNC. For the user, this approach has the advantage that the packet delay is reduced. From a network point of view, the advantage is that the SGSN requires fewer processing resources per active user, which helps to reduce equipment costs. This is especially important as the amount of data traversing the packet-switched core network is rising significantly.

A scenario where the one-tunnel option is not applicable is international roaming. Here, the SGSN has to be in the loop in order to count the traffic for inter-operator billing purposes. Another case where the one-tunnel option cannot be used is when the SGSN is asked by a prepaid system to monitor the traffic flow. This is only a small limitation, however, since in practice it is also possible to perform prepaid billing via the GGSN.

Proprietary enhancements even aim to terminate the user data tunnel at the NodeB, bypassing the RNC as well. However, this has not found the widespread support of companies in 3GPP and is not likely to be compatible with some HSPA+ extensions, like the enhanced mobile device states.

2.2.6   Competition for LTE in 5 MHz

As has been shown in the earlier sections, there have been many significant enhancements of HSDPA, HSUPA, and the system in general over the past years in various 3GPP releases. In practice, many of those enhancements such as 64QAM and dual-carrier operation are used in life networks today. As a consequence, enhanced HSPA networks have become a viable alternative to LTE deployments in the short and medium terms, as the spectral efficiency of both the systems is similar. As will be described in the next section, LTE scores over HSPA when more bandwidth is available, as it is not limited to 5 MHz channels. In practice, most network operators thus continue to improve their HSPA networks and many have started the rollout of an LTE network in parallel.

2.3   LTE

2.3.1   Introduction

For several years, there has been an ongoing trend in fixed line networks to migrate all circuit-switched services to a packet-switched IP infrastructure. In practice, it can be observed that fixed line network operators are migrating their telephony services to a packet-switched architecture offering both telephony and Internet access either via DSL or a cable modem. This means, that circuit-switched technology is replaced by VoIP-based solutions, as will be described in more detail in Chapter 4. In wireless networks, this trend is still in its infancy. This is mostly due to the fact that current 3.5G network architectures are still optimized for circuit-switched telephony in the radio network. In the core network, however, many network operators have switched to the Bearer Independent Core Network (BICN) architecture that separates the circuit-switched MSC into an MSC Server and a media gateway. While the protocols used are still from the circuit-switched era, many interfaces are now based on IP technology. One of the reasons for this slower adoption is, for example, that today's VoIP telephony implementations significantly increase the amount of data that has to be transferred over the air interface, which means fewer voice calls can be handled simultaneously. Also, interworking between a VoIP solution and the circuit-switched core network that has to be used in areas where no broadband network is available is difficult to achieve in practice. Besides such challenges, however, migrating voice telephony to IP offers a number of significant benefits such as cheaper networks and integration with other IP-based applications as discussed in more detail in Chapters 5 and 6.

At the same time, the general trend of ever increasing transmission bandwidths is highlighting the limits of current 3G and 3.5G networks. It was therefore decided in 2005 by the 3GPP standardization body to start work on a next generation wireless network design that is only based on packet-switched data transmission. This research was performed in two study programs. The LTE program focused on the design of a new radio network and air interface architecture. Slightly afterwards, work started on the design of a new core network infrastructure with the Service Architecture Evolution (SAE) program. Later, they were combined into a single work program, the EPS program. By that time, however, the abbreviation "LTE" was already dominant in literature and most documents still refer to LTE rather than EPS.

Besides being fully packet-based, the following design goals were set for the new network:

- **Reduced time for state changes** — in HSPA networks today, the time it takes a mobile device to connect to the network and start communication on a high-speed bearer is relatively long. This has a negative impact on usability, as the user can feel this delay when accessing a service on the Internet after a period of inactivity. It was thus decided that with a new network design it should be possible to move from idle state to being fully connected in less than 100 ms.
- **Reduced user plane latency** — another downside of cellular networks at that time was the much higher transmission delay compared with fixed line networks. While one-way delay between a user's computer at the edge of a DSL network to the Internet is around 15 ms today, HSPA networks had a delay of around 50 ms. This is disadvantageous for applications such as telephony and real-time gaming. For LTE, it was decided that air interface delay should be in the order of 5 ms to reach end-to-end delays equaling fixed line networks.
- **Scalable bandwidth** — HSPA networks are currently limited to a bandwidth of 5 MHz. At some point, higher throughput can only be reasonably achieved by increasing the bandwidth of the carrier. For certain applications, a carrier of 5 MHz is too large and it was thus decided that the air interface should also be scalable in the other direction.
- **Throughput increase** — for the new system, a maximum throughput under ideal conditions of 100 Mbit/s should be achieved.

The following sections now describe how these design goals are met in practice.

2.3.2   Network Architecture

2.3.2.1   Enhanced Base Stations

Figure 2.14 shows the main components of an LTE core and radio access network as described in [22]. Compared with UMTS, the radio network is less complex. It was decided that central RNCs should be removed and their functionality has been partly moved to the base stations and partly to the core network gateway. To differentiate UMTS base stations from LTE base stations, they are referred to as Enhanced NodeB (eNodeB). As there is no central controlling element in the radio network any more, the

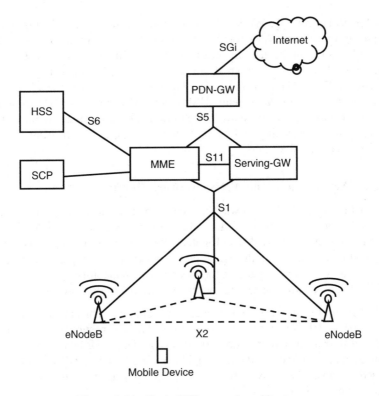

Figure 2.14 Basic LTE network architecture.

base stations now perform air interface traffic management autonomously and ensure QoS. This was already partly the case in UMTS with the introduction of HSPA, as discussed in the previous section. Control over bearers for circuit-switched voice telephony, however, rested with the RNC.

In addition, base stations are now also responsible for performing handovers for active mobiles. For this purpose, the eNodeBs can now communicate directly with each other via the X2 interface. The interface is used to prepare a handover and can also be used to forward user data (IP packets) from the current base station to the new base station to minimize the amount of user data lost during the handover. As the X2 interface is optional, base stations can also communicate with each other via the access gateway to prepare a handover. In this case, however, user data is not forwarded during the handover. This means that some of the data already sent from the network to the current base station might be lost, as once a handover decision has been made, it has to be executed as quickly as possible before radio contact is lost. Unlike in UMTS, LTE radio networks only perform hard handovers, that is, only one cell communicates with a mobile device at a time.

The interface that connects the eNodeB to the gateway nodes between the radio network and the core network is the S1 interface. It is fully based on the IP protocol and is therefore transport technology agnostic. This is a big difference to UMTS. Here, the interfaces between the NodeB, the RNCs, and the SGSN were firmly based on the ATM protocol for the lower protocol layers. Between the RNC and the NodeB, IP was not used at all for packet routing. While allowing for easy time synchronization between the nodes, requiring the use of ATM for data transport on lower protocol layers makes the setup inflexible and complicated. In recent years, the situation has worsened as rising bandwidth demands cannot be satisfied any more with ATM connections over 2 Mbit/s E-1 connections. The UMTS standard was thus enhanced to also use IP as a transport protocol to the base station and many HSPA networks make use of this today. LTE, however, is fully based on IP transport in the radio network from day 1. Base stations are either equipped with 100 Mbit/s or 1 Gbit/s Ethernet ports, as known from the PC world, or with gigabit Ethernet fiber ports.

2.3.2.2 Core Network to Radio Access Network Interface

As shown in Figure 2.14, the gateway between the radio access network and the core network is split into two logical entities, the Serving Gateway (Serving-GW) and the Mobility Management Entity (MME). Together, they fulfill similar tasks as the SGSN in UMTS networks. In practice, both logical components can be implemented on the same physical hardware or can be separated for independent scalability.

The MME is the "control plane" entity responsible for the following tasks:

- Subscriber mobility and session management signaling. This includes tasks such as authentication, establishment of radio bearers, handover support between different eNodeB and to/from different radio networks (e.g., GSM, UMTS).
- Location tracking for mobile devices in idle mode, that is, while no radio bearer is established because they have not exchanged data packets with the network for a prolonged amount of time.
- Selection of a gateway to the Internet when the mobile requests the establishment of a session, that is, when it requests an IP address from the network.

The Serving-GW is responsible for the "user plane," that is, for forwarding IP packets between mobile devices and the Internet. As already discussed in the section on UMTS, IP tunnels are used in the radio access network and core network to flexibly change the route of IP packets when the user is handed over from one cell to another while moving. The GTP is reused for this purpose and the mechanism is the same as shown for UMTS in Figure 2.3. The difference from UMTS is that the tunnel for a user in the radio network is terminated directly in the eNodeB itself and no longer on an intermediate component such as a RNC. This means that the Base Transceiver Station (BTS) is directly connected via an IP interface to the Serving-GW and that different transport network technologies such as Ethernet over fiber or optical cable, DSL, microwave, and so on, can be used. In addition, the S1 interface design is much simpler than similar interfaces of previous radio networks, which relied heavily on services of complex lower layer protocols.

As the S1 interface is used for both user data (to the Serving-GW) and signaling data to the MME, the higher layer protocol architecture is split into two different protocol sets: the S1-C (control) interface is used for exchanging control messaging between a mobile device and the MME. As will be shown later, these messages are exchanged over special "non-IP" channels over the air interface and then put into IP packets by the NodeB before they are forwarded to the MME. User data, however, is already transferred as IP packets over the air interface and these are forwarded via the S1-U (user) protocol to the Serving-GW. The S1-U protocol is an adaptation of the GTP from GPRS and UMTS (cf. Figure 2.3).

If the MME and the Serving-GW are implemented separately, the S11 interface is used to communicate between the two entities. Communication between the two entities is required, for example, for the creation of tunnels, when the user attaches to the network, or for the modification of a tunnel, when a user moves from one cell to another.

Unlike in previous wireless radio networks, where one access network gateway (SGSN) was responsible for a certain number of RNCs and each RNC in turn for a certain number of base stations, the S1 interface supports a meshed architecture. This means that not only one but several MMEs and Serving-GWs can communicate with each eNodeB and the number of MMEs and Serving-GWs can be different. This reduces the number of inter-MME handovers when users are moving and allows the number of MMEs to evolve independently from the number of Serving-GWs, as the MME's capacity depends on the signaling load and the capacity of the Serving-GW depends on the user traffic load. These can evolve differently over time, which makes separation of these entities interesting. A meshed architecture of the S1 interface also adds redundancy to the network. If, for example, one MME fails, a second one can take over automatically if it is configured to serve the same cells. The only impact of such an automatic failure recovery is that users served by the failed MME have to register to the network again. How the meshed capabilities of the S1 interface are used in practice depends on the policies of the network operator and on the architecture of the underlying transport network architecture.

2.3.2.3   Gateway to the Internet

As in previous network architectures, a router at the edge of the wireless core network hides the mobility of the users from the Internet. In LTE, this router is referred to as the Packet Data Network (PDN)-Gateway and fulfills the same tasks as the GGSN in UMTS. In addition to hiding the mobility of the users, it also administers an IP address pool and assigns IP addresses to mobiles registering to the network. Depending on the number of users, a network has several PDN-Gateways. The number depends on the capabilities of the hardware, the number of users and the average amount of data traffic per user. As shown in Figure 2.14, the interface between the PDN-GW and the MME/Serving-GWs is referred to as S5. Like the interface between the SGSN and the GGSN in UMTS, it uses the GTP-U (user) protocol to tunnel user data from and to the Serving-GWs and the GTP-S (signaling) protocol for the initial establishment of a user data tunnel and subsequent tunnel modifications when the user moves between cells that are managed by different Serving-GWs.

2.3.2.4   Interface to the User Database

Another essential interface in LTE core networks is the S6 interface between the MMEs and the database that stores subscription information. In UMTS, this database is referred to as the Home Location Register. In LTE, the HLR is reused and has been renamed the Home Subscriber Server (HSS). Essentially, the HSS is an enhanced HLR and contains subscription information for GSM, GPRS, UMTS, LTE, and the IMS, which is discussed in Chapter 4. Unlike in UMTS, however, the S6 interface does not use the SS-7-based MAP (Mobile Application Part) protocol, but the IP-based Diameter protocol. The HSS is a combined database and it is used simultaneously by GSM, UMTS, and LTE networks belonging to the same operator. It therefore continues to support the traditional MAP interface in addition to the S6 interface for LTE and also the interfaces required for the IMS as discussed in Chapter 4.

2.3.2.5 Moving between Radio Technologies

In practice, most network operators deploying an LTE network already have a GSM and UMTS network in place. As the coverage area of a new LTE network is likely to be very limited at first, it is essential that subscribers can move back and forth between the different access network technologies without losing their connection and assigned IP address. Figure 2.15 shows how this is done in practice when a user roams out of the coverage area of an LTE network and into the coverage area of a UMTS network of the same network operator. When the user moves out of the LTE coverage area, the mobile device can be instructed by the eNodeB to report neighboring UMTS (or GSM) cells. This report is forwarded to the MME which contacts the responsible 3G (or 2G) SGSN and requests a handover procedure. The interface used for this purpose is referred to as S3 and is based on the protocol used for inter-SGSN relocation procedures. As a consequence, no software modifications are required on the 3G SGSN to support the procedure. Once the 3G radio network has been prepared for the handover, the MME sends a handover command to the mobile device via the eNodeB. After the handover has been executed, the user data tunnel between the Serving-GW and the eNodeB is re-routed to the SGSN. The MME is then released from the subscriber management, as this task is taken over by

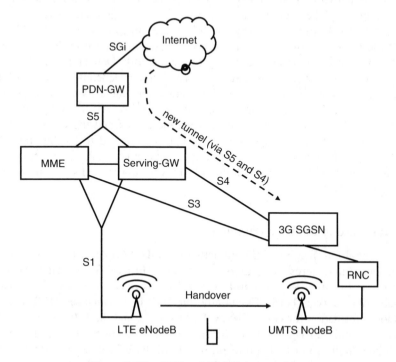

Figure 2.15   LTE and UMTS interworking.

the SGSN. The Serving-GW, however, remains in the user data path via the S4 interface and acts as a 3G GGSN from the point of view of the SGSN. From the SGSN's point of view, the S4 interface is therefore considered to be the 3G Gn interface between the SGSN and the GGSN.

Some early LTE networks deployed at the time of publication use simpler ways to transfer the user from one radio network technology to another. The simplest way is for the network to wait until the mobile device loses contact and then autonomously searches for a GSM or UMTS network. Once found, the mobile device then performs a location and routing area update, which moves the user's context including the assigned IP address to the radio network the device has picked up. A somewhat more sophisticated option is to aid the mobile device in the network and radio technology selection process by giving it details of GSM and UMTS cells in the neighborhood and instructing the device to go to those networks once the LTE coverage becomes very weak. This way, the mobile device does not have to try to regain contact with the LTE network first before it starts searching for alternative radio networks. Also, by giving it details of where and what to search for, the GSM and UMTS cell search time is significantly improved.

2.3.2.6   The Packet Call becomes History

A big difference of LTE from GSM and UMTS is that mobile devices will always be assigned an IP address as soon as they register to the network. This has not been the case with GSM and UMTS because 2G, 3G, and 3.5G devices are still mostly used for voice telephony and so it makes sense to attach to the network without requesting an IP address. In LTE networks, however, a device without an IP address is completely useless. Hence, the LTE network attach procedure already includes the assignment of an IP address. From the LAN (Local Area Network)/WLAN (Wireless LAN) point of view this is nothing new. From a cellular industry point of view, however, this is revolutionary. The GPRS and UMTS procedure of "establishing a packet call," a term coined with the old thinking of establishing a circuit-switched connection with a voice call in mind, has therefore become a thing of the past with LTE. Many people in the industry will have to change their picture of the mobile world to accommodate this.

2.3.3   Air Interface and Radio Network

While the general LTE network architecture is mainly a refinement of the 3G network architecture, the air interface and the radio network have been redesigned from scratch. In the 3GPP standards, a good place to start further research beyond what is covered below is TS 36.300 [23].

2.3.3.1   Downlink Data Transmission

For transmission of data over the air interface, it was decided to use a new transmission scheme in LTE which is completely different from the CDMA approach of UMTS. Instead of using only one carrier over the broad frequency band, it was decided to use a transmission scheme referred to as Orthogonal Frequency Division Multiple Access, or

OFDMA for short. OFDMA transmits a data stream by using several narrow-band sub-carriers simultaneously, for example, 512, 1024, or even more, depending on the overall available bandwidth of the channel (e.g., 5, 10, 20 MHz). As many bits are transported in parallel, the transmission speed on each subcarrier can be much lower than the overall resulting data rate. This is important in a practical radio environment in order to minimize the effect of multipath fading created by slightly different arrival times of the signal from different directions. The second reason this approach was selected was because the effect of multipath fading and delay spread becomes independent of the amount of bandwidth used for the channel. This is because the bandwidth of each subcarrier remains the same and only the number of subcarriers is changed. With the previously used CDMA modulation, using a 20 MHz carrier would have been impractical, as the time each bit was transmitted would have been so short that the interference due to the delay spread on different paths of the signal would have become dominant.

Figure 2.16 shows how the input bits are first grouped and assigned for transmission over different frequencies (subcarriers). In the example, 2 bits (representing a QPSK modulation) are sent per transmission step per subcarrier. A transmission step is also referred to as a symbol. With 16QAM modulation, 4 bits are encoded in a single symbol, while 64QAM modulation encodes 6 bits in a single symbol, thus raising the data rate even further. On the other hand, encoding more bits in a single symbol makes it harder for the receiver to decode the symbol if it was altered by interference. This is the reason why different modulation schemes are used depending on transmission conditions.

In theory, each subcarrier signal could be generated by a separate transmission chain hardware block. The output of these blocks would then have to be summed up and

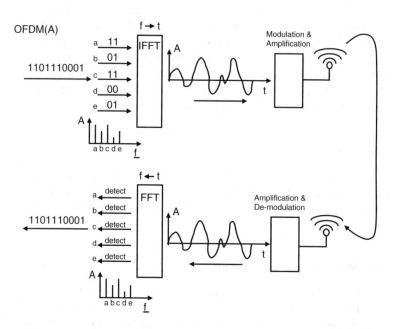

Figure 2.16 Principles of OFDMA for downlink transmissions.

the resulting signal could then be sent over the air. Because of the high number of subcarriers used, this approach is not feasible. Instead, a mathematical approach is taken as follows. As each subcarrier is transmitted on a different frequency, a graph which shows the frequency on the x-axis and the amplitude of each subcarrier on the y-axis can be constructed. Then, a mathematical function called Inverse Fast Fourier Transformation (IFFT) is applied, which transforms the diagram from the frequency domain to the time domain. This diagram has the time on the x-axis and represents the same signal as would have been generated by the separate transmission chains for each subcarrier when summed up. The IFFT thus does exactly the same job as the separate transmission chains for each subcarrier would do, including summing up the individual results.

On the receiver side, the signal is first demodulated and amplified. The result is then treated by a fast Fourier transformation (FFT) function which converts the time signal back into the frequency domain. This reconstructs the frequency/amplitude diagram created at the transmitter. At the center frequency of each subcarrier a detector function is then used to generate the bits originally used to create the subcarrier.

The explanation has so far covered the Orthogonal Frequency Division aspect of OFDMA transmissions. The Multiple Access (MA) part of the abbreviation refers to the fact that the data sent in the downlink is received by several users simultaneously. As will be discussed later, control messages inform mobile devices waiting for data which part of the transmission is addressed to them and which part they can ignore. This is, however, just a logical separation. On the physical layer, this only requires that modulation schemes ranging from QPSK over 16–64QAM can be quickly changed for different subcarriers in order to accommodate the different reception conditions of subscribers.

2.3.3.2 Uplink Data Transmission

For data transmission in the uplink direction, 3GPP has chosen a slightly different modulation scheme. OFDMA transmission suffers from a high Peak to Average Power Ratio (PAPR), which would have negative consequences for the design of an embedded mobile transmitter; that is, when transmitting data from the mobile terminal to the network, a power amplifier is required to boost the outgoing signal to a level high enough to be picked up by the network. The power amplifier is one of the biggest consumers of energy in a device and should therefore be as power-efficient as possible to increase the battery life of the device. The efficiency of a power amplifier depends on two factors:

• The amplifier must be able to amplify the highest peak value of the wave. Due to silicon constraints, the peak value determines the power consumption of the amplifier.
• The peaks of the wave, however, do not transport any more information than the average power of the signal over time. The transmission speed therefore does not depend on the power output required for the peak values of the wave but rather on the average power level.

As both power consumption and transmission speed are of importance for designers of mobile devices, the power amplifier should consume as little energy as possible. Thus, the lower the difference between the PAPR, the longer is the operating time of a mobile device

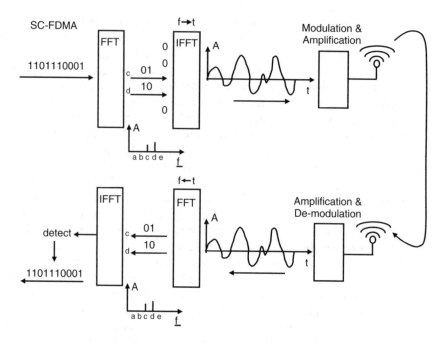

Figure 2.17   SC-FDMA modulation for uplink transmissions.

at a certain transmission speed compared with devices that use a modulation scheme with a higher PAPR.

A modulation scheme similar to basic OFDMA, but with a much better PAPR, is SC-FDMA (Single Carrier-Frequency Division Multiple Access). Due to its better PAPR, it was chosen by 3GPP for transmitting data in the uplink direction. Despite its name, SC-FDMA also transmits data over the air interface in many subcarriers, but adds an additional processing step as shown in Figure 2.17. Instead of putting 2, 4, or 6 bits together as in the Orthogonal Frequency Division Multiplexing (OFDM) example to form the signal for one subcarrier, the additional processing block in SC-FDMA spreads the information of each bit over all the subcarriers. This is done as follows: again, a number of bits (e.g., 4 representing a 16QAM modulation) are grouped together. In OFDM, these groups of bits would have been the input for the Inverse Discrete Fourier Transform (IDFT). In SC-FDMA, however, these bits are now piped into a FFT function first. The output of the process is the basis for the creation of the subcarriers for the following IFFT. As not all subcarriers are used by the mobile station; many of them are set to zero in the diagram. These may or may not be used by other mobile stations.

On the receiver side the signal is demodulated, amplified, and treated by the FFT function in the same way as in OFDMA. The resulting amplitude diagram, however, is not analyzed straight away to get the original data stream, but fed to the inverse FFT function to remove the effect of the additional signal processing originally done at the transmitter side. The result of the IFFT is again a time domain signal. The time domain signal is now fed to a single detector block which recreates the original bits. Therefore, instead of detecting the bits on many different subcarriers, only a single detector is used on a single carrier.

The differences between OFDM and SC-FDMA can be summarized as follows: OFDM takes groups of input bits (0s and 1s) to assemble the subcarriers which are then processed by the IDFT to get a time signal. SC-FDMA in contrast first runs an FFT over the groups of input bits to spread them over all subcarriers and then uses the result for the IDFT which creates the time signal. This is why SC-FDMA is sometimes also referred to as FFT spread OFDM.

2.3.3.3 Physical Parameters

For LTE, the following physical parameters have been selected:

- Subcarrier spacing, 15 kHz;
- OFDM symbol duration, 66.667 μs;
- Standard cyclic prefix: 4.7 μs. The cyclic prefix is transmitted before each OFDM symbol to prevent inter-symbol interference due to different lengths of several transmission paths. For difficult environments with highly diverse transmission paths a longer cyclic prefix of 16.67 μs has been specified as well. The downside of using a longer cyclic prefix, however, is a reduced user data speed since the symbol duration remains the same.

The selected subcarrier spacing and symbol duration compensate for detrimental effects on the signal such as the Doppler effect (frequency shift) due to the mobility of subscribers. The parameters have been chosen to allow speeds of beyond 350 km/h.

To be flexible with bandwidth allocations in different countries around the world, a number of different channel bandwidths have been defined for LTE. These range from 1.25 MHz on the low end to 20 MHz on the high end. Table 2.1 shows the standardized transmission bandwidths, the number of subcarriers used for each and the FFT size (the number of spectral lines) used at the receiver side to convert the signal from the time to the frequency domain.

In practice, network operators have two options to deploy LTE: In many countries, not all of the spectrum assigned for GSM and UMTS is used and can thus be used for LTE right away. This makes only sense for bandwidths of at least 10 MHz, since there is no speed advantage of using LTE in a 5 MHz band over HSPA+. The smaller bandwidths

Table 2.1 Defined bandwidths for LTE

Bandwidth (MHz)	Number of subcarriers	FFT size
1.25	76	128
2.5	151	256
5	301	512
10	601	1024
15	901	1536
20	1201	2048

of 1.25 and 2.5 MHz were specified for operators with little spectrum or for operators wishing to "re-farm" some of their GSM spectrum in the 900 MHz band. In practice, however, it is questionable if this would bring a great benefit since the achievable data rates in such a narrow band are lower than what can be achieved with HSPA today.

In addition to the use of already existing frequency bands, new bands are being made available by regulating authorities. In Europe, for example, the 2.6 GHz band, also referred to as the International Mobile Telecommunication (IMT) extension band, has been opened for LTE as well as 30 MHz of spectrum in the 800 MHz band previously used for terrestrial TV broadcasting.

At the time of publication, first LTE networks make use of several existing and newly allocated frequency bands. The following list gives some examples and it is expected that further frequency bands will be used in the future:

- Europe
  - The 800 MHz digital dividend band with 10 MHz LTE channels, not previously used by cellular networks.
  - The 1800 MHz band with 15–20 MHz LTE channels, also used by GSM.
  - The 2600 MHz band with 20 MHz LTE channel bandwidths, not previously used by cellular networks.
- North America
  - The 700 MHz band with 10 MHz channel bandwidths, not previously used by cellular networks.
- Japan
  - The 2100 MHz band also used for UMTS.

2.3.3.4 From Slots to Frames

Data is mapped to subcarriers and symbols, which are arranged in the time and frequency domain in a resource grid as shown in Figure 2.18. The smallest aggregation unit is referred to as a slot or a resource block and contains 12 subcarriers and 7 symbols on each subcarrier in case the default short cyclic prefix is used. The symbol time of 66.67 μs and the 4.7 μs cyclic prefix multiplied by 7 results in a slot length of 0.5 ms. In case the long cyclic prefix has to be used, the number of symbols per slot is reduced to six, again resulting in a slot length of 0.5 ms. The grouping of 12 subcarriers together results in a resource block bandwidth of 180 kHz. As the total carrier bandwidth used in LTE is much larger (e.g., 10 MHz), many resource blocks are transmitted in parallel.

Two slots are then grouped into a subframe, which is also referred to as a Transmit Time Interval (TTI). In case of Time Division Duplex (TDD) operation (uplink and downlink in the same band), a subframe can be used for either uplink or downlink transmission. It is up to the network to decide which subframes are used for which direction. Most networks, however, are likely to use Frequency Division Duplex (FDD), which means that there is a separate band for uplink and downlink transmission. Here, all subframes of the band are dedicated to downlink or to uplink transmissions.

Ten subframes are grouped together to form a single radio frame, which has a duration of 10 ms. Afterwards, the cycle repeats with the next frame. This is important for mobile

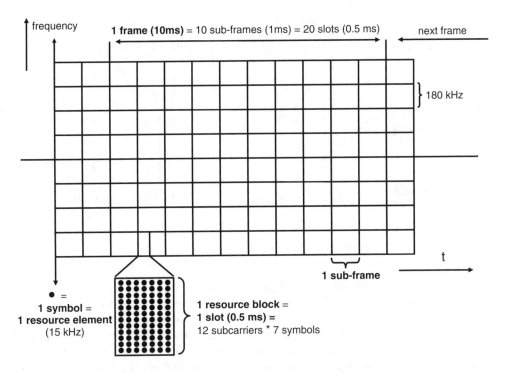

Figure 2.18 The LTE resource grid.

devices because broadcast information (e.g., uplink bandwidth assignments) is always transmitted at the beginning of a frame.

The smallest amount of resource elements (symbols) that can be allocated to a single mobile device at an instant in time is two resource blocks, which equals one subframe or one TTI. To increase the data rate for the mobile device, the scheduler in the network can concatenate several resource blocks in both the time and the frequency direction. Since there are many resource blocks being transmitted in parallel, it is also possible to schedule several mobile devices simultaneously, each listening to different subcarriers.

2.3.3.5 Reference Symbols, Signals, and Channels

Not all resource elements of a resource block are used for transmitting user data. Especially around the center frequency, some resource elements are used for other purposes, as described later.

To enable a mobile device to find the network after power on and when searching for neighboring cells, some resource elements are used for pilot or reference symbols in a predefined way. While data is transferred, pilot symbols are used by the mobile device for downlink channel quality measurements and, since the content of the resource element is known, to estimate how to recreate the original signal that was distorted during transmission.

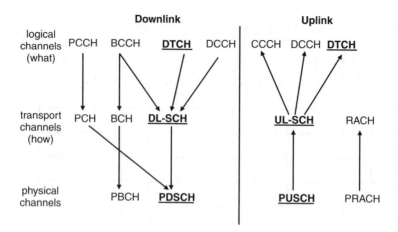

Figure 2.19   LTE uplink and downlink channels.

For the transmission of higher layer data, LTE re-uses the channel concept of UMTS as shown in Figures 2.8 and 2.9. Compared with UMTS, however, all devices use the shared channel on the physical layer. The LTE channel model is therefore much simpler than that of UMTS. LTE also re-uses the concept of logical channels (what is transmitted), transport channels (how is it transmitted) and physical channels (air interface) to separate data transmission over the air interface from the logical representation of data. Figure 2.19 shows the most important channels that are used in LTE and how they are mapped to each other.

2.3.3.6   Downlink: Broadcast Channel

While no data is transmitted and the mobile is in idle state, it listens to two logical channels. The logical BCCH is used by the network to transmit system information to mobile devices about the network and the cell, that is, in which resource blocks and resource elements to find other channels, how the network can be accessed, and so on. The basic parameters sent on the BCCH are mapped to the BCH transport channel and the Physical Broadcast Channel (PBCH). The PBCH is then mapped to dedicated resource elements in the subchannels of the inner 1.25 MHz of the band. Which resource elements are used for the PBCH is calculated with a mathematical formula which generates a certain pattern and thus distributes the broadcast information between different subcarriers over time [24]. In addition, a number of additional resource elements are used for a Synchronization Channel (SCH), which is not shown in Figure 2.19. As the name implies, these resource elements help mobile devices to synchronize to the cell and to find the resource elements on which the broadcast information can be found. In addition to basic cell parameters, the BCH also carries further information such as neighboring cell information, which is also necessary but not required for initial access to the network. To save bandwidth and to be flexible in the future, this information is not carried on the PBCH but on the Physical Downlink Shared Channel (PDSCH) instead, which is also used for transferring user data (IP packets). A pointer on the PBCH informs mobiles where to find the broadcast information on the PDSCH.

2.3.3.7 Downlink: Paging Channel

The paging channel is used to contact mobile devices in an idle state when a new IP packet arrives in the core network from the Internet and needs to be delivered to the mobile device. In idle state, which is usually entered after a prolonged period of inactivity, only the tracking area (i.e., the identity for a group of cells) where the mobile is located is known. The paging message is then sent into all cells of this group. When the mobile device receives the message, it establishes a connection with the network again, a bearer is set up and the packet is delivered. For services such as instant messaging, push e-mail and VoIP, paging for incoming IP packets is quite common. Such applications, if programed properly and no NAT firewalls are used, have a logical connection with a server in the network but are dormant until either the user invokes a new action or the application is contacted by the network-based server, for example, because of a new instant message coming in [25]. As can be seen in Figure 2.19, there is no physical channel dedicated for paging messages. Instead, paging messages are sent on the downlink shared channel and a pointer on the logical BCH indicates where and when the paging messages can be found on the shared channel. The broadcast cycle for paging messages a mobile device needs to listen to is usually in the order of 1–2 seconds. This is a good balance between quick delivery of an incoming packet and power consumption of a mobile device while not being actively used.

2.3.3.8 Downlink and Uplink: Dedicated Traffic and Control Channels and their Mapping to the Shared Channel

From a logical point of view, user data and RRC control messages are transferred via DTCHs and the DTCHs. Each mobile device has its own dedicated pair of these channels. As can be seen in Figure 2.19, both channels are multiplexed to a single PDSCH or physical uplink shared channel (PUSCH) which are used for all devices. Higher software layers are therefore independent of the physical implementation of the air interface.

2.3.3.9 Downlink: Physical Layer Control Channels

In addition to the previously mentioned channels, there are a number of additional physical layer control channels which are required to exchange physical layer feedback information. As these channels only carry lower layer control information and are originated by the base station and not the network behind them, they are not shown in Figure 2.19.

To inform mobile devices which resource blocks are assigned to them for transmitting in the uplink direction, the PDSCH is always accompanied by a Physical Downlink Control Channel (PDCCH). In addition, this channel informs mobile devices about the resource allocation of the PCH and the downlink shared channel.

Since the amount of data carried on the PDCCH varies, the number of OFDM symbols assigned to the physical control channel is broadcast via the Physical Control Format Indicator Channel (PCFICH). Finally, the Physical HARQ Indicator Channel (PHICH) carries ACKs for proper reception of uplink data blocks. The HARQ ACK functionality used in LTE is similar to that used in UMTS. For details, see Section 2.2.4.

2.3.3.10   Uplink: Physical Layer Control Channels

In the uplink direction there are two physical layer control channels: The physical uplink control channel (PUCCH) is a per device channel and carries the following information:

- HARQ ACKs for data frames received from the network (cf. Section 2.2.4).
- Scheduling requests from the mobile to inform the network that further uplink transmit opportunities should be scheduled, as there is more data in the output buffer.
- Channel Quality Indications (CQI) to the network, so the base station can determine which modulation and coding to use for data in the downlink direction. CQI information is also important for the scheduler in the base station, as it can decide to temporarily halt data transmission to users in a temporary deep signal fading situation where it is likely that data cannot be received correctly anyway.

The second control channel used in the uplink direction is the PRACH. It is used when no bearer is established in the uplink direction to request new uplink transmission opportunities. It is also used when the mobile wants to establish a bearer for the first time, after a long timeout or in response to a paging from the network.

2.3.3.11   Dynamic and Persistent Scheduling Grants

The packet scheduler in the base station decides which resource blocks of the physical downlink and uplink channels are used for which mobile device. This way, the base station controls both uplink and downlink transmissions for each mobile and is therefore able to determine how much bandwidth is available to a mobile device. Input parameters for the scheduler are, for example, the current radio conditions as seen from each device, so the data transfer rate can be increased or decreased to mobiles temporarily experiencing exceptionally good or bad radio conditions. Other input parameters are the QoS parameters for a connection and the maximum bandwidth granted by the operator to a mobile device, based on the user's subscription.

There are two types of capacity grants for uplink data transmissions: dynamic grants are announced once and are valid for one or more TTIs. Afterwards, the network has to issue a new grant for additional transmit opportunities or for the mobile to receiver further data in the downlink direction. Dynamic grants are useful for data that arrives in a bursty fashion, like during Web browsing, for example, and sporadic downloading of content.

Applications such as voice and video calls require a constant bandwidth and as little variation as possible in the time difference between two adjacent packets (jitter). For such applications, the base station can also issue persistent grants which are given once and are then valid for all subsequent TTIs. This way, no signaling resources are required to constantly re-assign air interface resources while a voice or video call is ongoing. This increases the overall efficiency of the cell and increases the amount of bandwidth that is available for user data. The issue arising with persistent grants is how the base station can know when to use this type of assignment. One way to achieve this is to base the decision on the connection's QoS requirements which are signaled to the network during bearer establishment. This works well for applications which require a constant bandwidth and are based on the IMS. In addition to the QoS signaling initiated by the mobile device, the

IMS has a connection to the transport network and can influence the bearer as well. This is discussed in more detail in Chapter 4. For Internet-based voice and other applications that are not using the IMS, however, using persistent grants is much more difficult. It is likely that such applications are used over a default bearer which has no guaranteed bandwidth and latency as it is used simultaneously for other applications on the same device such as Web browsing. In practice, it remains to be seen if schedulers will also take a look at the bandwidth usage of a mobile device over time and decide on this basis to use persistent or dynamic grants.

2.3.3.12 MIMO Transmission

So far, this chapter has focused on data transmission via a single spatial stream between a transmitter and receiver. Most wireless systems today operate in this mode and a second transmitter on the same frequency is seen as unwanted interference that degrades the channel. In practice, however, it can be observed that even a single signal is reflected and scattered by objects in the transmission path and that the other end receives several copies of the original signal from different angles at slightly different times. For simple wireless transmission technologies, these copies are also unwanted interference. LTE, however, makes use of scattering and reflection on the transmission path by transmitting several independent data streams via individual antennas. The antennas are spaced at least half a wavelength apart, which in itself creates individual transmissions which behave differently when they meet obstacles in the transmission path. On the receiver side, the different data streams are picked up by independent antenna and receiver chains. Transmitting several independent signals over the same frequency band is also referred to as Multiple Input Multiple Output, and Figure 2.20 shows a simplified graphical representation. In practice, this means that several LTE resource grids, as shown in Figure 2.18, are sent over the same frequency at the same time but via different antennas.

The LTE standard specifies two and four individual transmissions over the same band, which requires two or four antennas at both the transmitter and receiver side respectively. Consequently, such transmissions are referred to as 2 × 2 MIMO and 4 × 4 MIMO. In practice, only 2 × 2 MIMO is currently used, because of size constraints of mobile devices and due to the fact that antennas have to be spaced at least half a wavelength apart. Furthermore, most mobile devices support several frequency bands, each usually requiring its own set of antennas in case MIMO operation is supported in the band. More details on this topic will be discussed in Chapter 3 from a capacity point of view and in Chapter 5 from a mobile hardware point of view. On the network side, 2 × 2 MIMO transmissions can be achieved with a "single" cross polar antenna that combines two antennas in a way that each antenna transmits a separate data stream with a different polarization (horizontal and vertical).

While Figure 2.20 depicts the general concept of MIMO transmission, it is inaccurate at the receiver side, as each antenna receives not only a single signal but the combination of all signals as they overlap in space. It is therefore necessary for each receiver chain to calculate a channel propagation that takes all transmissions into account in order to separate the different transmissions from each other. The pilot carriers mentioned earlier are used for this purpose. The characteristics required for these calculations are the gain,

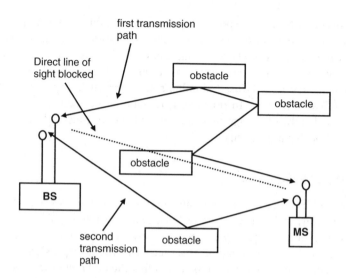

Figure 2.20 Principle of MIMO transmissions. (Reproduced from *Communication Systems for the Mobile Information Society*, Martin Sauter, 2006, John Wiley & Sons, Ltd, Ref. [1].)

phase, and multipath effects for each independent transmission path. A good mathematical introduction is given in [26].

As MIMO channels are separate from each other, 2×2 MIMO can increase the overall data rate by two and 4×4 MIMO by four. This is, however, only possible under ideal signal conditions. MIMO is thus only used for downlink transmissions since the base station transmitter is less power-constrained than the uplink transmitter. In less favorable transmission conditions, the system automatically falls back to single stream transmission and also reduces modulation from 64QAM, to 16QAM, or even QPSK. As has been shown in the previous section on HSPA+, there is also a tradeoff between higher order modulation and MIMO use. Under less than ideal signal conditions, MIMO transmission is therefore only used with 16QAM modulation, which increases but does not double the data rate compared with a single-stream transmission using 64QAM.

In the uplink direction, it is difficult for mobile devices to use MIMO because of their limited antenna size and output power. As a result, uplink MIMO is currently not used. The uplink channel itself, however, is still suitable for uplink MIMO transmissions. To fully use the channel, some companies are thinking about implementing collaborative MIMO in the future, also known as multiuser MIMO [27]. Here, two mobile devices use the same uplink channel for their resource grid. At the base station side, the two data streams are separated by the MIMO receiver and treated as two transmissions from independent devices rather than two transmissions from a single device that have to be combined. While this will not result in higher transmission speeds per device, the overall uplink capacity of the cell is significantly increased.

2.3.3.13 LTE Throughput Calculations

Based on the radio layer parameters introduced in this section, the physical layer throughput of an LTE radio cell can be calculated as follows: the transmission time per symbol is $73.167\,\mu s$ ($66.667\,\mu s$ for the symbol itself $+ 4.7\,\mu s$ for the cyclic prefix), the highest modulation order is 64QAM (6 bits per symbol) and there are 1201 subcarriers in a 20 MHz band:

$$\text{Physical speed} = (1/0.000\ 073\ 167) \times 6 \times 1201 = 98.487.022 \text{ bit/s}$$

$$\text{(i.e., about 100 Mbit/s)}$$

When 2×2 MIMO is used, the physical layer speed doubles to about 200 Mbit/s and in case 4×4 MIMO is used for transmission, the theoretical data transmission speed is 400 Mbit/s, based on a 20 MHz channel.

These values are usually quoted in press releases. However, as already discussed for HSPA, these raw physical layer transmission speeds are not reached in practice for a variety of reasons:

- 64QAM modulation can only be used very close to the base station. For the majority of users served by a cell, 16QAM (4 bits per symbol) or QPSK (2 bits per symbol) is more realistic.
- Error detection and correction bits (coding) are usually added to the data stream as otherwise the bit error rate over the air interface would become too high. Under average signal conditions it is common to see coding rates of 1/3. In practice, the coding overhead is thus in the range of 25–30%.
- Retransmissions — with a very conservative transmission strategy, the coding described earlier is sufficient to correct most transmission errors. In practice, however, more aggressive transmission strategies are used to make the best use of air interface resources. This usually results in air interface packet retransmission rates in the order of 20%.
- There is a significant overhead from pilot channels and control channels such as the BCH and the dedicated signaling channels per user to acknowledge the correct reception of data packets and to convey signal quality measurement results. Many of those channels are transmitted with a lower order modulation so even devices in very unfavorable signal conditions can receive the information.
- In many cases, less than 20 MHz of bandwidth is available for LTE.
- When using MIMO, the modulation order has to be reduced under less than ideal transmission conditions.
- The overall capacity of the cell has to be shared by all users.
- The interference caused by transmissions of neighboring cells on the same frequency band has a further detrimental effect.
- Mobile device capabilities: Current devices are either of LTE category 3 with enough processing power for a sustained data rate of 100 Mbit/s in the downlink direction or category 4 with processing power for 150 Mbit/s.

In practice, the peak throughput that can be achieved is only about 30–50% of the theoretical values given above for medium to good signal conditions. For a cell with a 10 MHz carrier and 2×2 MIMO, transmission speeds on the IP layer of 30 Mbit/s can be observed in practice and around 60 Mbit/s with a 20 MHz channel. A more detailed capacity analysis can be found in Chapter 3.

2.3.3.14 Radio Resource Control

As in UMTS and HSPA, the LTE network controls access to the air interface resources for both the uplink and the downlink. As there is no longer a central node in the radio network for the administration of resources, the base stations themselves are now responsible for the following tasks:

- Broadcasting of system information.
- Connection management — the mobile devices and the network use control channels such as the random access channel, the paging channel, and the DCCHs to exchange RRC messages. The first RRC messages exchanged when accessing the network for the first time, or after a long time of inactivity, are connection establishment messages. The eNodeB is then responsible for setting up a logical signaling bearer to the device via the shared uplink and downlink channel or denying the request where the system is overloaded. Connection management also includes the establishment of dedicated bearers, again over the shared physical channel, based on the QoS parameters of the user's subscription.
- Measurement control — as users move, the radio environment is very dynamic and devices therefore need to report signal strength measurements of the current and neighboring cells to the network.
- Mobility procedures — based on signal measurements of the mobile device, the eNodeB can initiate a handover procedure to another cell or even to another radio network such as UMTS or GSM/GPRS where the LTE coverage area is left.

2.3.3.15 RRC Connected State

To minimize the use of resources in the network and to conserve the battery power of mobile devices, there are several connection states. While data is exchanged between the network and a mobile device, the RRC connection is in the connected state. This means the network can assign resources to the device on the shared channel at any time and data can be instantly transmitted. The mobile remains in connected state even if no data is transferred for some time, for example, after the content of a Web page has been fully loaded. This ensures instant package transmission without any further resource control overhead, for example, when the user clicks on a link.

While in full connected state, the mobile has few opportunities to deactivate its receiver, which has a negative impact on the battery capacity. After some time of inactivity, the network can thus decide to activate a DRX Mode while the mobile is still in connected state. This means that the mobile only has to listen to downlink bandwidth assignments and control commands periodically and can switch off its receiver at all other times. The DRX interval is flexible and can range from milliseconds to seconds.

Even while in DRX mode, mobility is still controlled by the network. This means that the mobile device has to continue sending signal measurement results to the network when a defined high or low signal threshold for the current cell or a neighboring cell is met. The eNodeB can then at any time initiate a handover procedure to another cell if required.

2.3.3.16 RRC Idle State

If no packets have been transmitted for a prolonged amount of time, the eNodeB can put the connection to a user in RRC Idle state. This means that, while the logical connection to the network and the IP address is retained, the radio connection is removed. The MME is informed of this state change as well, as IP packets arriving from the Internet can no longer be delivered to the radio network. As a consequence, on receipt of IP packets the MME needs to send a paging message to the mobile device, which leads to the re-establishment of a radio bearer. In case the mobile device needs to send an IP packet while in RRC idle state, for example, because the user has clicked on a link on a Web page after a long time of inactivity, it also has to request the establishment of a new radio bearer before the packet can be transmitted. In practice, the return from RRC idle to RRC active is very fast and requires around 100 ms. This is significantly faster than to reestablish a bearer in HSPA+. Here, the corresponding operation takes between 700 ms to several seconds depending on the initial state.

Furthermore, the network no longer controls the mobility of a device in RRC idle state and the device can decide on its own to move from one cell to another. Several cells are grouped into a tracking area, which is similar to location and routing areas used in UMTS. The mobile only reports a cell change to the network if it selects a cell which belongs to a different tracking area. This means that the network, or more specifically the MME, has to send a paging message via all the cells that belong to the tracking area when a new packet for the device arrives from the Internet.

2.3.3.17 Treatment of Data Packets in the eNodeB

In addition to radio resource specific tasks, the eNodeB is also responsible for several tasks concerning the data packets themselves before they are transmitted over the air interface. To prevent data modification attacks, also referred to as man-in-the-middle attacks, an integrity checksum is calculated for each data packet before it is sent over the air interface. Input to the integrity checksum algorithm is not only the content of the packet but also an integrity checking key which is calculated from a unique secret key that is shared between the eNodeB and each mobile device. If a message is fraudulently modified on the air interface, it is not possible to append a valid integrity checksum due to the missing key and the message is not accepted by the recipient. Integrity checking applies to IP packets, to RRC messages exchanged with the eNodeB and also to mobility and session management messages exchanged with the MME.

In addition to integrity checking, data packets are encrypted before being transmitted over the air interface. Again, the subscriber's individual shared secret key, stored on the SIM card and the HSS, is used to calculate a ciphering key on both sides of the connection. Data intercepted on the air interface can thus not be decoded as the ciphering key is not known to an attacker. Ciphering applies to IP packets, to RRC messages and also to mobility and session management messages, the latter two not being based on IP.

An optional task only performed on user data IP packets before they are transmitted over the air interface is header compression. For LTE networks, this feature is very important, especially for real-time applications such as VoIP. As VoIP is very delay-sensitive, typically only 20 ms of speech data is accumulated in a single IP packet. With a data rate of around 12 kbit/s produced by sophisticated speech codecs, such as an Adaptive Multi-Rate (AMR) codec with a good voice quality, each IP packet carries around 32 bytes of data. In addition, there is an overhead of 40 bytes for an IPv4 header, the UDP header and the Real-time Transfer Protocol (RTP) header. With IPv6, the overhead is even larger due to the use of 128 bit IP addresses and additional header fields. This means that there is more overhead per packet than speech data. This greatly inflates the required data rate and therefore significantly reduces the potential number of simultaneous calls per base station. As voice calls are likely to be an important feature for LTE networks, it is necessary to compress IP packet headers before transmission. For LTE, the Robust Header Compression (ROHC) algorithm, originally specified in [28], was selected. Its advantages are:

- A very good compression ratio. The 40 bytes overhead of various encapsulated protocols are typically reduced to 6 bytes.
- A built-in feedback mechanism detects compression process corruptions as a result of air interface transmission errors. This allows an immediate restart of the compressor logic instead of letting the error propagate into the compression process of subsequent packets, as was the case with previously used header compression algorithms.
- The ROHC algorithm not only compresses the IP header but analyzes the IP packet and also compresses further encapsulated headers such as the UDP header and the RTP header where the data packet contains audio information.
- In order not to focus only on VoIP packets, ROHC is able to detect different header types in a packet and selects an appropriate overall header compression algorithm for each packet. The different compression algorithms are referred to as profiles. For VoIP packets, the RTP profile is used, which compresses the IP header, the UDP header and the RTP header of the packet. Further profiles are the UDP profile, which compresses IP and UDP headers (e.g., of Session Initiation Protocol (SIP) signaling messages, cf. Chapter 4), and the ESP (Encapsulated Security Payload) profile, which is used for compressing headers of IPsec-encrypted packets.

Integrity checking, ciphering, and compression are all part of the Packet Data Convergence Protocol (PDCP), which sits below the IP layer and thus encapsulates IP packets. The packet size, however, does not usually increase, as the additional PDCP header overhead is more than made up for by the header compression.

2.3.4 Basic Procedures

An important aspect of LTE, in addition to increasing the available bandwidth over the air interface, is to streamline signaling procedures to reduce delay for procedures such as setting up an initial connection and resuming data transfers from idle state. Figure 2.21 shows the message exchange of a device attaching to the network after it has been switched on until the point an IP address is assigned. Attaching to the network and getting an IP address is, as mentioned before, a single procedure in LTE as all services are based on IP. It does not make sense, therefore, to attach to the network without requesting an IP address as is the case today in GSM and UMTS networks.

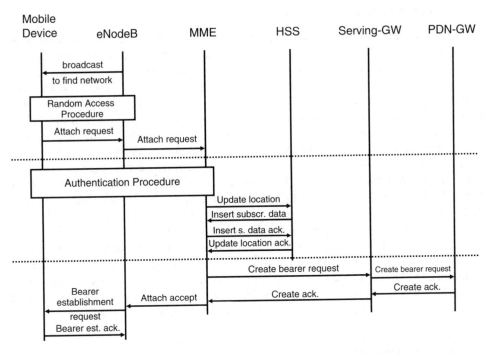

Figure 2.21 Attaching to the LTE network and requesting an IP address.

2.3.4.1 Network Search and Broadcasting System Information

The first step in attaching to the network after power on is to find all available networks and select an appropriate network to communicate with. For this, the mobile performs an initial scan in all frequency bands it supports and tries to find downlink synchronization signals. As most LTE-capable devices also support 2G and 3G networks such as GSM and UMTS, the network search procedure also includes such networks. As most bands are not dedicated to a single network technology, the mobile must be able to correlate downlink signals to different radio systems. In case of LTE, the mobile searches for synchronization signals which are placed at regular intervals in the center subchannels (1.25 MHz) of an LTE carrier. Once these are found and properly decoded, the BCH can be read and the mobile downloads the cell's complete system information. In most cases, the mobile device starts its search on the last used channel before it was switched off. If the device has not moved since it has powered off, the network is found very quickly. If a network is found, the registration process continues. If not, the process is repeated until the network is found or all supported frequency bands have been searched. Where the last used network before the device was powered down is not found, the mobile either selects a network on its own based on the preferences stored on the SIM card or presents the list of detected networks to the user who can then select the network of their choice (e.g., in case of roaming).

2.3.4.2 Initial Contact with the Network

After the broadcast information of a cell has been read and the decision has been made to use the network, the mobile device can attempt to establish an initial connection by sending a short message on the random access channel. The channel is referred to as a random access channel as the network cannot control access to this channel. There is thus a chance that several devices attempt to send a message simultaneously which results in a network access collision. If this happens, the base station will not be able to receive any of the messages properly. To minimize this possibility, the message itself is only very short and only contains a 5 bit random number. Furthermore, the network offers many random access slots per second to randomize access requests over time. When the network picks up the random access request, it assigns a C-RNTI (Cell-Radio Network Temporary Identifier) to the mobile and answers the message with a Random Access Response message. The message contains the 5 bit random number, so the mobile device can correlate the response to the initial message and the C-RNTI, which is used to identify the mobile device from now until the physical connection to the network is released (e.g., after a longer time of inactivity). In addition, the message contains an initial uplink bandwidth grant, that is, a set of resource blocks of the shared uplink channel that the mobile device can use in uplink direction. These resources are then used to send the RRC connection request message that encapsulates an initial attach request.

2.3.4.3 Authentication

When the eNodeB receives the connection request message, it forwards the contained attach request to the MME. The MME extracts the user identity from the message which is either the IMSI or a temporary identity that was assigned to the mobile device during a previous connection with the network. In most cases, a temporary identity is sent which is changed once the user has been authenticated to reduce the number of IMSIs that have to be transmitted before encryption can be activated. If the IMSI of the subscriber is not known to the MME, the Authentication Center in the HSS is queried for authentication information. If a temporary identity is sent that is unknown to the MME, a request is sent to the mobile to send its IMSI instead. Afterwards, the network and mobile authenticate each other using secret private keys which are stored both on the SIM cards and in the authentication center, which is part of the HSS. Once the subscriber is properly authenticated, the eNodeB and the mobile device activate air interface encryption. At the same time, the MME continues the attach process by informing the HSS with an Update Location message that the subscriber is now properly authenticated. The HSS in turn sends the user's subscription data, for example, what types of connections and services the user is allowed to use, to the MME in an Insert Subscriber Data Message. The MME confirms the reception of the message which in turn terminates the Update Location procedure with an ACK message from the HSS, as shown in Figure 2.21.

2.3.4.4 Requesting an IP Address

In the next step, the MME then requests an IP address for the subscriber from the PDN-GW via the Serving-GW with a Create Bearer Request message. When the PDN-GW receives the message it takes an IP address from its address pool, creates a subscriber tunnel endpoint and returns the IP address to the MME, again via the Serving-GW. Involving the Serving-GW in the process is required, as the user data tunnel is not established between the MME and the PDN-GW but between the Serving-GW and the PDN-GW. Once the MME receives the IP address, it forwards it to the eNodeB in an Attach Accept message, which is the reply to the initial Attach Request message. The eNodeB in turn forwards the Attach Accept message including the IP address as part of a Radio Bearer Establishment Request Message to the mobile device, which answers with a Radio Bearer Establishment Response message containing an Attach Complete message. The Attach Complete Message is then forwarded to the MME and the mobile device can now communicate with the Internet or the network operator's internal IP network (e.g., to connect to the IMS).

Despite the many messages being sent back and forth between the different functions in the network, the number of messages exchanged between the mobile device and the network has been reduced compared with GSM and UMTS by performing several tasks with a single message. This significantly speeds up the overall process. Excluding network detection and reading the BCH, the procedure only takes a few hundred milliseconds.

2.3.5 Summary and Comparison with HSPA

At its introduction, LTE competes with already deployed HSPA networks. In addition to deploying LTE, network operators also continue to upgrade their HSPA networks for the use of higher order modulation, MIMO, and Ethernet-based backhaul. In a 5 MHz band, the performance of LTE and HSPA+ is similar, so other reasons are required for operators to add LTE to their already existing GSM and HSPA infrastructure. One reason for adding LTE to a cell site is to increase the available bandwidth for a certain region by off-loading traffic from HSPA to LTE, for example, for LTE dongles, tablets, and other devices with a significant data demand. As network vendors are offering base stations capable of supporting several radio technologies simultaneously, the move to LTE could come as part of replacing aging base stations. By 2012, for example, many UMTS base stations will have been in the field for almost 10 years and therefore will have to be replaced anyway. As HSPA cannot be directly replaced by LTE because many devices still "only" being HSPA-capable, multiradio technology base stations are becoming very interesting for mobile network operators. If LTE is used in the same band as the other radio technologies, a single antenna can be used. Therefore, no additional antennas will be required for many base station sites. The existing antenna might be replaced with a new one, however, to enable MIMO transmissions for LTE or by a multiband antenna; so several radio technologies in different bands can use a single antenna simultaneously. When used with more than a 5 MHz bandwidth, which is usually the case, LTE can clearly show its advantages over previous technologies, as LTE radio channels can be easily extended to 10, 15, or even 20 MHz. In addition, the simpler radio and core network with fewer components and new technologies for backhaul transmission from the base station to the rest of the network will lower network operation costs. This will be an interesting driver for network operators, as bandwidth demands keep rising while the revenue per person is flat or even declining. Finally, due to the simplified air interface signaling, LTE is much more suitable for always-on IP connectivity and applications such as instant messaging and push e-mail, which frequently communicate with the network, even while the mobile device is not actively used. A current major downside of LTE compared to HSPA is the missing voice telephony integration. A practical solution has to be found for this before LTE can become a full replacement for GSM and HSPA. This is discussed in more detail in Chapter 4.

2.3.6 LTE-Advanced

As discussed in Section 1.5, LTE as per 3GPP Release 8 does not fully meet all requirements of the ITU for IMT-Advanced 4G wireless systems. While 3GPP Release 9 was mostly used for minor LTE enhancements and cleanup purposes after the introduction of LTE in Release 8, further significant new features have been introduced for LTE starting with 3GPP Release 10. The following section gives an overview of the main features and study items of 3GPP Release 10 and beyond from what is known in 2012 and as described in the 3GPP release overviews [16]. With these enhancements, the system is referred to as LTE-Advanced and has met the criteria as a full IMT-Advanced 4G technology [29].

2.3.6.1 Carrier Aggregation

The area that has received the most attention in the radio network for LTE-Advanced is how to further increase the average cell capacity and peak user data rates. One way to increase both the values is to increase the channel size (bandwidth) that can be used in a cell and that a mobile can receive. Up to 3GPP Release 9, the maximum carrier bandwidth was 20 MHz. In newer releases, bandwidths of up to 100 MHz are foreseen and a number of infrastructure manufactures have demonstrated the basic feasibility of such an ultralarge channel in early technology demonstrations [30]. To increase the total channel bandwidth while at the same time remaining backwards compatible, a CA scheme has been defined to bundle several standard LTE channels with a bandwidth up to 20 MHz in up to two different bands.

For a 100 MHz channel, five 20 MHz channels need to be bundled. In practice, this is difficult to achieve as mobile network operators do not have such a large amount of bandwidth available, even when refarming some of their already used spectrum for LTE. A typical network operator in Europe, for example, has the following amount of spectrum available today in the following bands (per transmission direction, i.e., per uplink and downlink):

- 10 MHz in the 800 MHz digital dividend band, used for LTE,
- 10–15 MHz in the 900 MHz, used for GSM and UMTS,
- 10–20 MHz in the 1800 MHz, used for GSM and LTE,
- 10–20 MHz in the 2100 MHz band used for UMTS,
- 20 MHz in the 2600 MHz for LTE.

The total combined spectrum of this list is far less than what would be required for a single 100 MHz channel even if all the spectrum currently used by GSM and UMTS would be refarmed. Such large channels can therefore just be created, for example, by assigning further spectrum in other bands, for example, the so far mostly unused 3.5 GHz band, which however has severe limitations in terms of range and is thus likely to be only suitable for small hotspot areas. Another option for larger channels would be for network operators to merge and be allowed by the regulator to keep all of their spectrum in the process. And even then, their spectrum would still be distributed in many bands, still making it difficult to get enough spectrum for 100 MHz channels when only two bands can be combined as currently defined in the specification. In the United States, current spectrum assignments to mobile network operators is in a similar order as described in [31].

In the short and medium terms, it is more likely that LTE-Advanced CA will be used to combine 20, 30, or 40 MHz of spectrum. The following list shows a number of potential combinations:

- Digital Dividend Band spectrum (10 MHz) + 20 MHz of additional spectrum in a higher band, such as the 1800 or 2600 MHz band in Europe or the 1700/2100 band (AWS — Advanced Wireless Services) in the United States for a total carrier bandwidth of 30 MHz.
- The combination of two 20 MHz channels, for example, one 20 MHz channel in the 1800 MHz band and 20 MHz in the 2600 MHz band for a total carrier bandwidth of 40 MHz.

Such combinations would fall far short of the potential of a 100 MHz channel but offers interesting possibilities to balance the data transfers across different bands in fading situations, especially when a channel on the low end of the spectrum (700 or 800 MHz band) and a channel at the upper end of the spectrum (1800 or 2600 MHz) are combined. Another scenario in which CA might become interesting is in situations where network operators want to refarm some of their spectrum, for example, in the 900 or 2100 MHz band so far used for UMTS or GSM. This would prevent a relatively narrow channel being on its own and resulting in lower data rates for users on this channel compared to others who are served on a broader channel in a different frequency band. At the time of publication, such scenarios are still only theoretical in nature and no network operators have yet announced such plans.

2.3.6.2 MIMO Enhancements

In today's LTE networks, MIMO is used only in the downlink direction and with two antennas at the base station site and two antennas at the mobile device. From a layer 1 point of view, this results in a theoretical peak data rate of 200 Mbit/s as described earlier in the chapter. With the launch of the first LTE network by Telia Sonera in Sweden in December 2009, LTE category 3 devices with a maximum processing capacity for receiving a data stream of 100 Mbit/s on a 20 MHz channel with 2×2 MIMO became available at first. In 2012, first LTE category 4 devices came on the market with the processing capacity for 150 Mbit/s on the same channel arrangement. In the uplink direction, the maximum transmission speed is defined at 50 Mbit/s.

While 4×4 MIMO transmission was already specified in LTE Release 8, it has so far not been made use of. This is because it is not only difficult to include four antennas in a small device but also because it is challenging to increase the number of antennas on cell towers. LTE-Advanced has in addition also specified 8×8 MIMO for the downlink direction. First technology demonstrations in 2011 and 2012 have shown that given enough space for the antennas and by using a receiver with sufficient computing power that data rates of 900 Mbit/s in a moving vehicle are achievable [32]. Table 2.2 gives a number of combinations of total channel bandwidth and number of MIMO channels and the resulting theoretical layer 1 peak throughput performance.

It is likely that first LTE-Advanced networks will make use of CA to combine a 10 MHz channel with another 10 MHz or perhaps a 20 MHz channel. An aggregation to obtain a 40 MHz channel might also be possible for some network operators. Owing to the limitations shown in Section 2.3.6.1, however, it is unlikely to see broader channels in the next few years.

Details on which LTE device categories are required to support higher speeds than what can be achieved in a 20 MHz channel with 2×2 MIMO can be found in 3GPP TS 36.306 [33]. It is interesting to note at this point that LTE category 3 and 4 devices found in practice today do not need to support 64QAM in the uplink direction. This also applies to future category 6 and 7 devices. 64QAM support is only required for future category 5 and 8 devices. This shows that in the uplink direction, transmission power is mostly the limiting factor rather than the modulation type. Studies about the use of MIMO techniques in the uplink direction have also been performed but the path forward in this domain is not quite clear at the time of writing.

2.3.6.3 CoMP—Coordinated Multi-Point Operation

Another important aspect in the evolution of LTE is how to improve the data rates of users in areas between cells and in areas where the signal strength of the macrocell is low. Measures to improve data rates in such locations also improve the overall network capacity, as data to such subscribers can be delivered faster and therefore, more time is available to send data to other subscribers that experience better signal conditions.

Before discussing how 3GPP Release 11 and beyond have set out to improve data rates for such users, it is interesting to understand how this is done in UMTS networks. In UMTS networks, voice calls are handed over from cell to cell with a soft-handover procedure. Two or more cells form an active set and each cell transmits the same data in

Table 2.2 Example of potential combinations of carrier aggregation and MIMO evolution and resulting theoretical peak data rates

2×2 MIMO, 20 MHz channel (LTE Release 8)	200 Mbit/s
2×2 MIMO, 30 MHz channel	300 Mbit/s
2×2 MIMO, 40 MHz channel	400 Mbit/s
4×4 MIMO, 20 MHz channel	400 Mbit/s
4×4 MIMO, 40 MHz channel	800 Mbit/s
8×8 MIMO, 40 MHz channel	1.6 Gbit/s
8×8 MIMO, 60 MHz channel	2.4 Gbit/s
8×8 MIMO, 100 MHz channel	4 Gbit/s

the downlink direction. The mobile device then decodes all incoming data streams simultaneously and reconstructs a single data stream. This scheme works for cells belonging to the same base station and also for cells belonging to different base stations and helps to reduce the transmission power spent by each cell to reach the subscriber. This significantly improves the network capacity and reduces the call drop rates during handovers. In the uplink direction, all cells try to decode the single transmission emitted by the mobile device and forward all data received to the RNC. If only one of the cells involved in a soft handover is able to capture an uplink voice packet correctly, the RNC can forward a correct voice packet to the MSC. In 3G networks, this mechanism works well for voice channels as the bearers for the soft handover are controlled by a central instance, the RNC. When HSPA was introduced and scheduling decisions were moved from the central instance to the base stations, a soft handover was no longer possible as coordination between the schedulers over backhaul links would have taken too much time. As a consequence, HSPA downlink data transmissions still use an active set but only one cell in the set transmits the data in the downlink direction. The RNC, however, can quickly switch from one cell in the active set to another to improve the overall handover performance.

As there is no central instance to control the radio bearers in LTE, the system was introduced without a soft-handover functionality. Instead, all handovers are performed as a hard handover, that is, the handover is prepared in the network by the originating cell communicating with the target cell to prepare the handover. Once data forwarding in the network is in place, the mobile device receives a RRC reconfiguration message and then starts communicating with the new cell. This is also referred to as a hard handover as the communication with one cell is stopped before it is resumed with another one. In other words, it is not possible to improve the cell edge data rates by improving the LTE handover mechanism.

For LTE-Advanced, a number of other mechanisms are therefore specified in a work item referred to as Coordinated Multi-Point (CoMP). Details are described in 3GPP TR 36.819 starting with 3GPP Release 11 [34]:

In the Joint Processing (JP) mode, the downlink data for a mobile device is transmitted from several locations simultaneously (Joint Transmission). A simpler alternative is Dynamic Point Selection (DPS), where data is also available at several locations but only sent from one location at any one time.

Another CoMP mode is Coordinated Scheduling/Beamforming (CS/CB). Here the downlink data for a mobile device is only available and transmitted from one point. The scheduling and optionally beamforming decisions are made among all cells in the CoMP set. Locations from which the transmission is performed can be changed semi-statically.

It is important to note that a CoMP transmission point is not necessarily a cell or an eNodeB as this would require coordination between autonomous entities. Here, the same problem occurs as in HSDPA, that is, each eNodeB has its own scheduler and it is not possible to synchronize them in real time because of the interconnection delays. Instead a number of CoMP scenarios rely on Remote Radio Heads (RRHs), which are connected to an eNodeB over an optical fiber link. Scheduling decisions including modulation and coding of data are performed in the eNodeB and then sent as a "ready-to-transmit" RF data stream to the RRH as an optical signal. At the RRH, the optical RF signal is then only transformed to an electromagnetic signal. A very high bandwidth connection is needed for this and in practice this can only be achieved with a fiber connection. In practice, such fiber connections are challenging to implement as a dedicated, and in some cases expensive, installation is required.

The scenarios studied by 3GPP, requiring different CoMP implementations are as follows:

- **Scenario 1, homogeneous network intrasite CoMP**: A single eNodeB base station site is usually comprised of three or more cells, each being responsible for a 120° sector. In this scenario, the eNodeB controls each of the three cell schedulers. This way it is possible to schedule a joint transmission by several cells of the eNodeB or blank out the resource blocks in one cell that are used in another cell for a subscriber located in the area between two cells to reduce interference. This CoMP method is easy to implement as no external communication to other entities is required. At the same time, this is also the major downside as there is no coordination with other eNodeBs. This means that data rates for mobile devices that are located between two cells of two different eNodeBs cannot be improved this way.
- **Scenario 2, high-power Transmit (TX) RRHs**: Owing to the inevitable delay on the backhaul link between different eNodeBs, it is not possible to define a CoMP scheme to synchronize the schedulers as described earlier. To improve on scenario 1, it was thus decided to study the use of many (nine or more) RRHs distributed over an area that would otherwise be covered by several independent eNodeBs. The RRHs are connected to a single eNodeB over fiber-optic links that transport a fully generated RF signal that the RRH only converts from an optical into an electromagnetic signal, which is then transmitted over the antenna. While this CoMP approach can coordinate transmission points in a much larger area than the first approach, its practical implementation is difficult as a fiber infrastructure must be put in place to connect the RRHs with the central eNodeB. A traditional copper-based infrastructure is insufficient for this purpose because of the very high data rates required by the RF signal and the length of the cabling.
- **Scenarios 3 and 4, heterogeneous networks**: Another CoMP approach is to have several low-power transmitters in the area of a macrocell to cover hotspots such as parts of buildings, different locations in shopping malls, and so on. The idea of this approach is to have a general coverage via a macrocell and offload localized traffic via local transmitters with a very limited range, reducing the interference elsewhere. This can be done in two ways. The localized transmissions could have their own cell IDs and thus act as independent cells from a mobile device's point of view. From a network point of view, those cells would be little more than RRHs with a lower power output instead of a high power output as in scenario 2. Another option would be to use RRHs as defined earlier with a low power output without a separate cell ID, which would make the local signal indistinguishable from the macrocell coverage for the mobile device. Again, fiber optical cabling would be required to connect the low-powered transmitter to a central eNodeB.

It should be noted at this point that the CoMP study item and first implementations focus on downlink transmissions. CoMP reception techniques to improve uplink transmissions have so far not been looked at.

Overall, the 3GPP CoMP study comes to the conclusion that data rates could be improved between 25% and 50% for mobiles at cell edges with neighboring interference, which is a significant enhancement. Except for scenario 1, however, fiber cable installations are required, which makes it unlikely that CoMP scenarios 2–4 are likely to be implemented on a broad scale in the next five years.

2.3.6.4 Hetnets and eICIC

Another complementary approach to increase the overall capacity of an LTE network is the use of small pico-eNodeB cells in places that are frequented by many users and where a high system capacity is required. Such places like airports or shopping centers are usually also covered by standard macrobase stations. Networks with macrobase stations and picobase stations are also referred to as a heterogenous network, as in contrast to homogeneous networks that only have macrocells that only overlap at their edges. The overlap between a large macrocell with a high power output and several small picocells with lower power output is desired to increase the total system capacity. The challenge arising out of such an overlap, however, is how to best mitigate the interference between the macrocell and pico cells as they use the same spectrum.

The solution specified in 3GPP is referred to as enhanced Inter-Cell Interference Coordination (eICIC) and basically separates transmissions of the macrocell from the picocell by reserving a number of subframes, which are not used by the macrobase station except for transmitting occasional control information to remain compatible with LTE Release 8. This way, there is less interference from the macrocell in these subframes, and subscribers in the picocell can thus be served with higher data rates. Also, the subframes reserved for the picocell can be used in another picocell for other subscribers. This means that the spectrum in the area of a macrocell can be reused several times. All other subframes are used by the picocells and the macrocells at the same time.

Which subframes are reserved for a particular picocell is negotiated between the macrocell and the pico cells, and the ratio between exclusive subframes of the picocell and the subframes used by all should adapt to the changing traffic patterns in the network. In [35], it is estimated that with four picocells in the area of one macrocell overall network throughput can be increased by a factor of 2.5.

Another option to increase the available network capacity in particular hotspots is to use Wi-Fi technology that is integrated into the overall cellular network with devices that can roam between the two technologies. While this is possible from a technical point of view, there are however, a number of practical advantages of picocells over tightly integrated Wi-Fi hotspots: one important aspect is security and encryption, which is the same in the picocell as in the rest of the network. Wi-Fi networks use different authentication and encryption methods and it is not clear at this point in time if all industry players can agree on how to integrate these procedures with the LTE network, which is crucial of the success of combined LTE/Wi-Fi deployments. Another advantage of picocells over Wi-Fi hotspots is the mobility management by the network, which allows to hand over connections from the macrocell to a picocell and vice versa without interruption of the connection. In this way, the IP address of the user is preserved, which is important in order to not break any ongoing connections such as for video streaming, VoIP, video calls, and so on. Furthermore, current-device-based Wi-Fi mobility management implementations would have to be significantly enhanced to ensure that the device decides early enough to leave a Wi-Fi network when the signal is getting too weak for proper service without causing a ping-pong effect between the Wi-Fi and the cellular network layers.

For completeness, it should be noted at this point that LTE Release 8 already contained a first Inter-Cell Interference Coordination (ICIC) method to mitigate interference between macro-eNodeBs. In this approach, the power of a certain number of subchannels in the frequency range of a channel can be lowered to reduce the interference in neighboring cells. These subchannels can then only be received close to the center of the cell, while all other parts of the overall channel could still be received at the cell edge. This concept is also referred to as Fractional Frequency Reuse (FFR). In contrast to this approach, eICIC introduced in LTE Release 10 as described earlier reduces the interference by blanking some subframes in the macrocell in the time domain entirely as described earlier [36]. This reduces the overall capacity of the macrocell which is why in contrast to Release 8 ICIC, close coordination between the macrocell and the picocells is required for eICIC to quickly change the ratio when less capacity is required in the picocells.

2.3.6.5 LTE-Advanced Summary

At the time of publication, the deployment of LTE networks as specified in 3GPP Release 8 is fully underway. The specification documents themselves are focused on the necessary features for making the overall system work. Release 9 then went on to specify corrections and minor enhancements. Also, significant time was invested in studies on how to further evolve the system beginning with 3GPP Release 10. Starting with Release 10, there was more time again in standardization to concentrate on nonessential features to increase the overall capacity of the radio network. As there was less time pressure for such features, it can be observed that features often have several options. Only practice will show which of these options will actually be implemented over time. While features such as CA might be picked up fairly quickly by network operators with small spectrum holdings in different bands, others such as advanced MIMO schemes will take much longer to implement because they are pushing the boundaries of what is technically feasible today. Functionalities to increase network capacity such as CoMP and Heterogeneous Networks are likely to be only considered once LTE networks are carrying significant load, which is not likely to be the case in the next couple of years. LTE-Advanced is therefore very much forward looking and offers the tools to develop the capabilities of cellular networks further once required in practice.

2.4 802.11 Wi-Fi

2.4.1 Introduction

At the end of the 1990s the first devices appeared on the market using a new wireless local area network technology that is commonly referred to today as Wireless LAN or Wi-Fi. Wi-Fi is specified by the IEEE in the 802.11 standard. It is very similar to the 802.3 fixed line Ethernet standard and reuses all protocol layers down to layer 2, as shown in Figure 2.22. The major difference between the two protocols is on layer 1,

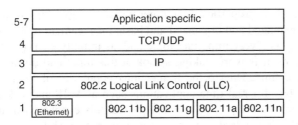

Figure 2.22 The 802.11 protocol stack. (Reproduced from *Communication Systems for the Mobile Information Society*, Martin Sauter, 2006, John Wiley & Sons, Ltd, Ref. [1].)

where the fixed line medium access has been replaced with several wireless variants. Furthermore, some additional management features were specified that address the specific needs of wireless transmissions that do not exist in fixed line networks, such as network announcements, automatic packet retransmission, authentication procedures, and encryption. Over the years, several physical layer standards were added to increase transmission speeds and to introduce additional features. Devices are usually backwards-compatible and support all previous standards to enable newer and older devices to communicate with each other.

Initially, Wi-Fi was not very popular or widely known as network interface cards were expensive and transmission speeds ranged between 1 and 2 Mbit/s. Things changed significantly with the introduction of 802.11b, which specified a physical layer for transmission speeds of up to 11 Mbit/s. Network interface cards became cheaper and devices appeared that could be connected to PCs and notebooks over the new high-speed USB (Universal Serial Bus) interface. Prices fell significantly and Intel decided to include Wi-Fi capabilities in their "Centrino" notebook chipsets. At the same time, the growing popularity of high-speed DSL and TV cable Internet connectivity made wireless networking more interesting to consumers, since the telephone or TV outlet was and still is often not close to where a PC or notebook is located. Wi-Fi was the ideal solution to this problem and Wi-Fi access points were soon integrated into DSL and cable modems. Likewise, Internet access in public places such as cafes, hotels, airports, and so on became popular, again enabled by Wi-Fi and cheap high-speed Internet access at the other end of the wireless connection via DSL. Today, Wi-Fi has become ubiquitous in notebooks and many other mobile and portable devices such as game consoles, mobile phones, and tablets.

Over time, two additional physical layer specifications were added to further increase transmission speeds. The 802.11g standard increased data transfer speeds to up to 54 Mbit/s on the air interface, and the 802.11n standard has the potential for up to 600 Mbit/s. It should be noted at this point that these speeds are only theoretical and not measured on the air interface. In practice, protocol overhead reduces the achievable speeds at the application layer to about half those values and even further in the case of 802.11n. This is discussed in more detail in the following sections. Standard 802.11a is another Wi-Fi air interface variant, but has never gained much popularity because it does not use the same standard frequency band as the other 802.11 variants.

The remainder of this chapter is structured as follows: as a first step, the Wi-Fi network infrastructure model is discussed. This is followed by an introduction to the different physical layers and their properties. Like other wireless networking technologies, the network needs to be managed and organized and basic management procedures are discussed next. Wi-Fi security is a very important topic and, as initially encryption algorithms were found to be insecure, it is worth taking a look at this topic and discussing how security was improved over time. Due to the tremendous popularity of Wi-Fi and the growing use of the technology for real-time applications such as VoIP and video streaming, QoS is becoming an important topic. As a consequence, this chapter then discusses the QoS extension of the Wi-Fi standard and how it can improve reliability for such applications.

2.4.2   Network Architecture

2.4.2.1   The Wireless Network in a Box

Unlike the network technologies described before, Wi-Fi is foremost a local area networking technology and most Wi-Fi networks are deployed as a "network in a box" as shown in Figure 2.23. In a typical home network, the Wi-Fi network bridges the final meters between the DSL modem and the devices using the fixed line Internet connection. The WLAN Access Point (AP) is usually combined with the DSL modem and also serves as the DHCP server, which provides network configuration parameters such as the IP address to notebooks and other wireless devices when they connect to the network. Most multipurpose DSL or cable routers also have a built-in Ethernet switch with several ports to connect PCs and other devices with a twisted-pair Ethernet cable.

2.4.2.2   Network Address Translation

Multipurpose routers such as the WLAN access point can connect many wireless and wired clients to the network; they usually also include NAT functionality, that separates

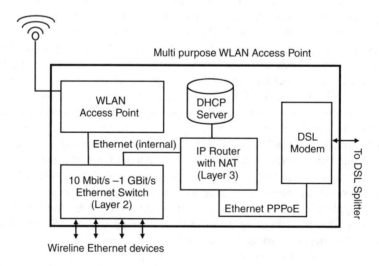

Figure 2.23   A DSL router with a wireless LAN interface. (Reproduced from *Communication Systems for the Mobile Information Society*, Martin Sauter, 2006, John Wiley & Sons, Ltd, Ref. [1].)

the local network from the DSL connection and translates IP addresses between the LAN and the Wide Area Network (WAN). This translation is required because the Internet service provider's network usually only assigns a single IP address per DSL connection. By using NAT, local IP addresses and TCP or UDP port numbers are mapped to the external IP address and the same or different TCP or UDP port numbers. This allows all local devices to communicate with servers on the Internet simultaneously via separate connections. From an external network point of view, all devices use the same IP address. While this works well for many applications, there are some for which this translation creates a problem, as incoming packets are discarded if they do not belong to a mapping that was created by an outgoing packet first. This can be solved by configuring static mappings, which forward incoming packets for a server (e.g., a Web server) to a specific internal IP address.

Static mappings, however, are only useful for servers that always use the same TCP or UDP port numbers. SIP, however, which is the dominant protocol for VoIP applications, uses dynamic port numbers. In addition, SIP applications use an IP address they can query from the local protocol stack and include it in application layer messages. As the local network stack is not aware of the external IP address assigned by the DSL network, the wrong IP address is used by the SIP client. Consequently, there are some applications that require more sophisticated solutions for traversing a NAT than static port mapping. Details of these solutions are discussed in Chapter 4.

2.4.2.3   Larger Wi-Fi Networks

Several access points can be used to extend the coverage area of a wireless network. In order that the Wi-Fi network provides a high throughput in each cell, each access point should use a different frequency. All access points broadcast the same network id, referred to as the SSID (Service Set ID) and wireless devices dynamically select which access point to connect to based on signal conditions. Mobile devices can change their association and select a different access point without losing their IP address. For this purpose, the WLAN adapter continuously scans the supported frequency bands to see if it can detect access points with a known SSID and then decides if it would be beneficial to change the association.

Today, most Wi-Fi networks use the Industrial, Scientific, and Medical (ISM) frequencies in the 2.4 GHz band. As the band has become very popular, it is becoming more and more crowded. Some 802.11n devices are therefore now also supporting an additional frequency band in the 5 GHz range.

All access points in the same network must be connected to be able to exchange data packets between devices being served by different access points and to have a single gateway to an external network (e.g., the Internet). One possibility is to connect all wireless access points via Ethernet cables to a common backhaul network infrastructure. In home environments, however, this is usually not possible so a better alternative in most cases is therefore to connect them wirelessly. The 802.11e specification contains an extension to the standard referred to as the Wireless Distribution System (WDS). WDS enables an access point to act as a standard access point for client devices and to transmit packets to a neighboring access point also via the air interface. A data packet of a device associated with an access point without a fixed line backhaul connection is therefore transmitted once over the air interface between two WDS access points and then once again from the access point to the client device. As access points usually have only one transmitter, all access points of a WDS have to operate on the same channel. This means that, in practice, devices communicating over one access point create interference for devices using another access point, which further reduces the overall throughput of the network. In home environments, this is often acceptable, as many applications work well even if only a fraction of the total bandwidth is available. Other applications such as high-definition video streaming, however, quickly run into a bandwidth bottleneck if WDS is used or if several geographically overlapping Wi-Fi networks with high traffic loads are operated on the same channel.

2.4.3 The Air Interface — From 802.11b to 802.11n

Over the last 10 years, the IEEE has specified a number of enhancements, also referred to amendments, to the original 802.11 air interface. A number of amendments were made to increase transmission speeds. All changes, however, have been specified in a backwards compatible manner, which means all devices can communicate with all networks, no matter which version of the standard they support.

2.4.3.1 802.11b — The Breakthrough

The breakthrough for Wi-Fi came after Wi-Fi chips became reasonably affordable and data rates became sufficient for the majority of applications with the 802.11b standard. Prior to the 11b amendment, data transmission rates were limited to 2 Mbit/s on the air interface, or about 1 Mbit/s at the application layer under ideal radio conditions with a channel bandwidth of 25 MHz. The 802.11b standard increased data rates on the air interface to up to 11 Mbit/s and about 7 Mbit/s at the application layer. At the time, this was more than sufficient for DSL and TV cable connections to the Internet, which were mostly in the range between 1 and 2 Mbit/s.

The air interface of both the initial 802.11 standard and the 802.11b amendment is based on a modulation scheme referred to as Direct Sequence Spread Spectrum (DSSS). In DSSS, each data bit is encoded in several chips, which are then transmitted over the air interface. This is similar to the spreading used in UMTS. Wi-Fi, however, always uses 11 chips per bit and only uses two-chip sequences, one for a "0" bit and another one for a "1" bit. This means that Wi-Fi only uses spreading to improve the robustness of the transmission but not for MA as in UMTS, which uses separate sequences for each client device.

In the slowest but most robust data transmission mode with 1 Mbit/s, Differential Binary Phase Shift Keying (DBPSK) modulation is used, which encodes one chip per transmission step (symbol). The 2 Mbit/s transmission mode uses QPSK modulation to transfer two chips per symbol.

To further increase data transfer rates, it was decided to reduce the amount of redundancy. Instead of encoding one bit into 11 chips, the High Rate Direct Sequence Spread Spectrum (HR-DSSS) physical layer introduced with 802.11b directly translates blocks of 8 bits into different chip sequences which are then transmitted over the air interface. This removes most of the redundancy, which reduces the reliability in less than ideal signal conditions. The amendment therefore also specifies a 5.5 Mbit/s data transfer mode.

To remain backwards-compatible to the original standard, the header of each data frame is transmitted with the original 1 Mbit/s DSSS modulation and DBPSK. Another reason for using this robust modulation and coding for all headers is that even the most distant devices can decode the headers of all frames and can thus decide if they have to receive and decode the rest of the frame.

The following list gives an overview of 802.11b air interface parameters, which are interesting to compare with those given for LTE in Section 2.3:

- **Bandwidth per channel** — 20 MHz.
- **Frame sizes** — 4–4095 bytes. Due to IP layer length limits, frames usually do not exceed 1500 bytes.
- **Frame transmission time** — depends on the modulation and coding used and the size of the frame. A frame with a payload of 1500 bytes requires a transmission time of 12 ms if sent with a speed of 1 Mbit/s. When signal conditions are good and the 11 Mbit/s HR-DSSS modulation is used, the same data packet is transmitted in only 1.1 ms. In addition to those times, each frame is acknowledged by a short ACK frame which, together with the gap between the frames, slightly increases the overall transmission time.
- **Retransmissions** — when a frame has not been received correctly, it is automatically repeated. Due to the decentralized medium access scheme, which is described in more detail below, a random timer is started before a retransmission occurs. In practice, it can be observed that a frame is usually retransmitted in around 0.5 ms. It is also quite common that a faulty packet is retransmitted more than once since some Wi-Fi implementations do not lower the transmission speed immediately to increase redundancy [37]. Under extreme circumstances in which the selected modulation and coding does not reflect the current signal conditions, more than five retransmissions can be observed before the frame is finally received correctly.

2.4.3.2   802.11g — The Mainstream

The 802.11g amendment to the Wi-Fi standard made a radical break in terms of modulation and coding as the spreading approach was replaced by OFDM.

OFDM is also used by LTE and for a basic introduction to OFDM, see Section 2.3.3. It is interesting to note that Wi-Fi was the first popular wireless technology to introduce OFDM. On the air interface, the maximum transmission speed specified is 54 Mbit/s, while at the application layer the highest achievable throughput is around 20–24 Mbit/s. This is, in most cases, sufficient for ADSL2+ and advanced cable modem connections.

To offer the best possible throughput for all signal conditions, 802.11g specifies a number of different modulation and coding modes that result in data transmission speeds between 6 and 54 Mbit/s, as shown in Table 2.3. To reach a transmission speed of 54 Mbit/s, 64QAM is used, which encodes 6 bits per transmission step. To be able to correct transmission errors to a certain degree, redundancy is added and a coding rate of 3/4 is used, that is, 1 extra bit is inserted for every 3 user data bits. As Wi-Fi uses 48 OFDM subchannels, 288 bits are transmitted per symbol. When the coding overhead is removed, 216 bits remain for user data.

The slowest transmission mode of 6 Mbit/s is foreseen for harsh signal conditions. Here, Binary Phase Shift Keying (BPSK) modulation is used, which encodes 1 bit per symbol. In addition, the much more robust 1/2 coding is used, which inserts one error correction bit for each user data bit (i.e., 50% overhead). In total, only 48 bits are transmitted using 48 subchannels, out of which only 24 bits are user data.

It is interesting to compare the 48 subchannels used by 802.11g in a 20 MHz band to the 1201 subchannels used by LTE in a similar bandwidth (cf. Table 2.1), as it reveals a lot about the different designs of the two systems. As LTE uses an order of a magnitude more subchannels, each symbol can be transmitted for a much longer time to counter the negative effects of long delay spreads that can appear when the signal travels over larger distances. Wi-Fi on the other hand does not require such equalization, as it is designed for short-range use where delay spread is not as pronounced. It thus uses fewer but broader channels which simplifies system design.

The 802.11g standard has been designed in a fully backwards-compatible manner. This means that older 802.11b devices can co-exist with newer devices in the same network.

Furthermore, 802.11g devices can also communicate with older 802.11b access points. This done as follows:

- Beacon frames, which broadcast system information, are modulated and encoded using the 802.11b standard.
- All frame headers, even those of 802.11g frames, are always modulated and encoded using the 802.11b standard. This means that even old devices can receive the beginning of 802.11g frames and see that they are not the recipient. Consequently, they ignore the rest of the frame, which they would not be able to decode anyway.
- When 802.11g devices detect 802.11b devices in the network, they automatically activate a protection mode and transmit short 802.11b-modulated Ready To Send (RTS) frames, which reserve the air interface for the time required to send the 802.11g-encoded frame. Devices using 802.11b decode the RTS frames and do not attempt to transmit or receive data for time specified in the RTS frame.

As these measures reduce performance, even if no 802.11b devices are in the network, most access points and client devices can be set into an 802.11g only mode.

Table 2.3 The standard 802.11g modulation and coding modes

Speed (Mbit/s)	Modulation and coding	Coded bits per subcarrier	Coded bits in 48 subcarriers per symbol	User data bits per symbol
6	BPSK, R = 1/2	1	48	24
9	BPSK, R = 3/4	1	48	36
12	QPSK, R = 1/2	2	96	48
18	QPSK, R = 3/4	2	96	72
24	16QAM, R = 1/2	4	192	96
36	16QAM, R = 3/4	4	192	144
48	64QAM, R = 2/3	6	288	192
54	64QAM, R = 3/4	6	288	216

2.4.3.3   802.11a — The Forgotten Standard

Most of the amendments made by 802.11g to the standard were already included in the earlier 802.11a amendment. The standard 802.11a, however, never became popular, since it was specified for the 5 GHz band. Therefore, it is not backwards-compatible with 802.11b, which exclusively uses the 2.4 GHz band. Devices supporting 802.11a, therefore, had to have two transceivers, one for the 2.4 GHz band and one for the 5 GHz band, which made them more expensive than single transceiver devices. As a result, there were only a few access points and devices available on the market that supported the standard.

2.4.3.4   802.11n — Breaking the Speed Barrier

A further amendment to the standard is 802.11n, which can raise data transmission speeds by an order of a magnitude compared with the 802.11g standard, if devices that communicate with each other implement all of the options of the standard. In practice, it can be observed today that 802.11n devices are capable of speeds between 100 and 150 Mbit/s at the application layer under good signal conditions. As technology progresses, it can be expected that still higher data rates will be reached with more sensitive receivers, additional antennas, and better noise cancellation techniques.

The IEEE 802.11n working group has specified a number of enhancements that all need to be implemented by a device to reach the transmission speeds quoted earlier:

- **Channel bundling** — two 20 MHz channels can be bundled to form a 40 MHz channel. This measure alone can more than double transmission speeds, as the guard band that is normally unused between two 20 MHz channels can be used for data transmission as well.
- **Support of 2.4 and 5 GHz** — since there are only three non-overlapping channels available in the 2.4 GHz band, it becomes very unlikely that a 40 MHz channel can be used in this band without interfering with other Wi-Fi networks in the same area. As a consequence, the 802.11n standard supports both the 2.4 GHz and the 5 GHz band. In the higher band, up to nine non-overlapping 40 MHz channels are available.
- **Shorter guard time** — in many environments, the guard time required to prevent the transmission of a symbol interfering with the next due refraction can be lowered from 800 to 400 ns. This reduces the symbol transmission time from 4 to 3.6 μs.
- **New coding schemes** — for excellent signal conditions, a 5/6 coding is specified which only inserts one error detection and correction bit for every five user data bits.
- **MIMO** — like LTE, 802.11n introduces the use of MIMO for Wi-Fi. More details on MIMO can be found in the section on LTE. Dual frequency (2.4 and 5 GHz) capable access points and other devices can have six antennas [38] or even more.
- **Frame aggregation** — in the default transmission mode, each transmitted frame has to be immediately acknowledged by the receiver. For transmissions of large chunks of data (e.g., a file transfer to or from a server), 802.11n aggregates several frames together. The receiver then only returns a single acknowledgement once all aggregated frames have been received.

In addition to these speed enhancements, the following additional features are also part of 802.11n:

- **Options to save cost** — since implementing all options described earlier is costly in terms of hardware and power requirements, most of them are optional. When a device connects to a network for the first time, the access point and the device exchange their capability information and then only use the options that they both support. This way, it is possible to transfer data with some devices with a very high rate, while other devices such as mobile phones, Internet tablets, and other small devices that do not require the full data transmission speeds offered by channel bundling, MIMO, and so on, operate in a single band and without MIMO to conserve battery power and to reduce hardware costs.
- **MIMO power-save mode** — if a MIMO-capable device has only a small amount of data to transfer, it can agree with the access point to switch off MIMO to reduce power consumption.
- **MIMO beamforming** — this option uses several antennas to transmit the same data stream and direct the transmission toward one device. This does not increase theoretical transmission speeds beyond the speed possible with a single antenna, but increases the range and the practical throughput for distant devices compared with a standard single stream transmission.
- **5 GHz backwards-compatibility** — 802.11n is backwards-compatible to 802.11a. In practice, however, there are only few 802.11a devices left that benefit from this as most of them have already been replaced by 802.11g or 802.11n devices.
- **2.4 GHz backwards-compatibility** — 802.11n is backwards-compatible in the 2.4 GHz band to all previous Wi-Fi standards. This means that 802.11n devices can be operated together with 802.11g and 802.11b devices in the same network. Only 802.11n devices, however, benefit from new features such as MIMO, channel bundling, and so on. As in 802.11g, new devices automatically react to older devices joining the network and start using CTS (Clear To Send) frames before transmitting 802.11n frames that cannot be detected by older devices.
- **Overlapping Base Station Subsystem (BSS) protection** — the standard also ensures that channel bundling has no negative effects on other Wi-Fi networks operating in the same area on one of the two 20 MHz channels. For this purpose, access points automatically scan their bands for beacon frames of other access points. If other beacon frames are detected, the use of two channels is discontinued immediately until both bands are clear again. This is required, as older devices cannot properly detect partly overlapping double channel networks and therefore cannot refrain from transmitting frames on the same channel while it is used by the other networks. In practice, it can be observed that some access points offer to deactivate this protection method.
- **Greenfield mode** — many access point vendors also implement a greenfield mode with no backwards-compatibility and no scans for neighboring networks. This slightly increases performance at the expense of older devices no longer being able to join the network and will potentially cause interference with neighboring networks.
- **QoS** — 802.11n devices should support QoS measures such as giving preference to frames carrying real-time data, as introduced with 802.11e, which is discussed in more detail below.

2.4.4   Air Interface and Resource Management

2.4.4.1   Medium Access

The major difference between Wi-Fi and the cellular wireless systems described earlier in this chapter is the way devices use the air interface. While LTE networks strictly control access to the network to stay in control over QoS and to prevent network overload, Wi-Fi uses a random medium access scheme which is referred to as the Distributed Coordination Function (DCF).

With DCF, access points do not assign timeslots or transmission opportunities. Instead, client devices autonomously listen to the air interface and transmit frames waiting in their output queue once they detect that no other device is currently transmitting a frame. To avoid simultaneous transmission attempts, each device uses a random backoff time. If after this random time the air interface is still unused, they transmit their frame. While highly unlikely, it is still possible, however, for two devices to start transmitting at the same time. In this case, both frames are lost and both devices have to retransmit their frames. For retransmissions, the time frame for the backoff increases to make it even less likely that devices will interfere with each other a second time.

Each frame also contains a field that informs all other devices of the duration of the transmission. This Network Allocation Vector (NAV) is analyzed by all devices and can be used to switch off the transceiver while the air interface is in use.

To ensure proper delivery of frames, the receiver has to acknowledge the proper reception of each frame, unless frame aggregation is used. Frames are confirmed by immediately returning an ACK frame. The time between a data frame and an acknowledgement frame is shorter than the shortest possible backoff time between two standard frames. This ensures that the ACK frame is always sent before any other device has a chance to send a new data frame.

Another major difference between Wi-Fi and cellular systems is that no logical channels are used, as in LTE, and that a single frame from the access point only contains data for a single device. Furthermore, management messages between the access point and a client device are sent in the same way as user data frames. Only the header of the frame marks them as management frames and Wi-Fi chips treat such packets internally instead of forwarding them to higher layers of the protocol stack. This makes the air interface very simple but much less efficient than the air interface of cellular systems.

Figure 2.24 shows how data is transmitted in practice. For each data frame, an acknowledgement is sent by the receiver after a short waiting time (the Short Interframe Space or SIFS), which is required to allow the receiver to decode the frame to check its integrity and for the transmitter to switch back into receive mode. Afterwards, the transmitter has to wait for a random time (the DCF Interframe Space or DIFS) before it can send the next frame. This gives other devices the chance to send their frames in case their random timer was initialized with a lower value. The lower part of Figure 2.24 shows how data is transmitted when frame aggregation is used and only a single ACK frame is sent to acknowledge reception.

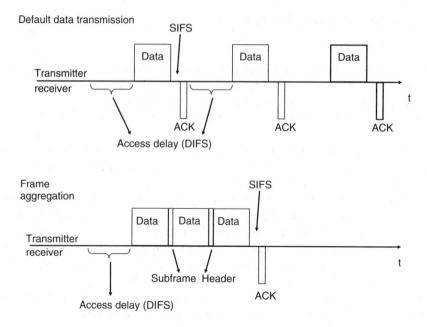

Figure 2.24   Data transmission in a Wi-Fi network.

2.4.4.2   Access Point Centric Operation

The default Wi-Fi operating mode is access point-centric. This means that client devices only communicate with the access point, even if they want to exchange data with each other. This works well for most applications, since the majority of data is exchanged between a wireless device and a server on the Internet. As devices at home or in the office become more and more connected, however, this quickly becomes an issue, as data being streamed from a notebook to a TV screen needs to traverse the air interface twice if both are wirelessly connected. The available bandwidth is thus cut in half. While the Wi-Fi standard also supports an ad-hoc mode in which no access point is required, this mode is not widely used in practice, as most home and office networks require an access point to connect the local network to the Internet.

To improve efficiency when two local wireless devices communicate with each other, the 802.11e amendment introduces the Direct Link Protocol (DLP) that enables two wireless devices to communicate directly with each other. A DLP session is initiated by the two devices exchanging DLP management frames via the access point in the network. If both devices are DLP capable they start to communicate directly with each other once the DLP negotiation is successful. In practice, however, only few devices currently support DLP.

2.4.4.3   An Example Frame

Figure 2.25 shows a typical Wi-Fi user data frame when data was sent from a client device to an access point. The data session to which this frame belongs has been traced with Wireshark (www.wireshark.org). As Wireshark is freely available at no cost, it is an ideal tool for obtaining hands-on experience with the technology [39]. The upper part of the

Figure 2.25 A typical Wi-Fi user data frame. (Reproduced from Wireshark, by courtesy of Gerald Combs, USA.)

figure shows a number of frames that have been received by the trace software (frames 778–782). Frame 778, for example, is acknowledged by the recipient with frame 779. The trace also shows that the CTS protection frames are sent before a data frame (e.g., frame 780). This indicates that older 802.11b devices are used in an 802.11g network and that the legacy protection mode has been enabled.

The main part of the window shows the header part of frame 778. The frame control field indicates that the frame "type" is a data frame; that is, it carries user data and not a Wi-Fi control message. Other important pieces of radio layer information contained in the header are the flags that indicate if the frame is a retransmission of a previous frame which was not correctly acknowledged, whether the data part of the frame is encrypted (protected), whether the originator of the frame intends to enter sleep mode after transmitting the frame and the NAV (duration) for the frame.

Unlike fixed line Ethernet frames which only contain the MAC hardware address of the source and the destination of the frame, a Wi-Fi frame additionally contains the MAC hardware address of the Wi-Fi access point. This is necessary, as several independent Wi-Fi networks can be operated in the same area. As an access point receives all frames of all networks, it relies on this information to process only frames of its own clients. At the end of the MAC header, a TKIP (Temporal Key Integrity Protocol) vector is included, which is used as an input parameter for the encryption and decryption algorithm, as discussed below.

2.4.4.4 Sleep Mode

With the increased use of mobile devices it is important to be as power-efficient as possible, especially for battery-driven Wi-Fi devices such as Internet tablets and smartphones. Many of these devices are continuously connected to the network while they are in reach but do not transmit data during most of that time. However, they must be reachable by the network for incoming phone calls, instant messages, and so on. A good balance therefore had to be found between the times during which the receiver is powered down and the time during which the device is reachable. Over the years, a number of different power-saving features have been standardized, but it is still the original sleep mode that is used today by battery-driven devices to reduce energy consumption. This sleep mode works as follows:

- When a device first connects to the network, it agrees the duration of a sleep period with the access point. The access point then assigns the device a bit in the Traffic Indication Map (TIM), which is broadcast in beacon frames. This bit is later set to 1 by the access point whenever frames have been buffered while the device was in sleep mode.
- When a device wants to activate the sleep mode, it sends an empty data frame and sets the Power Management (PWR MGT) bit in the frame header to 1. The access point acknowledges the frame and then buffers all frames that are received for the device.
- After the agreed sleep period has elapsed, the client device's receiver is switched on to receive a beacon frame. If the bit assigned to the device is still 0, no data frames are buffered and the device goes back into sleep mode. If, however, the bit is set to 1, the device usually returns to the fully active mode, powers on the transmitter and sends a request to the access point to forward the buffered frames.
- If at least one client device is in power-save mode, multicast and broadcast are buffered as well.

In practice, it can be observed that battery-driven devices enter sleep mode quickly after all data in the output buffer has been sent. A Nokia N95 smartphone, for example, enters sleep mode after no frame has been sent or received for 100 ms.

2.4.5 Basic Procedures

As in other wireless systems, Wi-Fi devices need to perform a number of management steps before access to the network is granted. This process is a similar but much simpler than the processes shown for LTE in Figure 2.21. In the first step, the client device scans all possible channels in the 2.4 GHz band and the 5 GHz band (if supported) to detect beacon frames of nearby access points. If beacon frames are received that contain a known SSID (i.e., the name of the network), the device then proceeds to the next step and performs a pseudo-authentication with the network. In practice, this step is only maintained for backwards-compatibility and is no longer used for authentication purposes as the concept was found to be flawed. This is discussed in more detail in Section 2.4.5. Afterwards, the device sends an association management frame to request the access point to accept it as its client device and to inform the access point of its capabilities, such as supported modulation schemes and supported authentication and encryption methods. The access point accepts the request with an association acknowledgement frame in which it in turn informs the client device of its capabilities. Depending on the type of authentication and encryption methods used, the device is then granted access to the network immediately or the access point enforces an authentication and ciphering key exchange message flow before access to the network is finally granted.

From the point of view of the access point, admission to the network means that it forwards Ethernet frames to and from the MAC hardware address of the device. It is therefore possible to use any kind of higher-layer protocol in the network. In practice, however, the IP protocol is dominant and other protocols are rarely used anymore.

Once the device is granted full access to the network, it usually requests an IP address from the DHCP server. This process is identical to a DHCP request in a fixed-line Ethernet network and does not contain any Wi-Fi specific elements.

2.4.6   Wi-Fi Security

2.4.6.1   Early Wi-Fi Security

On the security side, Wi-Fi had a difficult start, as the initial Wired Equivalent Privacy (WEP) authentication and encryption scheme proved easy to break in practice. While first attacks required a considerable amount of time and effort, it is now possible to break into a WEP-secured Wi-Fi network within minutes [40]. As a consequence, WEP has been superseded by more modern encryption techniques.

2.4.6.2   Wi-Fi Security in Home Networks today

When the vulnerabilities of WEP became apparent, both the IEEE and the Wi-Fi Alliance started programs to improve the situation. These parallel activities resulted in the following authentication and encryption features, which are widely used today:

- **Wireless Protected Access (WPA)** — this authentication and encryption algorithm builds on a draft IEEE security amendment. The WPA personal mode enforces an authentication procedure and a ciphering key exchange immediately after a device has associated with an access point. During the authentication phase, the client device and access point exchange random values which are used in combination with a secret password known on both sides to authenticate each other and to generate encryption keys on both ends. While there is only one secret password that is used with all client devices, the keys that are generated during this procedure are unique to each connection. This means that devices are not able to decode frames destined for other devices, despite using the same secret password. The algorithm used by WPA for authentication and ciphering is referred to as the Temporal Key Integrity Protocol. It is based on the initial algorithm used by WEP and in addition fixes all known weaknesses. At the time, this implementation was preferred over the more thorough approach proposed by the IEEE to speed up market availability. Nevertheless, WPA is still considered to be highly secure. Today, no attacks are known that could break WPA authentication and encryption, given that the password length is sufficient and that the password used cannot be broken with dictionary attacks.

- **Wireless Protected Access 2 (WPA2)** — this authentication and encryption algorithm conforms to the IEEE 802.11i security amendment and uses AES (Advanced Encryption Standard) for encrypting the data flow. All new devices offer both WPA and WPA2 authentication and encryption. Many access points can be configured to allow both WPA and WPA2 or only WPA2. To inform client devices which authentication and ciphering method they should use, a number of new information elements were added in the beacon frames.

The only known vulnerability of WPA and WPA2 personal mode is that all devices have to use the same secret password. For home networks, this is usually acceptable and also a pragmatic solution as only a few devices and a limited number of trusted people use the network. For Wi-Fi use in corporate environments, however, using a single password creates a security risk as it is much more difficult to keep the password secret.

2.4.6.3 Security for Large Office Networks

For professional use, WPA and WPA2 also have an enterprise mode that uses a standalone authentication server that is not included in the access point. This is necessary, as companies often deploy several access points, which necessitates the storage of authentication information in a central location. To authenticate devices and to be able to revoke network access for a user, individual certificates are used that need to be installed on each device. The public part of the certificate is also stored in the authentication server. When a device associates with an access point, authentication is initiated by the access point just like in personal mode. Instead of verifying the credentials itself, however, it transparently forwards all authentication frames to the authentication server in the network. Several protocols exist for this purpose and one that is often used and certified for WPA and WPA2 is EAP-TLS (Extensible Authentication Protocol — Transport Layer Security) as specified in [41]. Full network access is only given to the client device once the authentication server authorizes the access point to do so and once it supplies the encryption keys required to encrypt the traffic on the air interface.

2.4.6.4 Wi-Fi Security in Public Hotspots

A major security issue that remains to this day is public Wi-Fi hotspot deployments. For easy access to public Wi-Fi hotspots in hotels, airports, and other public places, no authentication and encryption is used on the air interface. This allows a number of different attacks of which two of the most common are described below:

If no encryption is used, data frames can easily be intercepted by anyone in range of a public hotspot. While some Web-based communication over the Internet such as online banking is transported via encrypted Hypertext Transport Protocol Secure (HTTPS) connections, many other applications such as Web mail, VoIP as well as the standard POP3 (Post Office Protocol), and SMTP (Simple Mail Transfer Protocol) e-mail are often transported without encryption on the application layer. As a consequence, passwords and Hypertext Transport Protocol (HTTP) cookies can easily be intercepted and used by an attacker either immediately or later on. A possible countermeasure against such attacks is to use software that encrypts all traffic and sends it through a tunnel to a gateway on the Internet.

An equally serious attack that has been reported from various locations is hackers cloning the start page of public hotspot operators and deploying false access points with network names of public operators. The user is redirected to the clone start page when they first access the net. The clone start page is often secured via HTTPS and looks and acts like the real landing page of the targeted operator. Using this method, credit card information can be stolen and used for other purposes without the user being able to detect the fraud. In practice, it is difficult or even impossible to protect users against such attacks, as only the URL (i.e., the Web address) of the landing page potentially reveals such an attack, as it does not belong to the operator.

2.4.7   Quality of Service: 802.11e

When the network load is low, real-time and streaming applications such as VoIP and video streaming work well over Wi-Fi networks. As soon as the network becomes loaded, however, it is necessary to prioritize the data packets of such applications to ensure a steady stream of data. This is not possible with the default DCF approach, as it treats all frames equally. The 802.11e amendment introduces several features to prioritize packets of real-time and streaming applications. In practice, only the Enhanced Distributed Channel Access (EDCA) is likely to gain widespread acceptance, as it is the only 802.11e feature that has been included by the Wi-Fi Alliance in its Wireless MultiMedia (WMM) certification program. Today, many devices already support WMM and it is likely that in the future the majority of devices will support it, as Microsoft's Windows certification program for wireless network adapters requires support for WMM. In principle, WMM works as follows: When a device wants to send a data frame, it is required to wait for a certain time after which it has to start a random timer. Only once this timer has expired can the frame be sent where no other device that also wanted to send a frame selected a smaller random time and started its transmission earlier. WMM extends this method and defines four priority classes: voice, video, background, and best effort. For each priority class, WMM specifies the maximum value of the random timer. The smaller the value, the higher the likelihood that a device will win the race for accessing the air interface. In addition, the standard also defines the maximum time a frame of a certain class is allowed to block the air interface. A voice frame, for example, is put into the voice priority class and given the highest priority, which translates into the shortest random timer value. As voice packets are small, WMM also restricts the time on the air interface for this class to prevent misuse. The video priority class gets a slightly higher random timer value, which means it is less likely to gain access to the air interface before a voice packet can be sent by another device.

The different priority queues in a device are also useful to prioritize packets of certain applications over others on a single device. VoIP data frames, for example, should always be sent before data packets in the best effort queue. This leads to the question of how applications can inform the lower protocol layer of the network stack which priority queue to put the data into. One possibility is the "diffserv" field in the header of an IP packet, which can be set by an application when it opens a connection.

It should be noted at this point that WMM only ensures QoS on the air interface. QoS on the backhaul link to the Internet via an ADSL or cable modem must be ensured by other means. In the uplink direction, the QoS can be controlled by the access point/DSL modem which can also analyze the "diffserv" field of IP packets and expedite the transmission of time critical packets over the backhaul interface. In the downlink direction, the access point/DSL modem has no control over the packet order. It is therefore the network side that should ensure the expedited forwarding of time critical packets. In practice, however, this is rarely done. Despite this lack of QoS control in the downlink direction, most real-time services such as VoIP still work well even under high network load since the downlink capacity is usually much higher than the uplink capacity. QoS control is therefore much more important in the uplink direction, which is controlled by the access point/DSL modem and not the network.

2.4.8   Gigabit Speeds with 802.11ac and 802.11ad

At the time of publication, the core features of 802.11n such as several MIMO streams, 40 MHz channels, and use of the 5 GHz band are widely available in products in the market. While the standard still contains many features, which have not yet been implemented such as, for example, an even higher number of MIMO streams, the industry has continued to develop additional enhancements to further increase the transmission speeds and pave the way for new applications.

The 802.11ac specification is the direct successor to the 802.11n specification and is expected to be finalized in 2013. A draft of the specification is available in [42] and contains the following new features.

The most straightforward way to further increase the theoretical and practical data rates is to use an even wider channel than the 2×20 MHz bonded channel currently used in practice with 802.11n. The first possibility is to use an 80 MHz channel, in effect doubling the data rate. The specification also has an option to combine two 80 MHz channels to a 160 MHz channel and has a further option to separate the two 80 MHz channels in the band to avoid reserved regions. In some countries, applications such as weather radars split the 5 GHz bands into two areas. As only the 5 GHz band offers enough spectrum for such wide channels (around 350 MHz), 802.11ac is exclusively specified for this band. In practice, it is likely that new Wi-Fi chips will support 802.11b, g, and n in the 2.4 GHz band and 802.11a, n, and ac in the 5 GHz band. While the 80 MHz channel configuration is likely to be supported by early chipsets, other channel options are likely to be added only later on.

To profit from exceptionally good signal conditions at very close range, a new modulation scheme was specified. While 802.11n supports the transmission of up to 6 bits per transmission step with 64QAM modulation, 802.11ac supports up to 8 bits per transmission step with 256QAM modulation. This, however, requires a signal-to-noise ratio of the channel that is at least twice as good as that required for reliable transmission of data with 64QAM.

At least a similar signal-to-noise ratio increase is also required for the optional transmission mode that uses 8×8 MIMO, up from four simultaneous spatial transmissions in 802.11n. As access points might have more sophisticated hardware than client devices, an option allows the use of the eight spatial streams with up to four different devices simultaneously. The access point could, for example, therefore simultaneously transmits data to four devices each using two MIMO streams, or to one device with four MIMO streams and in addition to two devices with two MIMO streams. Beamforming has also been enhanced from the previous version of the standard to increase the data rates.

The theoretical top speed that 802.11ac can potentially deliver is 6.93 Gbit/s, which however requires the combination of a 160 MHz channel, eight spatial streams, 256QAM modulation with little error correction and a short guard interval between the packets. This is likely far away from what can be achieved in practice in most scenarios. Nevertheless, 802.11ac will be a significant improvement over 802.11n. This will mostly be due to the four times wider channel bandwidth. In combination with improved antenna designs and increasing sensitivity of new Wi-Fi chips compared to previous generations, it is fair to assume that 160-MHz channel capable Wi-Fi devices will be able to reach speeds between four and six times faster than 802.11n devices. Further details can be found in [43]. A first demonstration of 802.11ac with an 80 MHz channel and a single stream data transmission from a smartphone at close range resulted in a data rate of 230 Mbit/s [44]. A conducted demo with 3×3 MIMO and an 80 MHz channel resulted in a data rate of over 800 Mbit/s [45].

2.4.9 Summary

This chapter has shown that, from a technical point of view, Wi-Fi does not compete with HSPA or LTE, as these are cellular network technologies designed to cover large geographical areas, while Wi-Fi is a local area network technology for covering hotspot areas, homes, and offices. From a commercial point of view there is a slight overlap between the two kinds of technologies since Wi-Fi is not only used in homes and offices but also for hotspot coverage in public places such as hotels, train stations, airports, and so on. Here, Wi-Fi directly competes with cellular network coverage for Internet access.

As will be discussed in the next chapter, the small cell sizes of Wi-Fi and high adoption rates are significant advantages of the technology, as many access points can be operated closely alongside each other compared with the distances required for cellular network base stations. The resulting overall bandwidth is at least 1–2 orders of a magnitude higher than the bandwidths that can be achieved with cellular networks in the same geographical area. They will therefore be a key element of future converged access network architectures that use cellular technology in combination with personal and business Wi-Fi networks that connect wireless devices to the Internet via a DSL or cable modem connections.

References

1. Sauter, M. (2006) *Communication Systems for the Mobile Information Society*, Table 3.6, John Wiley & Sons, Ltd, Chichester.
2. Ericsson (2011) Microwave Capacity Evolution–New Technologies and Additional Frequencies. Ericsson Review (June 2011), http://www.ericsson.com/news/110621_microwave_capacity_evolution _244188810_c (accessed 2012).
3. Donegan, M. (2008) T-Mobile Busts the Backhaul Bottleneck. Lightreading (Jan 2008), http://www.light reading.com/document.asp?doc_id=143211 (accessed 2012).
4. Sauter, M. (2008) The Dangers of Going SIM-Less, WirelessMoves, March 2008, http://mobilesociety .typepad./com/mobile_life/2008/03/the-danger-of-g.html (accessed 2012).

5. 3GPP (2005) Customized Applications for Mobile Network Enhanced Logic (CAMEL); Service Description; Stage 1. 3GPP TS 22.078, December 19, 2005, http://www.3gpp.org/ftp/Specs/html-info/22078.htm (accessed 2012).

6. Freescale Semiconductor. 3G Radio Network Controller (2008) http://cache.freescale.com/files/32bit/doc /white_paper/NIC3G8360EWP.pdf (accessed 2012).

7. Sauter, M. (2007) How File Sharing of Others Drains your Battery, WirelessMoves, May 2007, http://mobile society.typepad.com/mobile_life/2007/05/how_file_sharin.html (accessed 2012).

8. Holma, H. and Toskala, A. (2006) *HSDPA/HSUPA for UMTS: High Speed Radio Access for Mobile Communications*, John Wiley & Sons, Ltd, Chichester.

9. 3GPP (2008) High Speed Downlink Packet Access (HSDPA). Overall Description. TS 25.308, June 25, 2008.

10. Ferrus, R., Alonso, L., Umbert, A. *et al*. (2005) Cross layer scheduling strategy for UMTS downlink enhancement. *IEEE Radio Communications*, **43** (6), S24–S28.

11. Caponi, L., Chiti, F., and Fantacci, R. (2004) A dynamic rate allocation technique for wireless communication systems. *IEEE International Conference on Communications*, **7**, 20–24.

12. Rohde and Schwarz (2007) HSPA+ Technology Introduction. Application Note 1MA121, July 2007.

13. Sauter, M. (2011) Sleepless in Cologne–But At 16Mbit/s, June 2011, http://mobilesociety.typepad.com /mobile_life/2011/06/sleepless-in-cologne-but-at-16-Mbits.html (accessed 2012).

14. Sauter, M. (2012) Dual-Carrier HSPA+: 30Mbit/s and Counting, February 2012, http://mobilesociety .typepad.com/mobile_life/2012/02/dual-carrier-hspa-30-Mbits-and-counting.html (accessed 2012).

15. 3GPP (2011) UE Radio Access Capabilities. 3GPP TS 25.306 version 10.5.0, December 2011.

16. 3GPP (2012) Overview of 3GPP Release 11, version 0.0.9, January 2012, http://www.3gpp.org/ftp /Information/WORK_PLAN/Description_Releases (accessed 2012).

17. 3GPP (2007) Continuous Connectivity for Packet Data Users. 3GPP TR 25.903 version 7.0.0.

18. Rao, A.M. (2007) HSPA+: Extending the HSPA Roadmap, Alcatel Lucent, http://3gamericas.com/PDFs/ Lucent_RAO-MBA-Nov14-2007.pdf (accessed 2012).

19. Qualcomm (2006) System Level Analysis for HS-PDSCH with Higher Order Modulation. 3GPP TSG-RAN WG1 #47, R1-063415, November 2006.

20. 3GPP (2006) One Tunnel Solution for Optimisation of Packet Data Traffic. TR 23.809.

21. 3GPP (2007) General Packet Radio Service (GPRS); Service Description; Stage. TS 23.060 v.7.6.0.

22. Lescuyer, P. and Lucidarme, T. (2008) *Evolved Packet System (EPS)*, John Wiley & Sons, Ltd, Chichester.

23. 3GPP (2008) Evolved Universal Terrestrial Radio Access (E-UTRA) and Evolved Universal Terrestrial Radio Access Network (E-UTRAN); Overall Description; Stage 2. 3GPP TS 36.300 version 8.4.0.

24. 3GPP (2008) Evolved Universal Terrestrial Radio Access (E-UTRA); Physical Channels and Modulation. TS 36.211, version 8.2.0.

25. Sauter, M. (2007) Why IPv6 will be Good for Mobile Battery Life, WirelessMoves.com, Mar 2007, http://mobilesociety.typepad.com/mobile_life/2008/03/why-ipv6-will-b.html (accessed 2012).

26. National Instruments (2006) Addressing the Test Challenges of MIMO Communications Systems, http://zone.ni.com/devzone/cda/tut/p/id/5689 (accessed 2012).

27. Nortel (2008) Nortel MIMO Technology Provides up to Double the Existing Access Network Capacity to Serve More Customers at Less Cost. Embedded Technology Journal (Oct 25, 2008).

28. IETF (2001) RFC3095. *Robust Header Compression (ROHC): Framework and Four Profiles: RTP, UDP, ESP, and Uncompressed*, June 2001, http://www.ietf.org/rfc/rfc3095.txt (accessed 2012).

29. ITU (2012) IMT-Advanced Standards Announced for Next-Generation Mobile Technology, January 18, 2012, http://www.itu.int/net/pressoffice/press_releases/2012/02.aspx (accessed 2012).

30. Nokia Siemens Networks (2012) 4G Speed Record Smashed with 1.4 Gigabits-per-second Mobile Call #MWC12, February 2012, http://www.nokiasiemensnetworks.com/news-events/press-room/press-releases/4g-speed-record-smashed-with-14-gigabits-per-second-mobile-call-mwc12 (accessed 2012).

31. Sauter, M. (2011) Spectrum Usage Comparison, March 2011, http://mobilesociety.typepad.com/mobile _life/2011/03/spectrum-usage-comparison.html (accessed 2012).

32. Ericsson (2011) LTE Advanced: Mobile Broadband up to 10 Times Faster, June 28, 2011, http://www .ericsson.com/thecompany/press/releases/2011/06/1526485 (accessed 2012).

33. 3GPP (2011) UE Radio Access Capabilities. 3GPP TS 36.306 version 10.4.0, December 22, 2011.

34. 3GPP (2011) Coordinated Multi-Point Operation for LTE Physical Layer Aspects. 3GPP TR 36.819 version 11.1.0, December 22, 2011.

35. Qualcomm (2011) A Comparison of LTE Advanced hetNets and Wi-Fi. Whitepaper (Oct 2011).
36. ZTE (2012) Enhanced ICIC for LTE-A HetNet, February 6, 2012, http://wwwen.zte.com.cn/endata /magazine/ztetechnologies/2012/no1/articles/201202/t20120206_283266.html (accessed 2012).
37. Sauter, M. (2008) Sniffing Wifi Packets and Exploring Retransmission Behavior, May 2008, http:// mobilesociety.typepad.com/mobile_life/2008/05/sniffing-wifi-p.html (accessed 2012).
38. Higgins, T. (2007) Slideshow: Linksys WRT600N Dual-band Wireless-N Gigabit Router. SmallNetBuilder (Oct 2007), http://www.smallnetbuilder.com/content/view/30204/187/1/5 (accessed 2012).
39. Sauter, M. (2008) Wifi Tracing with an eeePC, April 2008, http://mobilesociety.typepad.com/mobile _life/2008/04/wifi-tracing-wi.html (accessed 2012).
40. Tews, E., Pychkine, A., and Weinmann, R.-P. (April 2007)aircrack-ptw/ Aircrack-ptw. http://wireless defence.org/Contents/Aircrack-ptw.htm (accessed 2012).
41. Simon, D., Aboba, B., and Hurst, R. (2008) RFC 5216. *The EAP-TLS Authentication Protocol*, March 2008, http://tools.ietf.org/html/rfc521 (accessed 2012).
42. IEEE (2011) Proposed TGac Draft Amendment, January 18, 2011, http://mentor.ieee.org/802.11/dcn/10 /11-10-1361-03-00ac-proposed-tgac-draft-amendment.docx
43. Leung, I. (2012) Moving to IEEE 802.11ac Wireless Standard. Electronics News (Mar 5, 2012), http://www.electronicsnews.com.au/features/moving-to-ieee-802-11ac-wireless-standard (accessed 2012).
44. Klug, B. (2012) Qualcomm Atheros Demos 802.11ac on MSM8960. AnandTech (Feb 27, 2012), http://www.anandtech.com/show/5594/qualcomm-atheros-demos-80211ac-on-msm8960 (accessed 2012).
45. Klug, B. (2012) 802.11ac RF Hands-On with Buffalo AirStation WZR-1750H. AnandTech (Jan 11, 2012), http://www.anandtech.com/show/5390/80211ac-rf-handson-with-buffalo-airstation-wzr1750h (accessed 2012).

3

Network Capacity and Usage Scenarios

The way in which mobile networks will be used in the future depends on many factors. This chapter discusses the current state of GSM, UMTS, and LTE networks and how a rise in capacity could affect usage in the future. Since capacity is limited this chapter also takes a look at how to steer the use of mobile network resources from a financial point of view and if it is still possible to link profitability with how much a user spends per month for using a network.

3.1 Usage in Developed Markets and Emerging Economies

In developed markets, the use of the Internet to communicate is moving away from use in specific places where Digital Subscriber Line (DSL), cable, or other broadband connections are available. Wi-Fi has become very popular in recent years due to its ability to un-tether users and allow them to move with their devices through their offices and homes. Small portable devices such as smartphones and tablets with built in wireless connectivity are also becoming widespread. Wi-Fi has thus created a virtual Internet bubble around people. During his time at Nokia, Anssi Vanjoki referred to this phenomenon as "[the] broadband Internet is no longer a socket in the wall." Due to the rise of smartphones and ubiquitous 3.5G coverage in cities, people can now leave the private Internet bubbles in their homes and offices without loosing connectivity as UMTS and LTE networks can take over and have thus become the natural extension of the personal Internet bubble. Over time, connectivity may get even more seamless as converged devices will learn which application can use which networks due to the cost associated with the expected network usage and availability of services. Music downloads are a good example for such a behavior. If a Wi-Fi connection is available, converged 3.5G/Wi-Fi devices can automatically use this type of network as it offers ample capacity, high throughput, and cheap connectivity. Users, however, will also want to browse their favorite music store's catalog when they are underway. Given adequate pricing for cellular data connectivity and sufficient cellular network capacity the experience will be almost seamless. How much capacity

3G, 4G and Beyond–Bringing Networks, Devices and the Web Together, Second Edition. Martin Sauter.
© 2013 John Wiley & Sons, Ltd. Published 2013 by John Wiley & Sons, Ltd.

is available with B3G networks will be discussed throughout this chapter. As fixed line and wireless networks together provide seamless connectivity for people, cellular network capacity is just one parameter in an overall capacity equation which also takes the fixed broadband access made available via private and public Wi-Fi access points into account.

In emerging economies the picture is quite different. Fixed line telecommunications infrastructure is not very well developed and it is unlikely that fixed line DSL or cable connections made available to mobile devices via Wi-Fi will be able to significantly reduce the load on cellular networks. Wi-Fi, however, might turn out to be a great solution to create wireless mesh networks as pioneer projects such as the MIT's (Massachusetts Institute of Technology) One Laptop Per Child (OLPC) initiative has demonstrated. Here, individual computers can be connected to the Internet by using other mobile devices to relay data packets from and to the Internet connectivity hub. The Internet connectivity hub can then use either a fixed line connection or a cellular wireless connection to route data packets to and from the Internet. While less capacity is available it is also likely that usage of the Internet will be much lower than in developed markets. This is mainly due to devices that have to be much cheaper than those in developed markets to be affordable. This in turn limits screen resolution, processing power, and on device storage capacity. Also, it is unlikely that low cost phones will be capable of supporting bandwidth intensive applications such as video streaming in the foreseeable future. Things are different with notebooks and desktop computers but it is unlikely that within the next decade use of such devices will become widespread in the developing world.

3.2   How to Control Mobile Usage

As will be shown further down in this chapter, capacity in wireless networks is limited. In fact, capacity on all types of networks is limited and to prevent overload, steering mechanisms are required. This is also important from a financial point of view since operators have a certain amount of capacity in their network which they need to sell for a price that is high enough to recover the initial Capital Expenditure (CAPEX) for acquiring licenses and for buying and installing base stations and the infrastructure behind them. A network also creates recurring costs, the so called Operational Expenditure (OPEX). These consist among other things of rental costs for properties where equipment such as base stations are installed, costs for leasing transmission lines to backhaul traffic, the monthly power bill, staff for maintaining the network, marketing, customer acquisition, and support, and so on. When wireless networks where mainly voice centric, the main instrument to control usage was the price per voice minute. Flat rate packages which include unlimited minutes seem to indicate that the pricing per minute is no longer important. This is not the case, however, since prices of such packages are based on an estimation of the average number of voice minutes spent by users to calculate the amount to charge for such "unlimited" use. As it is an average, some flat rate users will use more than the average number of minutes, while some use less. Furthermore, the fine print in many contracts still limits the maximum number of minutes per month. "Flat rates" and "unlimited" are thus in most cases anything but flat and unlimited.

3.2.1 Per Minute Charging

For mobile Internet access, charging per minute does not usually make sense. While voice calls create a fixed amount of data has to be transported through the network per minute, the amount of data generated by accessing the Internet depends highly on the application. A minute of streaming video produces an amount of data which is an order of a magnitude higher than browsing the web on a small handheld device. Another application that requires even less throughput per minute is mobile email. Devices specifically tailored for this application are usually always connected to the network or re-connect in short intervals to receive incoming messages. The amount of data transferred over time, however, is very low as most of the time, the device is just idle despite being connected to the network. For packet data, it therefore makes more sense to charge for the amount of data that has been transferred regardless of the time the device was connected to the network.

3.2.2 Volume Charging

The problem with this approach, from the point of view of many mobile operators, is that users paying for the use of a certain data volume per month does not allow for billing based on services. In the example above, a minute of video streaming is likely to generate more data than a push email generates over the course of a full month. If pricing for mobile data is based on push email consumption, downloading videos would not be affordable and most of the capacity of wireless networks would be unused. Vice versa, if data tariffs were based purely on video downloading, the price for transferring mobile email would be almost zero.

3.2.3 Split Charging

In recent years, network operators have started to adopt a dual strategy. On the one hand they are now offering services such as push email by offering dedicated and in many cases preconfigured devices together with a service contract that includes access to the network. On the other hand, operators have also started to sell transparent access to the Internet their subscribers can use with notebooks and other devices. Such offers, even though often called "unlimited," usually come with an upper volume limit to ensure network integrity. Prices for such offers with a monthly usage cap of around 5 GB have initially started at around €50 a month in many countries but have declined over time and are now available for around €10–35. In this regard it is also interesting to see differences in price and usage per country. Especially in Europe, there is a huge price difference when comparing offers in different countries mostly dependant on how much competition is present between the different network operators.

3.2.4 Small Screen Flat Rates

Some mobile network operators have also started to offer application specific "flat rates" such as unlimited web and email access from a mobile device. Based on the fact that mobile devices have smaller screen resolutions and limited storage capacity these offers are made in the hope that overall consumption per month will be much lower than if the network was used with notebooks. Pricing of such offers is in the range of €5–10. The issue with such offers is that it is difficult to ensure and control that users will only use such offers with small devices. Many schemes have been invented to prevent the use with notebooks but most of these are easy to circumvent by savvy users. Some operators are therefore offering different volume bundles between 200 and 1000 MB per month at different pricing levels. The price per megabyte of lower volume bundles is usually significantly higher than that of higher volume bundles. This way, mobile device only Internet access can be sold at a higher price and thus with a higher margin.

3.2.5   Strategies to Inform Users when their Subscribed Data Volume is Used Up

Today, several strategies exist in practice of what to do once a user exceeds the monthly volume limit. Some operators still charge a certain amount per megabyte when the limit is exceeded. If this is done without notifying the user it leads to high customer dissatisfaction. Another approach is to monitor usage and terminating the service when subscribers exceed the limit. While this has the advantage that there are no bad surprises with the next invoice, the approach is equally problematic for the user. Some operators have therefore started to introduce soft boundaries by tolerating the higher use for some time (e.g., three months) and informing their customers in various ways. Extra charges will only be applied if the user does not reduce his use in the following months. Yet another approach which is used by many network operators is to throttle transmission speed (e.g., to 64 or 128 kbit/s) once the limit is exceeded. Some operators then allow users to remove the throttle by paying for an additional data package. This is done via the web and the throttle is removed once additional data volume has been bought.

Another way to control mobile usage is to use different pricing strategies for prepaid and postpaid users. Newtork operators in some countries only offer competitive data prices to their postpaid subscribers in combination with a minimum service duration of 12–24 months. In many cases this significantly inhibits uptake of data services. Whether this is the desired effect or an unfortunate side effect is up for discussion.

In many countries, operators have started to also offer mobile Internet access to prepaid customers. Depending on the country, offers range from high priced mobile device web access to prices identical to those offered to postpaid customers. Such offers can usually also be terminated on a monthly basis as operators cannot bind prepaid customers for a longer period of time. The advantage of such an approach is that it attracts young people and students. As this is the main user group for Internet services, such offers are paving the way for mobilizing the web.

3.2.6   Mobile Internet Access and Prepaid

Prepaid offers are also interesting for tourists and international business travelers because data roaming prices are often excessive, especially when traveling overseas. A wiki on the Internet has more information on this topic [1].

In recent years, however, the European Union has undertaken steps to limit roaming prices for voice, Short Message Service (SMS), and Internet access in the EU. As a result many network operators are now offering data bundles on a daily or weekly basis to their subscribers which cost a few euros a day and are sufficient for moderate web browsing and use of email on smartphones. As an example, at the time of publication one network operator offers a 10 MB roaming data bundle for €2 a day, 50 MB for €5 a day, and 100 MB for €15 per week in the EU. For Internet access with notebooks this is usually not sufficient and as a result local Subscriber Identity Module (SIM) cards remain the better alternative.

One international mobile network operator goes even further. Hutchsion with their "3" brand and subsidiaries in some European and Asian countries as well as their network in Australia allow their prepaid and postpaid subscribers to roam into any other "3" network without paying roaming charges.

From a technical point of view, it should be noted that mobile Internet access while roaming is only slightly more expensive due to the extra expense incurred in backhauling the data to the home network before forwarding it to the Internet.

In the future it is likely that prices for mobile Internet access nationally and while roaming will continue to decline to levels attractive to more users. As discussed in more detail below many mobile operators are in the process of acquiring fixed line assets or are reuniting with their fixed line divisions to offer mobile Internet access together with fixed line DSL access at home. This will also be an important instrument to offload data traffic from cellular B3G networks to personal Wi-Fi networks at home or at the office which can carry large amounts of data at lower cost.

Future international data roaming scenarios are difficult to predict. In the EU regulatory intervention will continue in an effort to bring about more affordable prices for cellular wireless Internet access while roaming by further reducing the maximum price for voice, SMS, and data traffic and measures to increase competition in providing roaming services. The model being favored by the EU at present is to mandate a separation between a home network operator and virtual network operators, which a user can choose when roaming in the EU [2].

3.3 Measuring Mobile Usage from a Financial Point of View

Today, mobile operators report their Average Revenue Per User (ARPU) as one of the main indicators of the success of their network operation and marketing. This term is quite adequate for voice centric and non fractured markets in which one user has a single SIM card and only uses voice and SMS. When looking at markets with rising non voice use of wireless networks, however, the ARPU has become an irrelevant key figure for a number of reasons.

First, people in many countries have started using several SIM cards because each SIM card offers an advantage the other doesn't. The ARPU is now split between two SIM cards. The mobile network is not run less profitably due to this but the revenue of such a user is now split over two SIM cards. Business users are a good example of split SIM card use: Many business travelers today have one SIM card for their mobile phone and a second SIM card for the 3G data card that connects their notebooks to the Internet. In addition, many people also carry their private smartphone with them and perhaps an additional portable device such as a pad with cellular connectivity, also requiring its own SIM card. To calculate a proper ARPU the revenues of all SIM cards should be added together and only the combined value should then be put into the ARPU calculation. In practice, this is not done as there is usually no way of correlating all SIM cards to a single user, especially if some of the SIM cards were bought by a company and also because in many cases the SIM cards belong to different network operators.

Second, MVNO's (Mobile Virtual Network Operators) in some countries have started to offer cheap voice minutes but sell SIM cards without phones. This raises the question which of the following two ARPU's would be preferable:

- An ARPU of €30 a month generated with a contract which required an initial €300 subsidy for an expensive phone which is then spread over 24 months.
- An ARPU of €20 a month generated via a prepaid SIM without subsidies.

On paper the first ARPU value sounds more appealing but it's likely that the operator makes more money with the prepaid SIM despite the lower ARPU figure.

Third, mobile networks offer a wide range of services today from voice calls to high speed Internet access. This raises the question which of the following two customers is more valuable for a network operator:

- A customer that spends €30 a month on voice calls.
- A customer that spends €30 a month for Internet access.

In most cases the voice ARPU is probably more profitable than the data ARPU. However, prices for voice minutes keep declining so in the end the data customer could eventually become more profitable.

As a consequence the ARPU should be replaced by some other, more meaningful key figure adapted to the continuing changes. The following list shows a number approaches that could be used in the future to better measure mobile use from a financial point of view:

- Average revenue for a voice minute, based on all voice minutes sold in the network over the period of a month.
- Average revenue per megabyte for mobile services, that is, web surfing and other Internet activities from mobile phones.
- Average revenue per megabyte achieved with high speed Internet access from notebooks.
- SMS and MMS should also be treated in the same manner as from a price per megabyte point of view, MMS is no longer more expensive to be transported over the network than SMS messages.

3.4 Cell Capacity in Downlink

The data rate in the downlink direction of a cell (network to the user) is often referred to as the capacity of a cell. The theoretical maximum of this value is often used by network manufacturers and network operators to demonstrate the capabilities of network technologies and compare them with each other. The capacity is measured in how many bits a system

Table 3.1 Theoretical peak data rates, channel bandwidths, frequency reuse, and spectral efficiency of different wireless network technologies

Network type	Theoretical peak data rate	Channel bandwidth	Frequency reuse	Spectral efficiency
GSM	14.4 kbit/s	200 kHz	4	0.032
GPRS	171 kbit/s	200 kHz	4	0.07
EDGE	474 kbit/s	200 kHz	4	0.2
Cdma2000	307 kbit/s	1.25 MHz	1	0.25
1xEV-DO Rev.A	3.1 Mbit/s	1.25 MHz	1	2.4
UMTS	2 Mbit/s	5 MHz	1	0.4
HSDPA	14 Mbit/s	5 MHz	1	2.8
HSPA+ (2 × 2 MIMO)	42 Mbit/s	5 MHz	1	8.4
LTE	100 Mbit/s	20 MHz	1	5
LTE 2 × 2 MIMO	172.8 Mbit/s	20 MHz	1	8.6
LTE 4 × 4 MIMO	326.4 Mbit/s	20 MHz	1	16.3

can transmit per Hertz of bandwidth per second (bits per second per Hertz). Table 3.1 shows the peak data rates, channel bandwidth, frequency re-use, and spectral efficiencies for current and future cellular wireless technologies under ideal circumstances [3].

The values given in the table represent the theoretical limit of each technology. Typical speeds experienced in practice are much lower and are discussed further below. The table nevertheless demonstrates that newer technologies make better use of the available spectrum when signal conditions are ideal. This is due to the following reasons:

- **Higher order modulation**: While GSM, for example, uses GMSK (Gaussian Minimum Shift Keying) modulation that encodes one data bit per transmission step, 64 QAM (Quadrature Amplitude Modulation) is used by HSPA+ and LTE under ideal radio conditions to encode 6 data bits per transmission step.
- **Reduced coding**: Wireless Systems usually protect data transmissions by adding error detection and correction bits to the data stream. The more redundancy that is added, the more likely it is that the receiver can reconstruct the original data stream in case of a transmission error. In good radio conditions, however, less error coding is required since the likelihood of transmission errors is lower.
- **MIMO**: Multiple Input Multiple Output techniques exploit the fact that radio signals scatter on their way from the transmitter to the receiver. A 2 × 2 MIMO system uses two antennas and signal processing chains at both transmitter and receiver to transmit two independent streams of data over two independent radio paths. Under ideal circumstances the data rate doubles without using additional spectrum. With 4 × 4 MIMO, the data rate increases fourfold.

- **Beamforming**: This method exploits the fact that cells usually cover large areas and mobile devices are located at different angles. A cell then forms individual radio beams by using several antennas and transfers an individual data stream over each beam. Again no additional spectrum is required to increase the data rate. Presently, however, MIMO seems to be the preferred way of increasing spectrum efficiency.
- **Use of larger frequency bands**: Using larger frequency bands makes data transmission in the cell faster but does not increase efficiency.

In practice, cell capacity is much lower than these theoretical values which are only applicable under the most ideal circumstances. Achievable cell capacity depends on the following factors:

- **Backhaul connection**: For cost reasons, some operators prefer to use a lower capacity backhaul link to the cell than the capacity supported by the cell over the air interface.
- **Inter-cell interference**: Beginning with 3G radio technologies, all cells of a network use the same frequency band for communication. Thus, each cell interferes with its neighboring cells and the more active users a cell has to handle the more it interferes with neighboring cells. Another approach is to use different frequencies in neighboring cells. In practice, however, this is no longer feasible with frequency bands of 5 MHz or more as an operator usually has no more than two of these bands available.
- **Network capabilities and technology**: Depending on the radio technology, a given amount of bandwidth is used more or less efficiently as described above.
- **Mobile device capabilities**: Similar to networks evolving over time, mobile devices also undergo changes as radio technology improves. As a consequence, a mix of devices is used in networks with different capabilities. By improving mobile device capabilities such as antenna performance, sensitivity, signal processing, higher order modulation support, maximum number of simultaneous codes in case of WCDMA systems, processing power, and so on, mobile devices are able to better cope with a given radio environment and receive data more quickly. This increases overall capacity of the cell as the network has more opportunities to use higher order modulation and coding schemes and thus transports more data during a certain timeframe.
- **Mobile device locations**: Networks which are designed for the use with devices using antennas installed on rooftops can have a much higher overall throughput per cell. This is because the average reception conditions are much better than in the case of systems which allow mobile devices with built in antennas that are used indoors and thus experience worse signal conditions. This in turn reduces the capacity of the cell as the time spent sending data to a mobile which experiences less favorable radio conditions can not be used to send data much faster to devices with better reception conditions.
- **Frequency band used**: Wireless networks that use lower frequencies have a much better in-house coverage than those using higher frequencies. This difference can be observed today between GSM networks that use the 900 MHz band and UMTS networks which use the 2100 MHz band in Europe and Asia. When entering a building, UMTS coverage is lost much sooner than 900 MHz GSM coverage. This is why many wireless network operators are interested in getting permission to re-use lower frequency band allocations currently used for 2G systems for their 3.5G networks.

Once all these influences are taken into consideration the overall achievable spectral efficiency of the data transmission in a cell is much lower than what is given as peak values in Table 3.1.

While new radio interface technologies are specified with ever higher and higher spectral efficiencies in mind it has to be taken into account that higher spectral efficiency for top performance requires a higher the signal to noise ratio. The signal to noise ratio is determined by the Shannon-Hartley capacity equation:

$$C = B \times \log2\ (1\ +\ SNR)$$

In this equation C represents the channel capacity, B the channel bandwidth in Hertz and signal to noise ratio (SNR) is the instantaneous linear Signal to Noise Ratio. Table 3.2 shows typical spectral efficiency values for a single channel (i.e., excluding the MIMO entries in Table 3.1) and the corresponding required linear and logarithmic SNR.

The following examples of the evolution of UMTS show that an increasing signal to noise ratio is required to reach the theoretical top speed of each new step. As a consequence this means that the area in which the specified theoretical top speeds are available is shrinking from step to step:

Example 3.1: UMTS

UMTS has a theoretical cell capacity of 2 Mbit/s if all users experience perfect conditions and there is no interference from neighboring cells. With a channel bandwidth of 5 MHz this requires a spectral efficiency of 0.4:

Peak Spectral Efficiency (UMTS) = 2 Mbit/s/5 MHz = 0.4 (cf. Table 3.1)

According to the Shannon-Hartley maximum theoretical capacity equation this requires a signal to noise ratio of **−5 dB or 0.316228** (linear) as shown in Table 3.2:

Peak Channel Capacity (UMTS) = 5 MHz × log2 (1 + 0.316228) = 1.9820 MHz

Table 3.2 Required signal to noise ratio (SNR) for different spectral efficiencies

Spectral efficiency (bit/s/Hz)	Required SNR (dB)	Required SNR (linear)
10	30	1000
5	15	31.6228
2.9	8	6.30957
2	5	3.16228
1	0	1
0.4	−5	0.316228
0.14	−10	0.1
0.04	−15	0.0316228

Example 3.2: HSDPA

With High Speed Data Packet Access (HSDPA) the cell capacity is increased to a theoretical peak of 14 Mbit/s. On the physical layer this is done by using a higher order modulation and by reducing the error coding rate. To reach this theoretical cell capacity the following peak spectral efficiency is required:

Peak Spectral Efficiency (HSDPA) $= 14$ Mbit/s/5 MHz $= 2.8$ (cf. Table 3.1)

According to the Shannon-Hartley maximum theoretical capacity equation this requires a much higher signal to noise ratio compared to UMTS of about **8 dB or 6.30957 (linear)** as shown in Table 3.2:

Peak Channel Capacity (HSDPA) $= 5$ MHz $\times \log 2 \ (1 + 6.30957) = 14.3$ Mbit/s.

Example 3.3: LTE

3GPP's Long Term Evolution raises the bar once again by specifying a transmission mode with a theoretical peak cell capacity of 100 Mbit/s in a 20 MHz channel. In a 5 MHz channel the peak cell capacity would be 25 Mbit/s, which is 11 Mbit/s faster than HSDPA. To be able to reach this theoretical cell capacity the following peak spectral efficiency would be required:

Peak Spectral Efficiency (LTE) $= 100$ Mbit/s/20 MHz $= 5$ (cf. Table 3.1)

Again according to the Hartley maximum theoretical capacity equation this requires once again a much higher signal to noise ratio compared to UMTS of about **15 dB or 31.6228 (linear)** as shown in Table 3.2:

Peak Channel Capacity (LTE) $= 20$ MHz $\times \log 2 \ (1 + 31.6228) = 100.55$ Mbit/s.

As a consequence, this means that the cited theoretical peak cell capacity becomes more and more unlikely since in practice all systems experience the same signal conditions. This means that all 3G, 3.5G, and B3G systems discussed in this book will have a similar cell capacity for a given signal to noise ratio at a certain location.

Thus, other means have to be used to increase cell capacity. B3G systems use the following:

- The use of MIMO technology will counter this effect to a certain extent. The effect however is limited by the interference of the independent data streams on the same channel with each other and the fact that the number of antennas in mobile devices and also on rooftops cannot be increased beyond a reasonable limit. This is also reflected in Table 3.1 where 4 × 4 LTE does not have four times the cell capacity compared to an LTE cell that only uses a single channel. Also, base station costs will increase due to the additional cables between the base station and additional antennas required for MIMO. Cables are quite expensive due to their low signal loss properties and thus already today make up a sizable proportion of the overall base station price.

- Advanced receivers in mobile devices that are able to filter out interference can significantly improve the signal to noise ratios at a given location compared to less capable devices. In practice, this results in capacity gains between 40% and 100% for HSDPA [4].
- Increasing the channel bandwidth. This is the only parameter that scales linearly, that is, doubling the bandwidth of a channel also doubles the capacity of the cell. While bandwidth scales linearly there are two limiting factors for this parameter as well. First, power consumption of battery driven devices keep rising the broader the channel is that has to be received and decoded. Increasing the bandwidth thus has a detrimental effect on autonomy on a battery charge. Second, bandwidth for wireless communication systems is in very short supply and it will be very difficult to assign bandwidths of 20 MHz per channel or more.

3.5   Current and Future Frequency Bands for Cellular Wireless

Not every frequency band is suitable for wireless communication with mobile devices. While lower frequencies are better for in-house coverage, they increase the antenna size in mobile devices. A good example are FM radios built into mobile phones. FM radio transmits on frequencies around 100 MHz and typical mobile devices are too small to have an internal antenna for this frequency range. As a consequence the headset cable is used as an antenna and the FM radio only works if the headset is plugged in. Efficient internal mobile device antennas are hard to design below 700–800 MHz. While lower frequencies offer better in-house coverage, they also propagate much better in free space and thus reduce overall network capacity due to the larger coverage area of a cell. On the upper end of the spectrum, 6 GHz is around the highest frequency which makes sense for cellular communication. At this end, however, in-house penetration is already quite poor. Thus, this frequency band is also not usable for all purposes. The optimal space for cellular communication is therefore the frequency range between 1 and 3 GHz. Here, however, most bands are already occupied.

Table 3.3 lists current and future frequency bands for terrestrial wireless communication in Europe as described in the European table of frequency allocations [5] and 3GPP TS 25.101 [6]. There are also some frequency bands reserved for IMT-2000 two way satellite communications to mobile devices which are however excluded from this discussion.

According to Table 3.3 there are around 500 MHz available for cellular wireless communication in downlink direction and about the same amount of spectrum for the uplink direction. At the date of publication of this book, only a few early LTE networks were making use of the IMT-2000 extension band and the 1800 MHz band, previously only used for GSM. The Broadband Wireless Access (BWA) band is still unused. In the IMT-2000 band there are 12 frequency blocks (channels) of 5 MHz each available for UMTS. On average there are four operators per country, each using two channels. Consequently, only 60% of this band is actively used at the moment. The 1800 MHz band for GSM is also only partly used at the moment. As a consequence, less than a quarter of the spectrum assigned to terrestrial cellular wireless communication is currently in use.

In other parts of the world the situation is similar but different frequency ranges are used. In the United States of America, for example, the 900, 1800, and 2100 MHz ranges are not available to cellular wireless communication. Instead frequency ranges in the 850,

Table 3.3 Current and future bands assigned to be used for cellular communication in Europe

Frequency band (MHz)	Name	Used for or are foreseen to be used for	Total bandwidth for one direction (uplink or downlink)
832–862 (Uplink) 791–821 (Downlink)	Digital dividend band	LTE	30 MHz
880–915 (Uplink) 925–960 (Downlink)	GSM-900, UMTS-900, and LTE-900	GSM, UMTS, LTE	35 MHz
1710–1785 (Uplink) 1805–1880 (Downlink)	GSM-1800, LTE-900	GSM, LTE	75 MHz
1920–1980 (Uplink) 2110–2170 (Downlink)	IMT-2000 band	UMTS UMTS	60 MHz
2500–2570 (Uplink) 2620–2690 (Downlink)	IMT-2000 extension band	LTE	70 MHz 25 MHz[a]
3400–3800	Broadband Wireless Access (BWA)	Possibly LTE	200 MHz[b]

[a]Due to rx/tx separation of 120 MHz there is a 50 MHz gap between 2570 and 2620. This could potentially be used by a time division duplex system such as TD-LTE. Therefore an extra 25 MHz are counted.

[b]The band is unlikely to be used in the short to medium term future. Therefore, it is not clear whether a potential network will use FDD or TDD technology. Consequently, 200 MHz are assumed as usable for downlink data transmissions.

1700, 1900, and 2500 MHz band are used. Furthermore, a frequency range in the 700 MHz band has been auctioned off for cellular wireless communication and first LTE networks have begun to use this spectrum.

In practice it is getting more and more difficult for mobile devices to support all frequency bands for all regions. This means that it is getting more and more difficult to develop radio chips that will work worldwide, which has far reaching consequences. Already today, it is difficult to find even high end mobile devices that can be used all over the world even if local networks use a cellular standard the device was built for. For mobile device manufacturers the situation is equally undesirable as designing individual chips for different parts of the world increases product prices due to a lower economy of scale.

3.6 Cell Capacity in Uplink

This chapter has so far focused on radio aspects in the downlink direction. In the uplink direction, an equal amount of spectrum is available to transport the increasing data that users send to the network. Until recently, uplink capacity was not in the spotlight of network designers or the public since a mobile user was mainly seen as a consumer of information who sent little data back to the network. As will be discussed in more detail in Chapter 6 this is changing. One the one hand the use of notebooks that are connected via a cellular network to the Internet is on the rise. Mainly used by business travelers, students who usually have high mobility requirements, and by the general public in countries where wireless access and DSL compete heavily [7], the uplink is beginning to carry significant traffic.

One main contributor is outgoing e-mail with large file attachments which is getting more and more common as documents are exchanged between colleagues and friends and presentation documents increasing in size. Increasing uplink capacity is also required for mobile multimedia devices which are no longer just used to consume content but are also used now for uploading pictures, videos, and so on, to social web services such as Facebook, web pages, and blogs. Sending pictures and videos to public broadcasters also requires uplink resources as it is no longer uncommon to see pictures and videos taken with mobile devices for breaking news [8].

Furthermore, the rise of Internet based applications on smartphones increases interference in the uplink direction indirectly, even when no data is transferred, control information continues to flow for some time to report signal conditions and enable the network to instantaneously transmit data at the best possible speed due to renewed activity by the user. Only after some time has passed without any data being exchanged, typically about 5–15 s, the air interface connection is put into a less active state that does not require signaling information to be exchanged. Although the few seconds between state changes might seem insignificant at first, in terms of creating unwanted noise for other transmissions, it has to be kept in mind that a single base station typically serves 2000 users. As a consequence, accumulation of mobile devices simultaneously transmitting signaling data for keeping the connection in place during a significant amount of their activity time, quickly sums up to non negligible levels. For details see Chapter 2.

It can also be observed that 3G networks are more and more used by semi-professional and professional TV and radio broadcasters. Instead of using specialized radio equipment or fixed line voice connectivity for event reporting, a 3G network is used for transmitting a live stream to a broadcasting studio which then broadcasts the content via radio, TV, or the Internet to a wider audience interested in the event. Figure 3.1 shows the recording and transmission equipment of a semi-professional radio station reporting from an event and broadcasting their information via FM radio. The mobile phone in the lower part of the picture is connected via a cable to the PC and used by the software to transmit the audio stream to the Internet. In the future these trends will continue to evolve and the increased uplink bandwidth requirements of voice calls transported over IP will additionally increase the amount of data being sent to the network.

Most B3G systems use two separate channels to transmit data in the uplink and the downlink direction. This is called Frequency Division Duplex (FDD). An exception is TD-LTE used by a network operator in China and likely to be used by one network operator in the US. With this technology, the two transmission directions are separated in time (Time Division Duplex, TDD).

In principle uplink data transmission speeds are restricted by the same rules which also apply to the downlink direction. There are, however, a number of uplink specific limitations to consider for mobile devices:

- **Small antennas**: While base stations use directional antennas of considerable size to project the available transmission power in certain directions and vertical angles, mobile devices only use small omnidirectional antennas. These antennas are far less efficient than their counterparts at the base station. Transmitting in all directions is necessary since the user can change his location at any time which means that the direction of the base station keeps changing.

- **Limited transmission power**: While base stations in a typical urban scenario use a transmission power of 10–20 W per sector the mobile device is limited for continuous transmissions to 0.25 W or less depending on the frequency band. To some degree this is counterbalanced by the larger directional antennas of the base stations and more sensitive signal amplifiers in the base station.
- **Battery capacity**: High data rate transmissions require a lot of energy and thus severely impact the operating times of mobile devices on a battery charge. Every mobile device chipset generation tries to reduce the energy required for signal processing and the efficiency of the mobile's power amplifier. It is difficult, however, to keep up with the rising lower layer processing and power requirements.

Comparing the average HSDPA downlink capacity of a cell in practice today of 2–3 Mbit/s with the average High-Speed Uplink Packet Access (HSUPA) uplink capacity of a cell of around 1.4 Mbit/s [9], uplink efficiency is about 70% of downlink capacity. It is likely, that this ratio will not improve with newer technologies and network upgrades in recent years to enable higher speeds under good ratio conditions have mostly concentrated on the downlink direction.

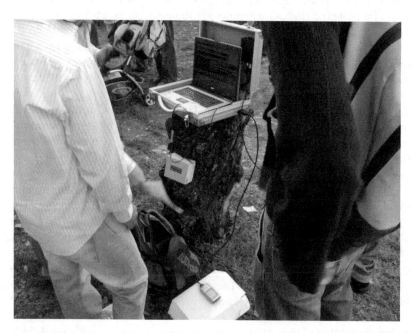

Figure 3.1 Recording equipment of a semi-professional radio station with 3G uplink.

3.7 Per-User Throughput in Downlink

As has been shown in this chapter, reaching the peak data rates of B3G systems by a mobile device is getting less and less likely. This is due to the continuously rising signal to noise ratio requirements for reaching these data rates.

In addition, the maximum throughput achieved by a mobile device depends on the following factors:

- The capabilities of the device itself.
- The number of other users and their activity (e.g., voice calls) in the cell.
- Maximum percentage of overall bandwidth that can be assigned to a single user: Previously, devices and networks were designed to use only a fraction of the cell's overall bandwidth for a single connection. B3G networks, however, can assign the majority of the bandwidth to a single mobile device if no other users are currently transmitting or receiving data in a cell.
- Operator defined bandwidth restrictions: B3G networks can restrict users to a certain bandwidth. Many network operators use this functionality to charge a premium for higher speeds.
- Amount of traffic on neighboring cells: Most 3.5G and B3G networks are based on radio access technologies in which the same frequency band is used by all cells of the network. The overall cell capacity thus also depends on the amount of traffic handled by neighboring cells. High activity in a cell increases interference for neighboring cells and as a consequence the maximum throughput that can be achieved over them.
- Backhaul capacity of the cell: To save cost, cells might be connected with a backhaul link which can carry less traffic than the air interface.

Table 3.4 shows typical per device throughput values in the downlink direction for different types of networks for average to good radio conditions. Values for GPRS, EDGE, UMTS, and HSDPA are shown for deployed networks. The first column shows throughput for a lightly loaded cell, that is, only few subscribers using the cell and a majority using applications such as web browsing with long inactivity periods as is typically the case today.

The second column shows values during an average cell load, that is, several subscribers are receiving data in downlink direction. These values are very subjective, since they depend on the number of simultaneous users and their bandwidth requirements. The column is nevertheless included as cells will get more loaded over time as usage picks up and applications require more bandwidth.

LTE is listed twice in the table, first with a channel bandwidth of 10 MHz, a typical bandwidth found in the 800 MHz digital dividend band in Europe and the 700 MHz band in the US. The second table entry gives the values for the maximum LTE channel bandwidth of 20 MHz, typically found in the 2.6 and 1.8 GHz bands used in Europe, which are significantly broader than the bands mentioned earlier and hence allow broader channel bandwidths.

All values given in the table are based on typical user speeds in networks today as reported, for example, in [10, 11]. The only exception is the speed limit given for average

Table 3.4 Typical end users data transfer rates in downlink for cells with light and average load

Technology	Typical speed with light cell load	Typical speed with average cell load	Download time of a 4.5 MB file during light cell load (s)
GPRS (5 timeslots)	60 kbit/s	40 kbit/s	614
EDGE (5 timeslots)	250 kbit/s	200 kbit/s	147
UMTS	384 kbit/s	128 kbit/s	96
HSDPA	1 Mbit/s (operator restriction)	1 Mbit/s	36
HSDPA	4 Mbit/s	2 Mbit/s	9
HSPA+ (64-QAM)	12 Mbit/s	3 Mbit/s	3
HSPA+ (64-QAM + Dual Carrier, i.e., 2 × 5 MHz)	25 Mbit/s	5 Mbit/s	1.4
LTE (10 MHz, 2 × 2 MIMO)	25 Mbit/s	5 Mbit/s	1.4
LTE (20 MHz, 2 × 2 MIMO)	50 Mbit/s	10 Mbit/s	0.7

cell loads of LTE as there were only few users in these networks at the time of publication of this book. Cells are thus only lightly loaded and values could only be measured for the light cell load scenario. Similar values also result from network simulations as, for example, described in [12].

In the last column of the table, download times for a transfer of a 4.5 MB file are shown for the given data rates under light cell load. This file size represents, for example, a 30 seconds MPEG (Moving Picture Experts Group)-4 encoded video clip with a resolution of 480p pixels and a frame rate of 30 frames per second. This is a typical video clip as encoded by Youtube in standard resolution [13].

In a wireless environment data rates are also subject to changing signal conditions. When used while being stationary, like for example, being connected to the Internet with a notebook while not moving, throughput during file downloads is stable given the number of other users in a cell and their behavior does not change.

In practice, putting the antenna or the mobile device used for connecting a notebook in a favorable position can have a significant impact on the achievable data rate. This can be demonstrated best with older HSDPA modems which only used a single antenna and had no mechanism to counter interference. Figure 3.2 shows an example of a file download in an HSDPA network with an old category 12 HSDPA modem with a theoretical top speed processing capability of 1.8 Mbit/s. At the beginning of the file transfer the antenna was in a very unfavorable position at a location with below average network coverage. The data transfer speed was around 60 000 bytes per second which equals 480 kbit/s. At about 50 seconds into the file transfer the antenna of the network card was slightly redirected while the notebook itself remained at the same place. Immediately data rates increased to about 150 000 bytes per second or around 1.2 Mbit/s. The two throughput drops during the remainder of the file download where caused by moving the antenna back and forth between the two positions to verify that it is the antenna change that caused

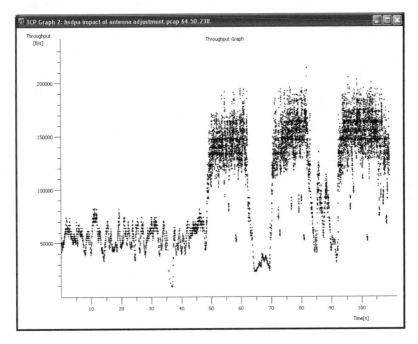

Figure 3.2 Impact of antenna position on transfer speed. (Reproduced from *Wireshark*, by courtesy of Gerald Combs, USA.)

this substantial change. From a user point of view, this means that for best performance in stationary use it is best to have an external antenna. In practice, many wireless cards thus have a connector for an additional external antenna connected to the network card via an extension cable.

Current HSDPA or LTE modems have much improved over the performance shown in Figure 3.2. On the one hand, their processing and radio capabilities allow much faster top speeds and on the other, dual antenna designs and interference cancellation algorithms make the device less susceptible to interference and dead spots. To improve reception further, many modems have a connector for a larger external antenna. By putting the antenna in a favorable location, reception can be further improved. A simpler and usually also more practical way of improving reception is to use a USB extension cable between the notebook and the USB HSPA or LTE modem. With allowed lengths of up to 3 m, the modem can be put closer to a place with better reception, for example, near a window or hung on a wall which usually results in a significantly better reception [14].

Mobility is another factor that can have a great impact on user data rates. Figure 3.3 shows the throughput of a file download while traveling on a train. Again an HSDPA network was used, this time in combination with an HSDPA category 6 mobile device (3.6 Mbit/s theoretical maximum speed). The signal picked up by the HSDPA device was received through the train windows since no repeaters were installed in the train for 3G signals. During the trace the train speed was around 160 km/h. This scenario is one of the most difficult encountered in practice since network coverage was not optimized for track coverage. Thus, signal levels varied by a great degree during the file transfer. Also, the

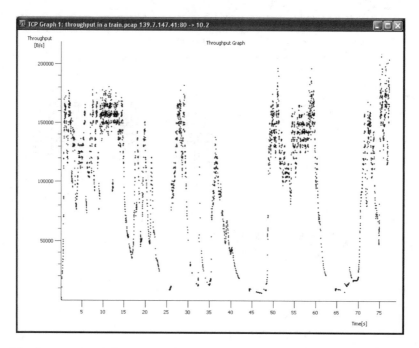

Figure 3.3 Impact of mobility on transfer speed. (Reproduced from *Wireshark*, by courtesy of Gerald Combs, USA.)

train itself and the sun and heat insulated windows have a high dampening effect on the radio signal. Handovers were another source of the throughput variations.

The impact of mobility on the user experience, that is, the resulting variable data rates, depends on the applications. For file downloads users will notice a lower average speed than while stationary. In the example above the maximum throughput achieved was around 1.5 Mbit/s while the average throughput of the 6 MB file transfer was around 850 kbit/s.

As long as the connection to the 3G network is not entirely lost, varying throughput has little impact on applications such as web browsing. This is due to the fact that web browsing, both in fixed line and wireless networks, does not usually take advantage of data transfer rates higher than 500 kbit/s, since relatively small amounts of data have to be transferred for a single web page. This will change over time as web page content keeps getting more complex as more pictures and other multimedia elements are added.

On voice over IP connections, varying signal conditions as those shown in Figure 3.3 have a great impact. Firstly, this is due to packet switched handovers from one cell to the next, which are not as optimized as those for dedicated circuit switched voice connections and thus cause a voice outage which is much longer than the optimized handover in a circuit switched voice call. Second, lost packets on the air interface are repeatedly sent until received correctly. While this approach is favorable for applications such as web browsing to prevent time intensive higher layer retransmissions, it is unsuitable for real time voice or video services. Voice codecs on higher layers have been designed to cope with packet loss to a certain extent since there is usually no time to wait for a repetition of the data. This is why data of circuit switched connections is not repeated when not received correctly but simply ignored. For IP sessions, doing the same is difficult, since a single session usually carries both real time services such as voice calls and best effort services such as web browsing simultaneously. In B3G networks, mechanisms such as "Secondary Packet Data Protocol (PDP) contexts" (UMTS) [15] and "Dedicated Bearers" (LTE) can be used to separate the real time data traffic from background or signaling traffic into different streams on the air interface while keeping a single IP address on the mobile device. This is done by an application providing the network with a list of IP addresses in a Traffic Flow Template [16]. The mobile device and the network will then screen all incoming packets and handle packets with the specified IP addresses differently, like not repeating them on the Radio Link Control (RLC) layer after an air interface transmission error. This is transparent to the IP stack and the applications on both ends of the connection. The IP Multimedia Subsystem (IMS) makes use of this functionality [17]. External providers of speech services such as Skype, however, do not have access to this functionality.

It is for further study how throughput would look in a train equipped with 3G repeaters in a similar fashion as is done today in some trains for 2G and an optimized network coverage along the railway track. The peak throughput values of 1.5 Mbit/s and Figure 3.3 suggest, that broadband wireless networks can handle the effects of high speed mobility quite well.

Data rates of individual users also depend on how many users are active in the same cell. If there are, for example, 10 users in a cell and all are browsing the net it is highly likely that when one of the 10 users loads a new page all others are reading a page and thus transfer no data. Thus, each user has the full capacity of the cell at his disposal at the moment the web page is transferred. On the other hand of the spectrum is a scenario in which all users are streaming video or downloading large files simultaneously. In this case the users have to share the available bandwidth of the cell while they are transmitting data and not in a statistical way as above. In practice, a typical scenario is in between these two extremes as different users use the network in different ways and due to the rising number of people with smartphones, tablets, and other devices connected to the network.

In addition, file transfers are becoming larger due to, for example, enhancements in camera resolution and video technology which further shifts the usage away from the previous example in which a single user has the full capacity of a cell available. Figure 3.4 shows how the data rate during a file download decreases when another user in the cell also starts a file transfer. The data rate of the file transfer is constant at around 120 000 bytes/s (around 1 Mbit/s) until the point where another user also starts a file transfer. The user's data rate then drops to around 80 000 bytes/s. The combined data rate of the two users in the cell is then on a level expected for an HSDPA enabled cell which not yet upgraded for 16QAM modulation or a cell which is only connected via a single E-1 (2 Mbit/s) to the network. In practice, such cells have become rare in most places at the time of publication of the second edition of this book. Most network operators have upgraded their UTMS cells for 16QAM and 64QAM operation, dual carrier operation, and backhaul capacity has been extended significantly and is now based on broadband backhaul technologies. This has increased overall cell capacity and top speeds but the general behavior of the speed being reduced for a user of a cell when another user also starts a data transfer has remained the same.

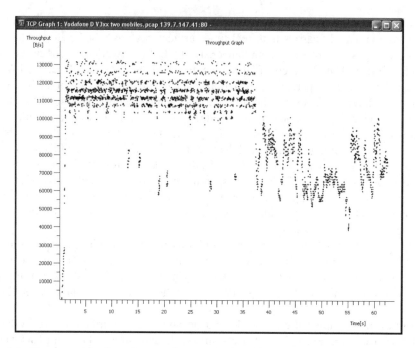

Figure 3.4 Data rate reduction when another user in the cell starts downloading a file. (Reproduced from *Wireshark*, by courtesy of Gerald Combs, USA.)

3.8 Per-User Throughput in Uplink

While most discussions around performance of beyond 3G networks focus on the downlink, the uplink is mostly neglected. With rising use of the uplink by mobile web 2.0 applications as well as growing use of notebooks with large file attachments in eMails and other applications generating large amounts of data to be sent to the network, it is worth taking a look at current and future uplink throughput per user. Table 3.5 gives an overview of typical upload speeds for some network technologies today and expected uplink performance of future systems. To translate throughput into time values the same file transfer example of a 4.5 MB file is used as previously for the downlink direction. Typical speeds were not divided into light and average cell load since uplink speeds are usually still restricted by the mobile's transmission power and not the overall cell capacity.

For GPRS and EDGE, only 3 timeslots are assumed in the uplink direction since this is the maximum number of timeslots current mobile phones support for uplink transmissions (GPRS multislot category 32) [18]. Despite virtually all 3G devices now being HSDPA enabled, there are still a significant number of new devices being sold without HSUPA capabilities. A reason for this could be the additional cost of intellectual property attached to HSUPA and the fact that even a 128 or 384 kbit/s dedicated bearers in the uplink direction result in a good user experience for applications that mainly transmit data in the downlink direction.

For LTE, standard single stream uplink transmissions were used for the estimation. While MIMO is also possible in uplink it is unlikely to be used in most situations since

Table 3.5 User data rates in uplink direction

Technology	Typical speed with light cell load	Download time of a 4.5 MB file during light cell load (s)
GPRS (3 timeslots)	36.6 kbit/s	1007
EDGE (3 timeslots)	150 kbit/s	245
HSDPA with legacy bearer in the uplink direction	384 kbit/s	96
HSDPA and HSUPA	1.4 Mbit/s	25
LTE (10 MHz)	4 Mbit/s	9
LTE (20 MHz)	4 Mbit/s	9

uplink transmission speed is usually limited by the mobile's transmission power rather than spectral efficiency. Dividing the signal energy in two traffic independent paths does therefore not make a lot of sense. From a standardization point of view, this has also been considered and in practice, mobile devices can and do concentrate their transmission power to a fraction of the available Orthogonal Frequency Division Multiplexing (OFDM) carriers to focus their signal energy in case signal conditions are not ideal. As a consequence, the data rates given in Table 3.5 for LTE do not change for a 10 or 20 MHz network deployment. Since mobiles only use a fraction of the OFDM carriers in power limited situations, the network nevertheless benefits from wider uplink channels since it allows more mobile devices to transmit their data simultaneously.

It is interesting to directly compare technologies used in practice today. With GPRS, which was introduced 12 years ago and still used today, the transmission of the 4.5 MB file takes over 16 minutes. With HSUPA, the same file can be transmitted in a mere 25 seconds. Even further advances have been made with LTE that can separate uplink transmissions on the frequency axis on a single channel and uses more sophisticated modulation schemes. For the transfer of the same file under average signal conditions, 9 seconds or even less is required. In theory, HSUPA has similar features, which, at the time of publication, however, have not yet been rolled out.

It should also be noted at this point, that file sizes are increasing as well. This counters the trend of rising per user data rates to some extent. The 30 seconds video file used as an example above was not likely to be sent when GPRS was first introduced. Only few mobile devices had built in cameras at the time with no or only very limited video capabilities. As mobile device processing power keeps increasing and camera technology in mobile devices matures, it is very likely that in the future, video file sizes will increase due to higher resolution and higher frame rates.

It remains to be seen if more advanced video compression algorithms can counter or slow down this trend. One of the best video MPEG-4 compression algorithms available today can encode a TV signal into a 1 Mbit/s data stream (which equals 3.7 MB for 30 seconds) in a quality similar to PAL (Phase Alternating Line), a color encoding system used for broadcast television. Starting with HSUPA, such a video stream could thus be sent from a mobile device to the network in real time, given the device has enough processing power to compress the input signal in real time and enough battery capacity to sustain this operation over a longer time. The sizes of other types of files users are transferring wirelessly are increasing as well. Thus, the video file example above is not only an isolated example but follows a general trend.

The current smartphone generation is able to record videos in 720p HD resolution and 30 frames a second but only compresses the data stream in real time to a stream rate of around 11 Mbit/s. This data rate is far too high for streaming over typical fixed (DSL, cable, etc.) and also wireless networks today, so a lower data rate such as 3.5 Mbit/s is required. This data rate offers a good compromise between stream quality and network capabilities and is used by video platforms such as Youtube for 720p HD video content today [13].

In summary it can be observed that while both per user uplink and downlink data rates are increasing, the amount of data transferred by users is increasing as well. It is therefore crucial that overall network capacity increases at least as fast as user demand.

3.9 Traffic Estimation Per User

Another factor with a major impact on the capacity required in today's and tomorrow's cellular wireless networks is how much data will be transferred per user per day or month. The cost for transferring data over a cellular wireless network will certainly be the main tool for network operators to steer usage to a certain extent as will be further discussed later on in this chapter. For current pricing strategies the following example shows the amount of data generated by typical notebook usage of an office worker.

When away from the office, a typical office worker generates about 70 MB of traffic in about 10 h, mostly in the downlink direction. This includes eMail, web browsing, and company database access. For this example, it is assumed that about 50 MB are received in the downlink and 20 MB are transmitted to the network. The following discussion focuses on the downlink only. The average data rate over time is 1.39 kB/s (50 000 kB/s/10 hour/60 minutes/60 seconds). Compared to the HSDPA cell bandwidth of about 300 kB/s (about 2.5 Mbit/s) as shown in Table 3.4, this is not a lot and more than 200 other subscribers with the same amount of traffic could use the same cell simultaneously. This calculation, however, is just as theoretical as assuming that all users will mostly use high bandwidth video streaming applications. Thus, a capacity requirement calculation needs to take a number of other variables into account:

- **Resource Handling**: Early 3G network technologies such as UMTS Release 99 assigned more bandwidth than required. During web browsing, for example, the channel was seldom fully used. After the web page had transferred, the network released the resources on the air interface after some time of inactivity. The channel was not released immediately to ensure a fast reaction to new data packets. This wasted a lot of capacity on the air interface, but was necessary to reduce delays. All 3.5G and beyond technologies use shared channels for data transmission in downlink for which no resource reservation is required. Also, it has been shown in Chapter 2 that systems are continuously optimized to use as little bandwidth as possible for the signaling required for the establishment a virtual channel between a mobile device and the network and maintaining it. Therefore, no additional overhead is taken into account in this example.

- **Busy Hour**: In most networks, there are certain hours of the day during which users are more active than during others. Let's assume that during busy hour, usage is three times higher than the daily average. This reduces the number of users per cell down to 72. On the other side, a single base station site is usually comprised of three cells, each covering a 120° sector. Thus, the number of high speed Internet users per base station per 5 MHz carrier increases to about 216. Note that not all 216 users would transfer data simultaneously due to the bursty nature of most of their data traffic.
- **Revenue**: Users who make use of the network in such a fashion on a daily basis are likely to accept a monthly charge of €30. These users would therefore generate the following revenue per month:

$$216 \text{ subscribers per site} \times € 30 \text{ per month} \times 12 \text{ months} = € 77.760/\text{year}$$

Over the lifetime of a base station, assumed to be 10 years, this amounts to €777.600. In addition, substantial additional revenue is generated via a base station by services requiring only little bandwidth compared to this intensive usage scenario such as voice calls, multimedia services, mobile web browsing, eMail, and other mobile device applications.

Holma and Toskala arrive at a similar number of people with a high monthly volume requirement that a base station can serve [19]. Instead of using the amount of data that users transmit per day, they base their calculation on the overall capacity of a cell per month. For a base station with two carriers per sector, compared to a single carrier used in the calculation above, they estimate that a single HSDPA cell can support up to 300 users with a monthly transmission volume between 2 and 4 GB. Furthermore, they come to the conclusion that when assuming a realistic base station price, a gigabyte of data could be delivered for around €2. This excludes all other cost such as site acquisition, backhaul transmission costs, marketing, customer acquisition, and so on.

The values for the calculations used above are certainly open for debate and will even change over time. A few years ago, the amount of data transmitted by the author while on business trips was around 40 MB per day. Lower network charges and faster transmission speeds, however, have encouraged higher use. In the future, it's likely that this trend will continue as prices decline and transmission speeds increase. At the same time, network capacity can increase as required as shown in Section 3.4. Furthermore, the cost of base station and network equipment keeps falling relative to the amount of bandwidth it is capable of transporting for reasons such as advances in technology and reduced space requirements. The deployment of small and cheaper cells to cover hotspot areas with a high bandwidth demand and the shared use of high speed fiber based backhaul networks by fixed and wireless networks also helps to reduce the cost relative to the amount of data transferred. These trends are important as revenue per user is unlikely to grow beyond the levels outlined in this chapter.

3.10   Overall Wireless Network Capacity

So far, this chapter has discussed the current and future per-cell throughput and the per-user throughput. For both, the noise generated by activity in neighboring cells was taken into account. Furthermore, it was discussed how many users can be served simultaneously by a single cell with acceptable quality of service and throughput today and how an evolution path in the future could look. These numbers were calculated based on cell capacity and expected amount of traffic generated per user. Based on these numbers, network operators can then determine how many cells are required for a certain area with a certain population density.

In cities cells are distributed with a site distance between 500 m and 2 km. Table 3.6 shows the downlink capacity per kilometer for different current and future network types for a single network with a base station inter distance of 1 km. The first row shows the capacity of a typical initial deployment. Once demand rises, operators usually install additional hardware in base stations to increase capacity. This enhanced throughput per kilometer is shown in a second column. In general the values in the table reflect typical network speeds in deployed networks rather than peak values often discussed but which are not achievable in practice due to the reasons discussed in this chapter.

For GPRS, two carriers per sector and three sectors per cell are assumed. Since the capacity of most GSM networks will no longer increase, no value is given for the low capacity deployment. This is due to the fact that it is more economical to increase the capacity by deploying B3G networks alongside a 2G network and migrating voice users to 3G. For two carriers per sector, 14 timeslots are available for voice and data traffic. Of these timeslots a varying number is used for voice calls. In Table 3.6, it is assumed that half are available for GPRS traffic. It is further assumed that the maximum GPRS speed per timeslot is 12.2 kbit/s. Higher speeds are possible with better coding schemes. In practice, however, these are only used in few networks.

Since EDGE is an upgrade for GPRS networks, the same assumptions were used as for GPRS. The maximum speed achieved per timeslot with EDGE is 59.2 kbit/s [20]. To take interference into account and users in areas with weak coverage, a speed per timeslot of 45 kbit/s is used for the estimation.

A single 5 MHz carrier is used per sector for UMTS in an initial low capacity deployment. Three sectors are assumed per base station. For low capacity deployments typical of early networks, a single E-1 was typically used per base station. This limited the total capacity of a base station to 2 Mbit/s without taking backhaul inefficiencies into

Table 3.6 Single network capacity per square kilometer with an inter cell distance of 1 km

Technology	Capacity per square kilometer with a low capacity deployment	Capacity per square kilometer with a high capacity deployment
GPRS	[a]	256 kbit/s
EDGE	[a]	945 kbit/s
UMTS	2 Mbit/s	[a]
HSDPA	7.5 Mbit/s	15 Mbit/s
HSPA+	15 Mbit/s	30 Mbit/s
LTE	60 Mbit/s	150 Mbit/s

[a] See text.

account. No value is given for a high capacity deployment since networks have increased their overall capacity by deploying HSDPA, various other enhancements and increased backhaul capacity.

A typical HSDPA low capacity deployment consists of cells with a single 5 MHz carrier per sector and three sectors per base station. When network operators are upgraded to HSDPA, they usually also install additional 2 Mbit/s E-1 connections to the base station. Thus, the overall base station capacity is assumed to be limited by the air interface capacity of 2.5 Mbit/s per sector. For a high capacity deployment, two carriers per sector are assumed and an IP based backhaul with capacity beyond the capabilities of several E-1 lines. This in effect doubles the available capacity.

It should be noted at this point that some of the base station's capacity is also used for voice calls which has to be subtracted from the capacity figures in the table. This is a growing effect as smartphones become ever more popular and users thus start avoiding the use of 2G networks because of their slow data speeds. From a network operator's point of view, the migration of voice calls to 3G can be also beneficial as this helps to to avoid further capacity upgrades to their 2G networks.

For HSPA+, no low capacity deployment is considered since the use of advanced features only makes sense to further increase HSPA capacity. The HSPA+ value takes advanced mobile receivers into account as well as other features mentioned in Chapter 2.

For LTE, capacity is calculated for a 10 and a 20 MHz carrier. For each variant, a three sector configuration and 2×2 MIMO is assumed for initial deployment. As the network expands, operators might increase the bandwidth of the carrier or start using two or even more carriers per sector, potentially in different bands. In Germany, for example, network operators have been assigned spectrum in the 800, 1800, and 2600 MHz for LTE and all bands are already used for first life network deployments in 2011 and 2012. At the time of publication, 800 MHz is mainly used in rural areas, and 1800 and 2600 MHz channels are used in cities. To increase capacity in the future, all three bands can be used simultaneously, with the 800 MHz spectrum significantly improving deep indoor coverage. In addition to these configurations, advanced receivers are assumed in mobile devices for high capacity deployments since by this time such devices are likely to be used in the network. To calculate capacity for a high capacity LTE deployment, three carriers with 10 MHz and two times 20 MHz is assumed.

It should be noted at this point that LTE specifies the use of 4×4 MIMO and LTE-Advanced even 8×8 MIMO. At present it is still not quite clear how such configurations can be used in practice and what kind of benefit they will bring. Therefore, such configurations are not taken into consideration for estimating the high capacity values for LTE given in Table 3.6.

As already discussed, the capacity increases shown in Table 3.6 beyond HSPA+ are mainly due to larger channel bandwidths. As capacity increases, achieving such high data rates on the backhaul link from the base station to the network becomes increasingly difficult. This is further discussed in Section 3.17.

The values discussed so far in this section are for a single network. In practice, most areas are typically covered by three or four operators. Thus, the values in Table 3.6 have to be multiplied by that number. Some network operators might also have more spectrum in bands mainly used for HSPA than indicated in Table 3.6 and less for LTE. As spectral efficiency of HSPA and LTE are comparable, the capacity as indicated for each technology would change while the overall capacity remains the same.

To further increase capacity per kilometer it is possible in theory to decrease the distance between the base stations further by adding additional sites. This is done, for example, around high traffic areas such as football stadiums, race tracks, downtown city streets, and other areas. Such deployments are exceptional and sometimes even only temporary since deployment costs for permanent installations is high. Increasing overall network capacity in this way is not feasible due to the cost involved and the limited number of suitable sites to install base stations. Even if suitable sites exist, the public is often opposed to installing ever more antennas in cities and resistance is likely to grow as network operators are weaving their networks ever denser.

Another option to increase overall network capacity without adding more base station sites and antennas is using more of the available bandwidth. This is what is typically done first when going from a low capacity to a high capacity network. As shown in Section 3.4 there is around 500 MHz of bandwidth available for cellular networks between 800 MHz and 4 GHz. In typical deployments in Europe today, the three to four B3G operators per country typically only use two 5 MHz downlink channels in the 2.1 GHz band in cities. This amounts to a use of 30–40 MHz of spectrum. This means that a significant amount of the capacity of this band is still unused. Even in the high capacity scenario described in Table 3.6 for UMTS, four operators only use two 5 MHz channels or 40 MHz in total which still leaves 33% of the capacity of this band unused. In addition, around 50 MHz is used by GSM networks in the lower bands today. In total, "only" around 85 MHz of the available spectrum is used today. This is up from around 70 MHz that was used on the date of publication of the first edition of this book. The increase is mainly due to a second UMTS 5 MHz channel now used by most network operators.

With the deployment of LTE by more than one network operator in cities over the next few years, this value is likely to increase. Used spectrum is likely to increase to 140 MHz once three LTE networks with 20 MHz carriers are taken into operation in the same place, a typical scenario, for example, in Europe. A further increase will be seen once network operators start using several frequency bands for LTE to increase their capacity and indoor coverage. This step, however, is unlikely to occur in the next few years due to the ample amount of capacity available offered by current networks UMTS networks and the significant addition of capacity with LTE.

Table 3.7 shows the amount of capacity per kilometer available today under the following assumptions: Four GSM operators have deployed their networks alongside each other. Each network operator uses base stations with three sectors and two carriers per sector. In a two carrier configuration an average of seven timeslots per carrier is used for data transfer (and voice calls). Two timeslots are used for signaling and are thus not counted. It is further assumed that two of the four operators use EDGE with an average data rate per timeslot of 50 kbit/s. The other two operators only use standard GPRS and the data rate per timeslot is 12.2 kbit/s. On average, the data rate per timeslot is thus 31.3 kbit/s. To account for voice calls, it is assumed that half the timeslots are used for this purpose. For UMTS the same number of network operators is assumed to have deployed base stations with three sectors and two 5 MHz channels per sector. They all use HSDPA+, which is a bit forward looking in 2011 but likely to be the case in many countries soon, and thus reach an average data rate of about 4 Mbit/s per carrier per base station. It is

Table 3.7 Downlink capacity per square kilometer with four GSM and four UMTS operators today in advanced markets

Operator	Capacity formula	Capacity per square kilometer
Four GSM operators, each operating with one base station per square kilometer, three sectors, two carriers per sector	Four operators × three sectors × two carriers/sectors × seven timeslots × 31.1 kbit/s/2	2.6 Mbit/s
Four UMTS/HSDPA operators, each operating with one base station per square kilometer, three sectors, and two 5 MHz carrier per sector with HSPA+	Four operators × three sectors × two carrier/sectors × 4 Mbit/s HSDPA throughput	96 Mbit/s
		Total capacity per square kilometer: 98.6 Mbit/s

further assumed that the majority of voice calls are still handled by the 2G network and thus there is no significant impact on 3G data capacity. As shown in Table 3.7 such a setup amounts to a capacity per kilometer of around 98.6 Mbit/s.

This value is up from 32.6 Mbit/s per km^2 calculated in the first edition of this book, as in 2008, most network operators only had a single 5 MHz UMTS channel deployed. Since then, many networks have upgraded their radio access networks to support higher HSDPA data rates and new smartphones now usually support either HSDPA category 10 (14.4 Mbit/s with 16 QAM modulation) or category 14 (21 Mbit/s, 64 QAM). High end USB data modems in the meantime also support dual-carrier operation, which, however, does not have a great effect on overall capacity as the functionality bundles the capacity of two 5 MHz carriers which is part of the overall calculation already.

For a long term estimation of capacity availability, the deployment of LTE by several network operators enables a significant increase in available capacity. First LTE deployments have already taken place and are now in regular service. It is expected that as capacity requirements increase more network operators will follow. Those that have already deployed LTE today are likely to use more spectrum with LTE as capacity requirements grow. Their capacity adds up to the existing networks today in the same geographical area.

Table 3.8 shows one potential combination and the resulting total capacity per square kilometer in Europe, where LTE can be deployed in the 800, 1800, and 2600 MHz bands. Note that since the table also considers the use of the 900 MHz band for B3G networks it is assumed that a major share of voice calls will also be handled by B3G networks either over legacy circuit switched connections or via Voice over IP. The impact of this is discussed in more detail in Chapter 3.10. In the example, even though showing an

Table 3.8 Potential downlink capacity per square kilometer with LTE and several operators

Operator	Bandwidth and band used	Capacity per square kilometer, based on an average spectral efficiency of 0.8 and a 3 sector BTS (Base Transciever Station) configuration
HSPA/LTE operator 1	1 × 10 MHz in 800 MHz 1 × 5 MHz in 900 MHz 1 × 20 MHz in 1800 MHz 2 × 5 MHz in 2100 MHz 1 × 20 MHz in 2600 MHz	156 Mbit/s
HSPA/LTE operator 2	1 × 10 MHz in 800 MHz 1 × 5 MHz in 900 MHz 3 × 5 MHz in 2100 MHz 1 × 20 MHz in 2600 MHz	120 Mbit/s
HSPA/LTE operator 3	1 × 10 MHz in 1800 MHz 4 × 5 MHz in 2100 MHz 1 × 20 MHz in 2600 MHz	120 Mbit/s
HSPA/LTE operator 4	1 × 5 MHz in 900 MHz 1 × 10 MHz in 1800 MHz 3 × 5 MHz in 2100 MHz 1 × 20 MHz in 2600 MHz	120 Mbit/s
LTE-Advanced extension by one network operator	1 × 40 MHz in 3.5GHz	96 Mbit/s
Sum:	Total bandwidth used: 255 MHz	Total capacity per kilometer: 612 Mbit/s

advanced use, only about half of the available 500 MHz of spectrum that is available is used. With such a setup, a total capacity per kilometer of over 600 Mbit/s can be reached. It is unlikely, however, that such a massive build out is to occur quickly but it can certainly be envisaged as a long term scenario.

As capacity is spread over many different bands, such a usage scenario requires mobile devices that can combine carriers in several frequency bands and networks that can intelligently distribute traffic over the available channels and switch devices quickly from one band to another depending on the changing signal conditions due to the mobility of the users.

The last line in Table 3.8 suggests the use of the 3.5 GHz band by mobile network operators. Spectrum has been reserved for this purpose but there are no efforts made at the time of publication to make use it. As the table describes a long term scenario, at least one network operator is assumed to make use of it eventually and the resulting additional capacity is included in the capacity estimation.

In Section 3.9, two approaches were used to estimate the number of users that could be served by a single network. The more conservative approach of [19] estimated 300 subscribers per base station site for HSDPA using two 5 MHz carriers. Applied to the current network deployment status in many countries as described in Table 3.7, a deployment of four UMTS networks each using two 5 MHz carrier results in 1200 users who could be served per kilometer with a monthly data volume between 2 and 4 GB. The second approach discussed in Section 3.9 resulted in 216 users per base station with 3 sectors and a single carrier per sector. Applied to the example in Table 3.7, 1728 people could be served per kilometer with a daily use of 70 MB.

When the two cell capacity calculation approaches of Section 3.9 are applied to the future scenario described in Table 3.8, the number of people per kilometer that could be served by HSPA and LTE with a high monthly data transfer volume can be calculated as follows: The more conservative approach estimated 300 subscribers per base station site for two 5 MHz carriers in each of the three sectors. Table 3.8 assumes four operators together using 255 MHz. Furthermore, advanced devices and 2×2 MIMO is assumed to be in widespread use which could double spectral efficiency from what was assumed in Section 3.8. The conservative approach would thus result in the following number of people that could be served with a monthly data volume of 2–4 GB:

$$\text{Number of users} = (300 \ \text{users}/10\,\text{MHz}) \times 255\,\text{MHz} \times 2 \ = 15\,300\,\text{users per km}^2$$

The second approach in Section 3.3 resulted in 216 subscribers for an HSDPA system and a bandwidth usage of 5 MHz. Based on this approach the following number of people could be served by the six assumed operators which together use a bandwidth of 165 MHz with a daily data volume of 70 MB:

$$\text{Number of users} = (216 \ \text{users}/5\,\text{MHz}) \times 255\,\text{MHz} \times 2 = 22\,032\,\text{users per km}^2$$

The data volume per user per day also includes data traffic generated by voice calls for this example, either via circuit switched connections over the HSPA network or voice over IP connections via LTE. This is discussed in detail in Section 3.11.

When discussing capacity in terms of supported users per kilometer, it is interesting to take a look at the population densities of some cities. The population density of Los Angeles is 3168 people per kilometer, Munich has 4316 people per kilometer, New York is densely populated with 10 456 people per kilometer and Manhattan has an astounding 24 846 people per kilometer. Smaller cities with less than 100 000 inhabitants usually have population densities between 500 and 1000 people per kilometer.

When comparing the number of users per kilometer calculated above for future network deployments with what today's networks are capable of it has to be kept in mind that it is likely that the monthly use per user will rise as well. If the average requirement has doubled by the time networks of this capacity are rolled out, the capacity in terms of number of users per kilometer is cut in half. Currently, the amount of data flowing through wireless networks doubles about once every 18 months, an average of reports discussed in [21]. How this fits into the equations above depends on how much of that growth is fueled by already existing subscribers using more data and how much by new subscribers having bought a smartphone for the first time.

It is also likely that data usage is growing in fixed line and wireless networks for different reasons. In fixed line DSL and cable networks, most of the growth today is from existing subscribers and the increasing use of online video services by existing users. While video is becoming an important service on mobile devices as well, the trend might be different here, as smartphones and tables are unlikely to be used for watching full length movies or television series episodes while roaming outside. Tablets with large screens are better suited for full length movie viewing but in most cases consumers are likely to do this at home where the private Wi-Fi network connected to their DSL or cable line is used either due to better coverage and data speeds or in order not to use a significant amount of their monthly wireless data allowance.

As discussed in Section 3.2 there is well over 500 MHz of available bandwidth for downlink transmission available for cellular networks. The total bandwidth used in the example above is well below that value. Thus, overall capacity could be further increased by adding more channels per base station site. Each channel, however, will add to the overall transmission power used at a base station site. At some point, the maximum allowed field strength per site might thus be reached. Today, values for most base station installations are far below the maximum allowed value. In some countries, the national body for telecommunication regulation performs regular field strength measurements. In Germany, for example, the federal network agency (Bundesnetzagentur) is responsible for this task and measurement results are published on a web page [22]. Even in densely populated areas with a high concentration of base stations, the field strength value close to base station sites today is in most cases still below 1% of the maximum allowed value.

3.11   Network Capacity for Train Routes, Highways, and Remote Areas

In practice there are some exceptions to the rule of requiring a certain amount of capacity depending on the population density. Areas around overland railway lines and highways, for example, have a much lower population density than city centers or residential areas. Users in such areas, however, might produce a much higher amount of data traffic in cellular networks on average since they are mobile and solely rely on cellular networks for connecting to the Internet. Base stations in those areas on average serve fewer users than base stations in city centers. Despite the smaller number of users, these cells nevertheless generate substantial revenue. Covering such areas can be a competitive advantage since users prefer using networks that cover most if not all areas where they are likely to use the network. In practice it can be observed that highways and to some degree railway lines are specifically covered by 2G networks to ensure coverage along their path to prevent call drops. This is especially important in trains since many people use their phones during train trips. Internet access also benefits from this since many network operators have upgraded their 2G networks with EDGE for higher data rates. In practice it is difficult to predict coverage along railway lines. Some lines are already covered by 3G networks (cf. Figure 3.3) and thus offer excellent connectivity. While this is still the exception today, it is expected that more and more railway lines will be covered by advanced cellular networks. Some train operators are also equipping their trains with Wi-Fi hotspots that use a satellite system and public B3G networks as backhaul connection. An example with mixed results can be found in [23, 24].

In rural areas, fast high-speed Internet access either via a fixed line or a wireless connection is still rare in some regions. This is due to the fact that the number of potential customers per square kilometer is low. Fixed-line high-speed Internet access thus suffers from the fact that the range of DSL for a bitrate of 1 Mbit/s is typically less than 8–10 km. This makes it financially unattractive for telecom companies to install the required equipment. B3G networks have thus become an interesting solution to this problem. Examples of companies with rural B3G deployments are T-Mobile, Vodafone, and O2 in Germany with LTE in the 800 MHz digital dividend band [25]. A coverage map of such a deployment can be found in [26]. Another example is Telstra in Australia where HSPA+ in the 850 MHz frequency band is used. An article published by Ericsson on the Australian HSPA+ deployment [22] gives further information on the technical and financial background. To make such a venture profitable, a single base station must cover as large an area as possible. One of the main factors in cell range is the frequency used. In the case of Telstra the 850 MHz band was used, which offers substantial increase in coverage range over the 2100 MHz band used for B3G networks in other regions. In addition rural deployments usually require a roof-mounted directional antenna for distant subscribers. In effect, the network is thus used as a mobile network by subscribers close to the base station and as a fixed wireless network rather than a truly mobile network at greater distances. Inside a house, a 3G/Wi-Fi router is used to connect all devices requiring Internet access to the B3G connection. Using a roof mounted directional antenna greatly improves the link budget (reception conditions) and thus the range over which data can be sent. Ericsson suggests in [27] that, compared with a handheld device used indoors, an indoor window-mounted omnidirectional antenna can improve the link budget by up to 12 dB. An omnidirectional antenna on the rooftop improves the link budget by up to 47 dB due to the greater height and the resulting vertical direction gain. A directional antenna mounted on a rooftop can increase the link budget by as much as 65 dB. In practice, a higher link budget both increases the range of the cell and also the data transfer rate that can be achieved by users and as a consequence overall network capacity. Both network operators and customers thus benefit from rooftop antennas.

The initial version of the UMTS/HSDPA standard allowed cell ranges of up to 60 km. In the case of Telstra, larger cell ranges were required for some regions. The limitation was due to base stations only being able to deal with propagation delays on the random access channel in the order of 768 chips, or a range of about 60 km. The 3GPP standards were thus enhanced and an extended cell range mode was introduced that can extend the cell range from a delay point of view to up to 200 km. While the enhancement requires software changes on the network side, no modifications are required on the mobile device. Further information can be found in the corresponding work item [28].

On the financial side, Ericsson estimates that a base station covering an area of 12 km^2 is financially viable in areas with population densities as low as 15 people km^2. This assumption is based on operator market share of 50%, a mobile penetration of 80% and a fixed mobile broadband penetration of 35% of the subscriber base among other values. The main service revenue was assumed to be €15 for mobile voice telephony and €30 for fixed mobile Internet access.

3.12 When will GSM be Switched Off?

It has been shown in this chapter that GSM-based GPRS and EDGE have difficulty keeping up with capacity enhancements of B3G networks and subscriber demands. Thus, switching off GSM networks and using the capacity in the 900 and 1800 MHz band for B3G networks would increase total capacity without requiring additional spectrum. It is expected that this will be slowly done over time as the specification of B3G network technologies such as HSPA and LTE now also allows operation in these frequency bands. An early example of this is O2 in the UK, who has begun using the 900 MHz frequency band for UMTS in central London [29].

Only a few years ago, most industry observers were predicting a rapid decline of GSM networks once 3G networks were in place. GSM had just celebrated its tenth birthday in terms of live network deployments and UMTS was already at the doorstep. Looking at lifetimes of analog wireless systems, it seemed certain that in another 10 years (2012) GSM would be a thing of the past. Today, 2012 it is certain that GSM will be used much beyond 2012, even in countries where B3G networks have been rolled out. This surprising development has several causes:

- **Equipment refresh** – in 2002, GSM equipment started to age as network vendors kept selling hardware that had been developed a decade previously. Since then, however, virtually all network vendors have completely refreshed their network equipment from base stations to core network routers. This was a necessity as the parts for aging designs (e.g., 486 processors) were no longer available at reasonable cost. Hardware evolution also meant lower prices. GSM base station controllers sold today, for example, are no less capable than the latest 3G radio network controllers in terms of processing power, memory, or storage capacity. GSM base station prices and sizes also keep decreasing, which in turn reduces CAPEX for network equipment.
- **New entrants** – another reason for refreshing aging hardware designs is the entry of new Asian companies like Huawei and ZTE into the GSM and 3G markets with new hardware and lower prices. Established vendors could no longer afford to continue selling expensive hardware with the new competition.
- **New markets** – back in 2002 it was not clear that GSM would have such tremendous success in emerging economies in Asia, India, and Africa. Compared with the 2.5 billion GSM subscribers today, the few (hundred million) 3G subscribers in 2008 almost seem like a drop in the ocean. This created economies of scale for GSM beyond anything imagined.
- **Continuous evolution** – back in 2002, it was assumed that most R&D would be put into the development of 3G networks. This has been true to a certain extent, but instead of being dormant, GSM has continued to evolve. Compared with 2002, GSM hardware is much more efficient due to the technical and economical hardware refresh. New features such as EDGE for higher packet-switched data rates have pushed the GSM standard far beyond the circuit-switched network it was once designed for.
- **Network refresh** – just like consumer IT equipment such as PCs and notebooks, network equipment such as base stations, controllers, switches, and routers have a limited lifetime and require replacement. The cycle is certainly longer than the 2 or 3 years for consumer PCs but after 10–12 years base stations have to be replaced because of aging components or due to their inability to support new features such as EDGE. Also, the power consumption of older systems is much higher than that of new base stations, so at some point the price of replacing a base station is absorbed quickly by reduced operational costs.
- **3G network coverage** – even in the most advanced 3G countries such as Italy, Austria, Germany, and the UK, 3G network coverage is nowhere near as ubiquitous as GSM coverage. This is different from the 1990s, where GSM coverage quickly approached the coverage levels of the analog networks.

- **Roaming** – with GSM, international roaming is a major benefit. For the foreseeable future, the majority of roamers will still have a GSM-only phone. Switching off GSM networks makes no sense as revenue from roaming customers is significant.

So where does that leave GSM in Europe and the USA in 2012? In five years, it is likely that the majority of subscribers in Europe and the USA will have 3G-compatible phones that are backwards-compatible to 2G. In urban areas, operators might decide to down scale their GSM deployments, as most people will now use the 3G network instead of the 2G network for voice calls. Cities will still be covered by GSM, but probably with fewer channels. Such a scenario could happen in combination with yet another equipment refresh, which will be required by some operators for both their 2G and 3G networks. At that time, base station equipment that integrates 2G, 3G, and B3G radios such as LTE could become very attractive.

3.13   Cellular Network VoIP Capacity

Even today, voice communication is one of the main applications and revenue generators in wireless networks. As voice communication generates a narrow band data stream, current networks such as GSM and UMTS have an optimized protocol stack for voice transmission in the radio network and on the air interface. As a result, voice transmission in those networks is very efficient in terms of required transmission capacity and hand over performance. Additionally, a dedicated core network infrastructure is used in such networks for transporting and managing voice calls, as discussed in Chapter 2. B3G networks on the other hand follow a trend that started in fixed-line networks to treat voice calls just as one of many communication applications that can be transported over IP. A detailed discussion on this topic will follow in Chapter 4. B3G networks can thus no longer integrate voice calls as tightly into the protocol stack. As a consequence voice transmissions over the air interface are less efficient in B3G networks. Before looking at B3G voice capacity, the following section discusses the voice capacity of GSM and UMTS base stations to act as a reference for the VoIP discussion that follows.

The Enhanced Full Rate (EFR) codec, often used in GSM and UMTS networks today requires a data rate of 12.2 kbit/s. For additional capacity, Adaptive Multi Rate (AMR) codecs were introduced in the standards some years ago. While the Full Rate AMR codec requires the same bandwidth as an EFR call, enhanced processing now allows to also use a 6.75 kbit/s half rate AMR codec that delivers almost the same speech quality but only requires half the resources on the air interface.

In practice, however, most GSM and UMTS networks in industrialized countries continue to use either EFR or Full Rate AMR which is transported in a single timeslot and requires a bandwidth of about 22.8 kbit/s on the air interface due to channel coding, which adds error detection and correction information. In addition, some networks have introduced AMR Wideband [30], a codec with a significant better sound quality but with almost the same bit rate required by a standard AMR full rate.

A 6.75 kbit/s AMR voice call is carried in half a timeslot and has a data rate of 11.4 kbit/s after channel coding. An average GSM base station with 3 carriers, 3 sectors, and 66 timeslots can therefore carry 66 EFR or 133 AMR 6.75 kbit/s voice calls. In practice, however, some timeslots are used for GPRS or EDGE data transmission, which reduces the number of simultaneous calls possible. The number of timeslots reserved for data transmission is an operator-defined value. As a general rule, it can be assumed that about 20–30% of the timeslots are reserved for data transmission. Voice timeslots can be used by GPRS and EDGE transmission while not required for voice calls.

A further voice capacity extension for GSM is VAMOS (Voice services over Adaptive Multi-user channels on One Slot) which is capable of multiplexing four voice calls in one timeslot [31]. In the medium and long term, VAMOS might be an interesting feature in the industrialized countries to reduce the amount of spectrum required for the remaining GSM capacity while in emerging countries, VAMOS might be used to increase capacity in existing networks without using additional channels.

UMTS uses the same codecs as GSM. On the air interface users are separated by spreading codes and the resulting data rate is 30–60 kbit/s depending on the spreading factor. Unlike GSM, where timeslots are used for voice calls, voice capacity in UMTS depends less on the raw data rate but more on the amount of transmit power required for each voice call. Users close to the base station require less transmission power in down link compared with more distant users. To calculate the number of voice calls per UMTS base station, an assumption has to be made about the distribution of users in the area covered by a cell and their reception conditions. In practice, a UMTS base station can carry 60–80 voice calls [32] per sector. A typical three-sector UMTS base station can thus carry around 240 voice calls. As in the GSM example, a UMTS cell also carries data traffic, which reduces the number of simultaneous voice calls.

In B3G networks such as HSPA and LTE, voice calls are no longer transported over dedicated circuit-switched channels and equipment in the core network. Instead, the voice data stream is packetized and sent over IP. To estimate the number of simultaneous VoIP channels, coding overhead has to be assessed differently since it is highly adaptive. The most important factor in this calculation is the average throughput, as discussed in Section 3.9, since this value includes the average channel coding used in a cell. Furthermore, it has to be taken into account how well a system can transport a high number of simultaneous low-speed connections. Usually it is more efficient to handle few users with high bandwidth requirements since there is a per user overhead for air interface management such as channel access, power control, and channel signal estimation. The percentage of this overhead is small for high-speed transmissions but grows for slow data streams such as voice.

When a voice call is transported over IP, the same voice codecs are used by optimized VoIP applications as those mentioned for GSM and UMTS above. In B3G networks voice calls no longer use dedicated and transparent channels but are transported over an IP network together with packets of other applications. One voice packet is usually sent every 20 ms and contains the voice data collected during this time. When the EFR codec with a data rate of 12.4 kbit/s is used, 31 bytes of voice data are sent in each packet. Nonoptimized VoIP applications make use of the G.711 codec, which is used in fixed-line analog voice telephony for compatibility reasons. The data rate of this codec is 64 kbit/s and thus much higher than the data rate of EFR. Sent in 20 ms intervals, a G.711 encoded voice call generates 160 bytes of voice data for each packet.

These 20 ms voice frames are then encapsulated in three protocol layers, each adding its overhead. Figure 3.5 shows the overhead for a 20 ms G.711 voice packet. In the lower part of the figure, the user data carried in the packet is marked in blue. The overhead in

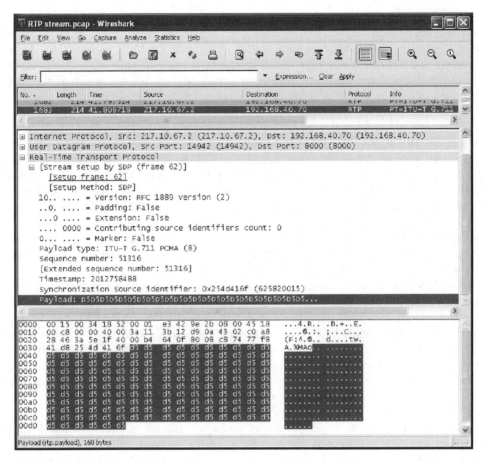

Figure 3.5 Voice transmission over IP, G.711 codec in an RTP packet. (Reproduced from *Wireshark*, by courtesy of Gerald Combs, USA.)

front is shown in white. The IPv4 layer adds 20 bytes to the overhead, the User Datagram Protocol (UDP) adds 8 and the Real-time Transfer Protocol (RTP) adds another 12 bytes. In total, the overhead for IPv4 is 40 bytes. If IPv6 is used in the network the overhead grows to 60 bytes. For a packet that carries a 20 ms EFR frame, the overhead exceeds the size of the payload. The resulting data rate is thus:

Datarate (EFR) over IPv4 = (31 voice bytes + 40 bytes overhead) /20 ms = 3.55 kbytes/s

In practice, the IP overhead would thus reduce the number of simultaneous voice calls per cell at least by two.

Several technical options exist to reduce the overhead. On the application level, more data could be collected before a packet is assembled and sent through the network. From a network perspective this looks appealing as the ratio between over head and user data could be influenced by the application. This unfortunately comes at a price. Collecting more voice data, that is waiting for a longer time before sending it over the network, would substantially increase the overall delay. Packetization delay is just one part of the overall delay, which additionally includes core and access network delay, air interface delay, and jitter buffer delay. A jitter buffer is required for VoIP transmissions at the receiving end since it is not guaranteed that packets arrive in time. As a consequence, the mouth-to-ear delay increases quickly to values beyond 200 ms, which is the limit described in [33] at which the delay becomes noticeable and distracting to the user.

Another possibility to reduce the overhead is to use header compression in the network between network elements that do not use the information in the IP header for forwarding the packet. This is possible since most fields of packets exchanged between two specific end points always contain the same values. For UMTS, HSDPA, and LTE radio access networks the Robust Header Compression (ROHC) algorithm [34] has been standardized for this purpose. In addition to compressing static fields like the source and destination IP addresses that never change, several profiles have been defined for ROHC so dynamic fields in different combinations of protocol layers used by IP applications can be compressed as well. One of these profiles is used when the ROHC compressor detects a VoIP transmission which uses IP, UDP, and RTP. This way an IPv4 or IPv6/UDP/RTP header can be reduced from 40 or 60 bytes down to 3 bytes.

In UMTS and HSDPA, header compression is performed between the mobile device and the Radio Network Controller (RNC), as described in Chapter 2. Thus, both the air interface and the backhaul connection between the base station and the RNC benefit from the compression. It should be noted at this point that header compression is optional and not yet widely used in wireless networks since most voice traffic is still carried over circuit-switched connections. In LTE, header compression is performed between the base station and the mobile device since the RNC has been removed from the architecture and most of its tasks are now performed by the base station.

Due to the almost complete removal of the IP protocol overhead by ROHC, VoIP transmissions over the air interface can be almost as efficient as circuit-switched transmissions. In [35] it is estimated that over 80 simultaneous voice calls can be transported over a 5 MHz HSDPA channel. The estimation is based on using the EFR voice codec and already includes the radio signaling overhead based on a Release 6 HSDPA implementation with a Fractional Dedicated Physical Control Channel (F-DPCH, cf. Chapter 2). This value is similar to the number of circuit-switched voice calls per UMTS sector given above.

For other B3G technologies, the number of voice calls over IP will be similar given the same amount of bandwidth used per sector. Radio network enhancements such as MIMO and advanced signal processing will further increase the number of simultaneous voice calls. In combination with ROHC, the number of VoIP calls per megahertz of bandwidth can thus exceed the number of circuit-switched voice calls per megahertz of bandwidth today.

3.14 Wi-Fi VoIP Capacity

IP over Wi-Fi shares the same basic technical background as discussed in the previous section for cellular network VoIP capacity. While VoIP capacity in cellular networks plays a significant role due to the amount of the network capacity being used for voice calls, it is likely that in the future only a small amount of capacity will be used for VoIP in private Wi-Fi networks. Wi-Fi VoIP capacity will therefore mainly play a role in office environments where Wi-Fi networks will be deployed for wireless telephony. This could come as part of office environment which relies exclusively on Wi-Fi for networking. While this has not been feasible in the past due to the speed of earlier Wi-Fi standards, 802.11n has the potential to overcome this limitation.

VoIP over Wi-Fi differs from what has been discussed for the cellular world in the previous section:

- Wi-Fi, unlike B3G cellular networks, does not have a centralized scheduler for packets (cf. Chapter 2). With a rising number of clients creating significant load, the number of collisions on the air interface is increasing. Thus, it is not possible to use the full capacity of a Wi-Fi network unlike in a scenario when only few devices are fully utilizing the network.
- The lack of a centralized scheduler makes it difficult to prefer small VoIP data packets to larger packets generated by other applications. The Wireless Multimedia extension discussed in Chapter 2 improves this behavior. Optionally, the Wi-Fi standard also defines a centralized scheduler. It is unlikely, however, that this option will be widely used due to economies of scale in both access point and client devices.
- Most cellular B3G technologies use FDD, that is different frequency bands for uplink and downlink. Wi-Fi on the other hand uses TDD. Thus, bandwidth requirements of the channel are twice as high as in the FDD system, since the uplink and downlink are sent in the same channel.
- The number of voice calls per access point is ultimately limited by the capacity available on the backhaul connection. For the examples below it is assumed that there is sufficient capacity in the backhaul in both the uplink and downlink direction.
- As in cellular networks, an average bandwidth of a Wi-Fi network has to take into account that some Wi-Fi/VoIP phones will be in unfavorable or distant positions from the access point and thus use a lower transmission speed. The transmission time for those packets is thus several times higher than the transmission time of packets for VoIP phones close to the access point.
- Unlike in cellular networks, no header compression schemes are currently defined to decrease the IP overhead in Wi-Fi networks.

Table 3.9 shows the number of concurrent voice calls as calculated in [36] in a Wi-Fi network under the following assumptions: all devices in the network are 802.11g capable and no protection schemes are required in the network for older devices. In the first column, the different voice codecs are listed and in the second column their bandwidth requirements. In the third column, the theoretical maximum number of simultaneous calls is given at the maximum speed of the network. In the second column the number of simultaneous VoIP calls is given for a lower average network speed, which is more realistic in practice since not all VoIP devices will be used under ideal network conditions. The table shows that, even when average conditions are assumed and the highest bandwidth codec (G.711) is used, the network supports up to 51 simultaneous VoIP calls. In practice, the number will be somewhat lower due to the negative effects of the decentralized scheduler. Nevertheless, given the short range of a typical Wi-Fi access point, it is unlikely that this limit will be reached in practice. Together with Wireless MultiMedia (WMM),

Table 3.9 Number of simultaneous voice calls in a 802.11g network excluding the effect of a decentralized scheduler

Voice codec	Bandwidth requirement over IP	Number of calls with 54 Mbit/s	Number of calls with 18 Mbit/s (averaged)
G.711	80 bit/s	78	51
GSM enhanced full Rate (EFR)	28.4 bit/s	92	71
iLBC 30 ms (Skype)	24 bit/s	133	101

which prioritizes small and bursty streams, using a combined Wi-Fi network for VoIP and other data is feasible in most environments. Should the overall traffic exceed the limits of a single network, it is also possible to deploy a dedicated Wi-Fi network for voice alongside a Wi-Fi network for general use.

3.15   Wi-Fi and Interference

Using Wi-Fi as part of a future converged fixed, nomadic, and mobile Internet access network substantially increases overall capacity per square kilometer. As applications such as TV broadcasting over IP gain in popularity, even Wi-Fi capacity limits are reached quickly in the unlicensed 2.4 GHz ISM (Industrial, Scientific, and Medical) band. In countries such as France, TV broadcasting over IP over DSL has already achieved great popularity today. Some DSL access providers have started to offer equipment to wirelessly connect a TV set top box to the DSL modem/router over Wi-Fi. An example is the Freebox of DSL provider Free [37]. Due to this and the general popularity of DSL/Wi-Fi routers for PC and notebook connectivity, it is not uncommon to be in the range of more than 20 Wi-Fi networks in a Paris apartment building. Since there is only room for three non-overlapping 20 MHz channels in the 2.4 GHz ISM band, there is a great partial and full overlap of these networks. While this does not have a big impact on the performance of a Wi-Fi network since the overlapping Wi-Fi networks are only broadcasting beacon frames in idle mode, performance quickly drops when use in overlapping networks increases. In the case of WiFi-connected set-top boxes, it can be observed in practice that a stream of 4–5 Mbit/s is continuously sent independently of whether or not the TV set is turned on. Consequently, this capacity is no longer available in the other overlapping networks.

Figure 3.6 shows a trace taken with a layer 1 tracer [38] under the conditions described. The lower graph in the figure shows the frequency range of the ISM band between 2400 and 2480 MHz. Instead of showing the frequency in MHz, the x-axis shows the 13 available Wi-Fi channels. On the y-axis the amplitude of the signal received over the band is shown. The color of the peak depends on the intensity of the signal received during 60 minutes. Bright colors indicate high activity level. The graph shows five partially overlapping networks with their center frequency on channel 1 (little traffic so the arch is not very visible) and channels 3, 5, 6, and 11. The most activity can be observed in the wireless network that is centered around channel 11.

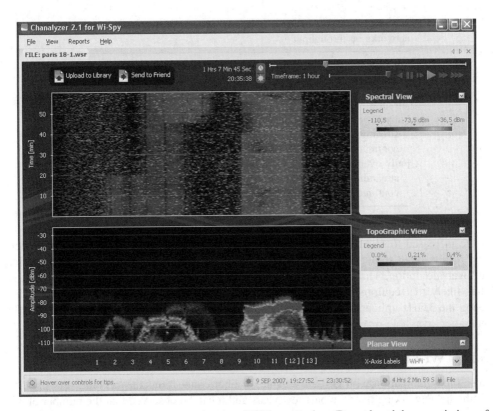

Figure 3.6 Layer 1 trace over overlapping Wi-Fi networks. (Reproduced by permission of MetaGeek LLC, 5465 Terra Linda Way, Nappa, ID 83687, USA.)

The upper graph in Figure 3.6 shows a time graph over the frequency range. On the y-axis a resolution of 60 minutes has been chosen to show the activity in the ISM band in the course of 1 h. The Wi-Fi networks on channels 5 and 11 were used for streaming as there was uninterrupted activity throughout the test period. The Wi-Fi networks on channels 3 and 6 were also used for streaming. Streaming was stopped on the Wi-Fi network on channel 6 after about 12 minutes while streaming was started on the Wi-Fi network on channel 3 about 40 min into the trace.

Table 3.10 shows the impact of partly and fully overlapping networks on the maximum throughput of a Wi-Fi network compared with a situation in which no overlapping occurs. The tests were performed with a Linksys WRT54 802.11g router and Iperf [39], a UDP and Transmission Control Protocol (TCP) throughput measurement tool. One of the notebooks for the test was connected to the Wi-Fi Access Point with an Ethernet cable, while the other one used the Wi-Fi network. With a fully overlapping Wi-Fi network, which is used for TV streaming, the capacity of the Wi-Fi network under test was reduced to 72%. Partial overlapping caused an even bigger speed penalty and performance was reduced to 59%.

In the future even more private Wi-Fi networks will be set up and TV and other multimedia streaming over Wi-Fi is likely to become even more popular. As discussed in Chapter 2, the 802.11n standard by default only allows the Wi-Fi channel bandwidth

Table 3.10 Effect of partly and fully overlapping Wi-Fi networks on throughput

Situation	Measured throughput (Mbit/s)
No interference	22.5
Fully overlapping Wi-Fi network with a streaming client and an estimated continuous stream of 4–5 Mbit/s	16.3
Fully overlapping Wi-Fi network with a streaming client and an estimated continuous stream of 4–5 Mbit/s	13.4

to be increased to 40 MHz in case there are no overlapping networks. As the presented test results have shown, this was a wise choice. In practice, many access points offer an override option. This will increase the problems for the ISM band even further. It is thus likely that equipment sold especially for multimedia streaming purposes will start using the 5 GHz unlicensed band in which there are 18 independent channels for 20 MHz operation available or nine for 40 MHz operation.

3.16   Wi-Fi Capacity in Combination with DSL, Cable, and Fiber

Wi-Fi in combination with DSL, TV cable, or fiber connections has become very popular for home networking over the past few years and it can be observed that the number of 3G and B3G mobile devices such as Personal Digital Assistant (PDAs) and mobile phones being equipped with a Wi-Fi interface is also rising. In addition, stationary devices such as set-top boxes, multimedia storage devices, TV, and satellite receivers and many other devices are beginning to be equipped with a Wi-Fi interface. As discussed in more detail in Section 3.14, the rising use of Wi-Fi has started to cause interference between networks. This is a trend that is likely to increase in the future as new devices and services are added. As a consequence, many 802.11n products are now additionally operating in the 5 GHz band, which is available for unlicensed use in many countries. Table 3.11 shows the frequency bands currently used by Wireless LAN. The exact bandwidths are country-dependent, but generally in the bands shown in the table. In total, there is around 500 MHz of bandwidth available. It should be noted at this point that the 2.4 GHz band is also used by other systems such as Bluetooth. Their use, however, is limited and narrowband and thus has no significant impact on the use of Wi-Fi, barring some exceptional cases.

Table 3.11 Frequency bands for Wi-Fi

Frequency range (MHz)	Total bandwidth available	Number of available 20 MHz channels	Number of available 40 MHz channels
2.410–2.480	70 MHz	3	1
5.150–5.350	455 MHz	18	9
5.470–5.725	—	—	—

From a capacity point of view current 802.11g networks provide a maximum practical throughput of around 23 Mbit/s in a channel with a bandwidth of 20 MHz. This requires that there is no interference by other overlapping networks, no legacy 802.11b devices in the network and only a short distance, in the range of several meters, between the Wi-Fi access point and the wireless device. The MIMO functionality of 802.11n, as shown in Chapter 2, increases this throughput to around 40 Mbit/s in a 20 MHz channel and to over 100 Mbit/s where a 40 MHz channel is used. With increasing distance or obstacles such as walls between the access point and the client device, data rates quickly decrease. In buildings, the range of a Wi-Fi network is thus typically less than 30 m with data rates of 3–5 Mbit/s at the coverage limit. For most applications this range is sufficient. One exception is VoIP over Wi-Fi. Due to the smaller area covered by Wi-Fi networks compared with that of a cordless phone system, Wi-Fi VoIP phones have a significant disadvantage [40].

Several options exist to increase the coverage for the use of VoIP over Wi-Fi or to enlarge a Wi-Fi network in general. Companies mostly prefer to deploy several access points and create a single network by using a fixed Ethernet infrastructure to tie them together. In a home network environment with a single access point connected to the Internet (i.e., no fixed Ethernet infrastructure is available), the Wireless Distribution System (WDS) functionality built into most access points can be used to replace the fixed-line backbone infrastructure with a wireless link. An access point can typically serve clients and forward packets over the WDS link simultaneously. Overall network capacity is significantly reduced, however, since packets traversing a WDS repeater are sent over the air once by the user and once by the repeater of the wireless link.

When making the capacity of fixed-line DSL connections available wirelessly via Wi-Fi, it is important to note that the throughput offered by a Wireless LAN is typically higher than that of the fixed-line Internet link. For the following estimation it is assumed that a home has on average an 8 Mbit/s DSL, cable, or fiber link which is shared by four people in a household via Wi-Fi. With a population density of 4000 inhabitants per km^2, which was used to set capacity estimations of cellular networks above into perspective, this would result in 1000 DSL connections per km^2, each delivering 8 Mbit/s. With an assumed DSL subscription rate of around 50% of the households, this would result in about 500 DSL connections per km^2. The line speed, however, is only available up to the DSLAM (DSL Access Multiplexer), which concentrates DSL lines and forwards the aggregated traffic to the wide area optical network. For an assumed concentration rate of 25 : 1 in 2011, the following capacity per square kilometer would result:

$$\text{Wi-Fi-DSL capacity per km}^2 = (4000/4) \times 50\% \times 8 \text{ Mbit/s}/25$$

$$= 160 \text{ Mbit/s per km}^2$$

It is important to note that the estimated value greatly depends on the concentration rate, which is also referred to as the oversubscription. With increasing load on networks due to TV and multimedia streaming, fixed-line operators can quickly change this to much lower values, if the broadcasting server is in their own network. This is done by adding capacity in the optical backbone network with no changes required between the subscriber and the DSLAM. For traffic from and to the Internet via an IP transit point, different rules apply. Here, traffic is typically charged based on required bandwidth. With prices around €4 per Mbit/s [41] in 2011 (down from around €20 in 2007), the use of 160 Mbit/s during peak hours would incur a cost of €640 a month. This would just over €1 a month per subscriber in the example given earlier. Even with a lower concentration rate and higher use, peering costs remain well in the single digit euro per month per user area. To this cost, the DSL operator has to add the capital and operation expenditure for his own backbone network to the IP transit point, the DSLAMs in the central offices, line rental to customer, and so on.

In the future, enhancements of DSL and cable will allow a further increase in fixed-line data rates. ADSL2+, which is already in widespread in many countries, allows subscriber data rates of up to 16 Mbit/s in practice. VDSL (Very-high-bit-rate digital subscriber line), which has also become very popular in recent years increases subscriber data rates to around 25–50 Mbit/s. In some regions, optical connections to subscriber homes are currently under deployment, pushing data rates even higher. From a financial perspective, it should be noted that such upgrades are mostly limited to cities, as discussed in Section 3.10. Also, to benefit from the rising speeds in the access for anything but TV and video streaming from a server in the operators network, IP interconnect prices at peering points must continue to fall.

Future bandwidth increases on the last mile to the subscriber come with an additional cost in comparison with standard DSL deployments. For ADSL2+ the DSLAM is usually installed in the telephone exchange and the cable length to the subscriber can be up to 8 km for a 1 Mbit/s service. For VDSL, which offers data rates of up to 50 Mbit/s in downlink, the cable length must not exceed 500 m. Thus, DSLAMs can no longer be only installed in central telephone exchanges but equipment has to be installed in street cabinets. This is a challenge since the cabinets are quite large, require power, and active cooling and create noise. Also, earthworks are necessary to lay the additional fiber and power cables required to backhaul the data traffic. Figure 3.7 shows a VDSL DSLAM cabinet that has been installed alongside a traditional small telecom cabinet. To connect a new subscriber, a technician is required to manually rewire the customer's line to one of the ports. Different sources currently specify the maximum capacity of such cabinets from about 50–120 VDSL ports [42]. To support 500 VDSL connections per km^2, several cabinets are thus required. If several network operators compete in the same area, the number of required cabinets will grow further.

With VDSL and fiber network deployment to the customer premises, data rates on the Internet link can come close to or even surpass what is currently offered on the wireless link by the 802.11n standard. Using the same oversubscription as in the previous example, the capacity of Internet connectivity via Wi-Fi and VDSL could reach the following level:

$$\text{Wi-Fi (VDSL/optical) capacity per km}^2 = (4000/4) \times 100\,\text{Mbit/s}/25 = 4\,\text{Gbit/km}^2$$

If both the 2.4 GHz band and the 5 GHz band were fully used by Wi-Fi and if it is further assumed that the per network peak throughput in 20 MHz would be 40 Mbit/s, 21 nonoverlapping networks could deliver around 800 Mbit/s in an area with a radius of 30 m. Even if undesired partial network overlapping, no MIMO, reduced cell edge data rates, interference from neighboring Wi-Fi cells and other radio systems using the band are considered, there is still enough capacity available on the wireless link for such data rates. It should be noted at this point that some Wi-Fi networks will also carry a substantial amount local traffic, for example, between a home media server and an IP-enabled television set, which will generate much more data traffic than what users request from the Internet.

Figure 3.7 A VDSL DSLAM cabinet alongside a traditional telecom wiring cabinet.

The following summary once more highlights the differences in how capacity can be increased per square kilometer in Wi-Fi/DSL networks compared with cellular wireless networks:

The capacity per square kilometer of Wi-Fi over DSL is mostly limited at the back end. As the price per Mbit/s capacity at the Internet peering point is expensive, an over-subscription factor is used to limit the maximum bandwidth available to all subscribers at a given time. As prices for interconnection to the Internet decline, the oversubscription can be lowered or the line speed can be increased while keeping the same oversubscription. An advantage of Wi-Fi in combination with DSL compared with cellular wireless networks is that TV and multimedia streaming from servers inside the network can be very cheap as the additional capacity used in the access does not increase the operator's cost.

The capacity per square kilometer of cellular networks is limited at the front end. To reach a throughput in the same order of magnitude compared with Wi-Fi over DSL, cable, or fiber, the air interface resources need to be fully utilized; that is, there is no oversubscription factor between the air interface speed and the speed at the Internet peering point. Thus, mobile operators cannot offer TV and multimedia streaming at the same price as fixed-line operators even if the streaming server is within their own network. To increase capacity in an area they cannot lower an oversubscription factor but have to add more capacity at the base stations and at the Internet peering point.

The following consequences can be deduced from the previous observations:

- The capacity of Wi-Fi/fixed line networks per square kilometer scales with the number of subscriptions per square kilometer. As each subscriber generates additional revenue, adding more subscribers only requires increased resources in the optical back-end network and increased capacity at the Internet peering point. Both increases are covered by the additional subscriber revenue.
- The capacity of cellular networks per square kilometer is lower but in the same order of magnitude as the capacity per square kilometer of Wi-Fi/DSL networks for the population density of a mid size city. Thus, cellular networks can compete for a sizable market share.
- Cellular network operators cannot offer TV and multimedia streaming at a competitive price compared with Wi-Fi/DSL. Their offers are thus limited to Internet access.
- As capacity per base station is increased, the individual backhaul connections to the network must be increased as well. This raises the question of how this can be done economically. This is discussed in more detail in the next section.
- To increase capacity for a Wi-Fi/DSL network, only centralized locations (the DSLAM) must be upgraded.
- Future fixed access technologies such as VDSL and fiber to the home require curb-side installations. The number of subscribers is limited by the number of ports available in the DSLAM at the curb-side.
- Wi-Fi/DSL automatically benefits from falling interconnection peering prices since there is ample capacity between the DSLAM and the subscriber.

3.17 Backhaul for Wireless Networks

It is interesting to note that in reality most links in wireless networks are not wireless. This starts with the connection between the base stations and the rest of the network. This link is also referred to as the base station backhaul or simply backhaul. With increasing air interface throughput, the capacity of backhaul links has to increase as well. Legacy backhaul equipment based on 2 Mbit/s E-1 links [43] does not scale with rising demand and has therefore been replaced in many networks offering high speed wireless Internet access. The following section describes the capabilities and limitations of E-1 based technologies to raise an understanding of why E-1 based backhaul technology has become the main reason for congestion in networks where they have not yet been replaced. This is followed by an overview of high speed IP and Ethernet based backhaul technologies used for HSPA+ and LTE networks.

There are two types of E-1 based technology used to backhaul the traffic from a base station:

- Wireless operators often choose microwave backhaul connections. This requires extra equipment at the base station and at the other end of the link but frees the operator from monthly fees to a fixed-line operator to lease a wired connection. For this reason, microwave backhaul has become very popular with alternative operators, especially in Europe. Figure 3.8 shows a typical base station setup that uses a microwave connection for the backhaul. The long antennas are used for the connection to the mobile devices while the round antenna creates the directional beam for the backhaul connection and receives the data stream from the other end of the link. Legacy E-1 based equipment can transport several E-1 connections over one microwave link, so data of several daisy-chained base stations can be backhauled at once. A typical maximum is 16 E-1 connections that equal an E-3 line with 32 Mbit/s.

• Other operators prefer fixed-line connections. This has the advantage that the backhaul network is managed by another company and thus offloads responsibility to an external third-party company.

In North America and Japan a Time Division Multiplexing technology referred to as T-1 is used for the backhaul links with 24 timeslots of 64 kbit/s. In the rest of the world E-1 links with 30 timeslots of 64 kbit/s are used. The bandwidth of a T-1 is thus 1.5 Mbit/s while the capacity of an E-1 is 2 Mbit/s. From the point of view of B3G networks the use of timeslots on a connection is a relic of voice-centric telephone networks. Here, T-1 and E-1 connections are used to transmit 25 or 31 individual telephone calls over the same line. For GSM and other voice-centric systems it made sense to use the same technology since base stations at the beginning were also exclusively used for voice calls. Due to compression used for voice calls in wireless networks, a single T-1 or E-1 timeslot carries four GSM voice calls. Since GSM is also a TDM (Time Division Multiplex)-based system, one timeslot in a T-1 or E-1 connection carries the content of four air interface timeslots. A base station with three sectors and two carriers per sector each having eight timeslots thus requires 12 of the 24 or 31 timeslots of a T-1 or E-1 connection, respectively. In addition, an additional timeslot is usually used for signaling purposes. A single GSM base station only uses a fraction of the timeslots of a T-1 or E-1. In practice several base stations are thus connected in a chain to the same T-1 or E-1. Note: in the remainder of the text, E-1s are used for further comparisons since T-1 and E-1 connections are essentially the same except for the different number of timeslots.

When the GPRS was first introduced in wireless networks, the same 1 : 4 mapping between E-1 timeslots and air interface timeslots kept being used. With the introduction

Figure 3.8 Base station antennas with a microwave dish for backhaul. (Reproduced from *Communication Systems for the Mobile Information Society*, Martin Sauter, 2006, John Wiley & Sons, Ltd.)

of EDGE, which increases packet-switched data transmission speeds, the fixed mapping between air interface timeslots and fractions of E-1 timeslots had to be abandoned. With EDGE the capacity of a single timeslot on the air interface is up to 58.2 kbit/s and thus much higher than the 16 kbit/s of a quarter of an E-1 timeslot. Mapping an EDGE air interface timeslot to a single E-1 timeslot also does not make sense since transmission conditions on the air interface are often not ideal and a lower transmission speed is used for the timeslot. Therefore, starting with EDGE, the timeslot concept which was invented for circuit-switched connections has for the first time become a problem for backhaul connections.

Initially and to some extent still today, some 3G and B3G networks based on HSDPA and code division multiple access (CDMA) EVDo (Evolution-Data only) technology use E-1 connections for backhaul. Since these systems use CDMA on the air interface, the time division multiplexing of E-1s is no longer required for transferring data. The decision of using this technology was mainly taken due to the fact that no other technologies were available at the time to transport data in both directions at the required speeds. Furthermore, UMTS/HSPA base stations must be synchronized tightly with each other, which is achieved in practice by locking the base station clock to the E-1 timing which is very precise due to the TDM nature of the technology.

Initially, UMTS/HSPA networks relied on the packet-switched Asynchronous Transfer Mode (ATM) protocol to transfer data between the base station and the RNC. Data exchanged between the base station and the RNC is packetized into ATM packets which have a fixed length of 53 bytes and is then sent over E-1 links. Instead of using timeslots individually the ITU has developed a standard of how ATM packets are logically transported over E-1 links by using all timeslots simultaneously [35]. Thus, while the timeslots still exist on the lower layers of an E-1 connection, higher layers at both ends of the connection are no longer affected by the timeslot structure. Since the timeslot structure on lower layers is maintained, it is not necessary to modify E-1 transmission equipment despite changing to a packet-switched transmission mode.

In practice, the transmission speed of approximately 2 Mbit/s of an E-1 or 1.5 Mbit/s of a T-1 is not sufficient even for a legacy low capacity three-sector UMTS/HSPA base station. As shown in this chapter such a base station offers a capacity of around 7.5 Mbit/s if one carrier is used per sector or around 15 Mbit/s if two carriers are used. Consequently, a UMTS base station requires four E-1 connections for a single carrier configuration and up to eight E-1 connections for a two-carrier configuration.

Since a single user can achieve speeds on the air interface exceeding the transmission speed of a single E-1 connection, ATM data packets have to be multiplexed over several E-1 links. In UMTS networks, this is done via Inverse Multiplexing over ATM (IMA). In essence IMA sends ATM packets in a round robin fashion over several E-1 links, as shown in Figure 3.9. Vendors usually included IMA multiplexers as part of the base station hardware so no additional hardware is required at the base station site.

As bandwidth capabilities on the air interface kept rising, it was not possible to increase the number of E-1 links per base station anymore to support the data rates now offered on the air interface with HSPA+ and LTE for two reasons. Firstly, there are only a limited number of copper cables available at a base station site. Secondly, network operators are paying line rental fees per link and not per base station site. Depending on the country, E-1 line rental prices per month are between €200 and 500 in 2008, the high time of

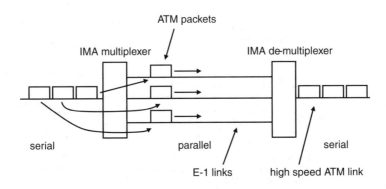

Figure 3.9 Inverse multiplexing over ATM.

the deployment of early (non-evolved) 3G systems. As a consequence, each additional E-1 link significantly increased the monthly operating cost of a base station. Revenue per user on the other hand has reached a ceiling in many countries or is even slightly declining, mostly due to declining prices for voice calls. Data services such as Internet access can compensate for this, but require significantly more bandwidth than voice calls and as a consequence more bandwidth in the backhaul as well. As a consequence, B3G technologies such as HSPA+ and LTE require backhaul bandwidths of 60 Mbit/s and more per base station. This would require more than 30 E-1 links, which is clearly not practical either from a financial or from a technical perspective.

As a consequence, a number of different technologies were developed to increase the bandwidth on the wireless backhaul and decrease transmission costs to counter the increasing prices for backhaul connectivity. An early idea to increase backhaul capacity beyond the limits of E-1 based access was to split backhaul traffic. Data for real-time voice calls continued to be sent over E-1 connections to ensure quality of service and to have a reliable link for synchronizing the base station with the network. All other types of data flows are sent over copper cables but using a different technology. In practice this could be ADSL (Asymmetric Digital Subscriber Line). Such an approach requires the installation of a device at the base station that can separate the two traffic classes and send them over the different connections. At the other end a similar device is used to combine them again before the next network element is reached. Figure 3.10 shows how such a setup looks for a UMTS/HSDPA network in which the ADSL network of a third-party company is used for the backhaul. The device to be installed at the base station site is usually small enough to fit in the base station cabinet. While this approach reduces the number of links required at the base station site, there are also a number of downsides. Current ADSL deployments are asymmetrical, which means that the bandwidth in the downlink is much higher then in the uplink direction. In a three sector base station configuration in which HSUPA is used there would not be enough uplink capacity in the backhaul to offload potential uplink traffic of all sectors. Another disadvantage is that network operators have to manage and monitor two types of backhaul networks, which creates additional overhead. Finally, this option is mainly interesting for mobile operators that have their own fixed-line ADSL networks, as other ADSL operators might not have a great interest in backhaul wireless traffic via their ADSL network. One of the reasons for this is the oversubscription factor per line discussed before, for which their networks have been dimensioned.

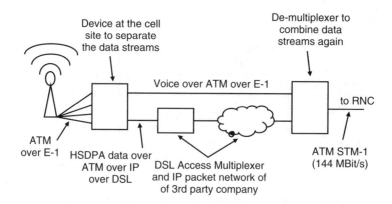

Figure 3.10   Use of DSL and pseudo-wires for backhaul.

A slightly different approach to the one shown in Figure 3.10 is for an operator to install their own DSLAM at a central site and terminate the ADSL links of base stations to their own equipment. In this scenario, no IP pseudo-wires are required since ADSL natively transports ATM packets. In this scenario, the demultiplexer would receive native ATM packets instead of IP-encapsulated ATM packets.

The use of two networks (E-1 and ADSL) for backhaul was mainly installed during a transition period in which accurate timing information to synchronize the base stations could not yet be sent over IP based links. This problem has been solved and there is a general trend toward fully IP-based packet backhaul solutions. This trend is driven by LTE, which natively uses IP over Ethernet instead of ATM. While early LTE deployments might have be made alongside and operated independently from already installed GSM and UMTS equipment, future LTE deployments by networks operators are likely to replace aging GSM or UMTS base stations with equipment supporting several radio technologies simultaneously. But at the same time, the opportunity to save cost of this approach is to use a single backhaul for all radio technologies, which have previously required three different types of backhaul connectivity: the GSM part of the base station required TDM, HSPA was initially based on ATM and the LTE part of the base station is based on IP. As a consequence, virtual connections will be used to backhaul traffic from all three radio technologies through a single connection and an IP metro network to the next node in the wireless hierarchy. Help is provided by either standardized or proprietary ways to connect GSM and UMTS base stations to the network over IP, thus in some cases replacing traditional backhaul connectivity with IP even before LTE is used. Another advantage of having all three radio technologies in a single base station instead of having three logically separate base stations at the same location is that a single operation and management system can be used to control and maintain the radio network instead of three. This scenario is shown in Figure 3.11. In the metro part of the network, IP based technologies over fiber are becoming more popular and metro Ethernet networks have in many cases already replaced or at least supplemented SDH (Synchronous Digital Hierarchy)-based network technology. On the backhaul link which connects the base station site to the optical metro network, several options exist:

From a technical perspective using an optical link to connect to the metro network is the best choice. Optical fibers offer very high bandwidths and thus offer scalability for the future. Unfortunately, only few base stations have fiber connectivity today and deploying new fibers to base station locations is expensive.

VDSL is a copper cable-based alternative to fiber deployment. Current VDSL standards allow data rates of up to 50 Mbit/s in downlink and 50 Mbit/s in uplink direction at cable lengths below 1 km. Several VDSL connections per base station site can be used to increase bandwidth if required. At the edge of the metro network the VDSL connection could be terminated by a DSLAM, which in addition to terminating wireless backhaul connections can also be used to terminate consumer or business VDSL connections.

For mobile operators without fixed-line metro network assets, packet-based micro wave backhaul solutions are an alternative. Ethernet microwave backhaul solutions support speeds of several hundred megabits/s today and it is likely that even higher bandwidths will be available in the future [44].

3.18   A Hybrid Cellular/Wi-Fi Network Today and in the Future

As has been shown in Section 3.9 for cellular wireless networks and in Section 3.15 for DSL/optical/cable networks in combination with WI-FI, there is sufficient wireless capacity available today to offer users a broadband connection to the Internet. Also, it has been shown that a significant amount of already existing and recently assigned spectrum for cellular and Wi-Fi communication is not used so far and hence capacity can be extended in the future as needed. Each network type has advantages over the other and hence, they need to be looked at together to form an overall picture on current and future possibilities and capacity requirements.

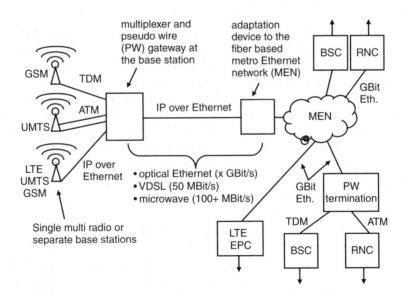

Figure 3.11   Packet-based backhaul options replacing today's E-1/T-1 links.

DSL/optical/cable Internet connectivity in combination with Wi-Fi is the technology of choice for most households in cities. The Wi-Fi access point is usually built into the DSL modem and thus the devices of all family members can be wirelessly connected within the home. From a financial point of view only one subscription is required to connect all members of the household. Furthermore, DSL/optical/cable network operators can offer TV and multimedia streaming to subscribers from a streaming server in their own networks due to the high capacity available on the last mile to the subscriber. In addition, the Wi-Fi network can also be used to connect devices at home with each other. This becomes more and more important as devices such as network enabled TV screens, multimedia servers, NAS (Network Attached Storage) servers, and PCs within the home communicate with each other. Streaming a recorded movie locally from a multimedia server to a TV screen in HDTV quality requires a large amount of bandwidth, which a Wi-Fi network connected to a fixed line Internet connection can support in addition to other simultaneous data traffic such as VoIP, Web browsing, and online gaming. The network thus creates a virtual local network bubble around the household and the people living in it. Many devices used in such a network, such as notebooks, netbooks, and tablets, are mobile to a certain degree and remain connected to the network even when moved through this bubble. The bubble, however, only has a limited size. When the first edition of this book was published, most of those devices were not yet in widespread use. Those devices that were, usually lost connectivity instantly when leaving this bubble. This has significantly changed when the second edition of this book was published as most of those devices now have a Wi-Fi and a B3G network adapter and thus continue to be connected when the user leaves the home and office. Many popular communication and social media services have become popular in recent years and as they run in the web browser and are delivered "from the cloud," that is, via servers located somewhere around the world, the transition from a home or office Wi-Fi to a B3G network and vice versa is almost seamless from a user's point of view.

Cellular networks show their strength outside a local Internet bubble. B3G networks offer an overall capacity that can by far be sufficient to ensure continued connectivity for people leaving a local Internet bubble. Available network capacity is not a spectrum issue as often suggested in the media but depends mainly on competition between network operators and willingness to offer an adequate service level. It is interesting to note that significant differences exist between countries which are often explained with a difference in population density and higher use. As this chapter has been shown, such views are usually unfounded as in those countries with excellent B3G capacity, the use of smartphones and other mobile devices is at least as widespread as in those countries from which problems are reported, and prices for connectivity are on a similar level. These countries and networks are therefore proof of what is possible with B3G technologies if their capabilities are appropriately used.

While many people today use cloud based services and store their personal information not at home but on a server on the Internet, more security conscious users, and businesses prefer to store their private data in their own home or company networks. With Virtual Private Network (VPN) tunnels, secure Hypertext Transport Protocol (HTTP) communication, and other methods they can access their data while not at home. From a cellular network capacity point of view, there is no difference between the two approaches as the amount of data that is exchanged over the wireless part of the network, which is the limiting factor, is the same.

Remote home control applications are also slowly becoming more popular. Little bandwidth is required for them, however, checking and changing the status of lights and windows, for example, requires little data to be exchanged.

Bandwidth requirements for streaming of stored contents such as music, videos, and movies from a multimedia server at home or from a server on the Internet to mobile devices on the other hand requires significant bandwidth and is one of the main drivers of B3G data volume growth. The limited battery capacity of mobile devices prevents the prolonged use of applications such as radio streaming, however, and small screens and mobility make it unlikely that full length movie viewing over B3G networks will become as popular as the same activity at home. Therefore, the effect of these applications on B3G networks is lower compared to their impact on fixed line networks.

For some users, such as students or business travelers, using the cellular network for high-speed Internet connectivity in combination with nomadic devices such as notebooks instead of a Wi-Fi/DSL connection is also appealing. A comparison of the values estimated in Sections 3.9 and 3.15 shows that cellular networks are able to meet these demands for a significant percentage of the population in addition to traffic generated by other applications such as voice calls and mobile access from small mobile devices. In rural areas and at special locations such as in the car or on the train, mobile networks will often be the only cost efficient way to offer broadband Internet access to subscribers.

As a consequence, cellular B3G networks have transformed into overlay networks to the private or business Internet bubble. Many users are likely to spend a significant time in their local Internet bubble. Mobile devices which include a Wi-Fi interface can thus use such private bubbles for a significant amount of timeto reduce the load on high speed cellular networks. Music downloads are a good example. While high-speed cellular networks deliver a similar experience when downloading music from a central server or via a secure connection from the user's database in a home network, using the local Wi-Fi network for this purpose makes more sense from a user's point of view as the amount of data per month on a cellular subscription is often limited.

A converged use of B3G networks in combination with personal Wi-Fi networks at home, at the office, and in hotspot locations also makes sense from a capacity perspective. Even if high-throughput streaming between local devices is taken out of the equation, cellular networks in cities undoubtedly depend on personal Internet bubbles to handle data traffic while users are at home or in the office. Plans and standards have been made to combine these different connectivity types and uses into a single overall network architecture. In practice, however, little to no practical effort has been undertaken on the network layer toward this goal.

Today, companies offering fixed-line high-speed Internet access and companies operating B3G networks are often separate entities. With the growing trend of using both network types for Internet access by a majority of the population, the former trend of splitting telecom companies into fixed-line and wireless divisions or even distinct companies is noticeably reverting. Network operators with both fixed-line and wireless assets have a competitive advantage since they can offer converged network access to their customers. From a backhaul perspective, having both wireline and wireless assets allows a telecom operator to use a single network infrastructure to backhaul both wireless and fixed-line

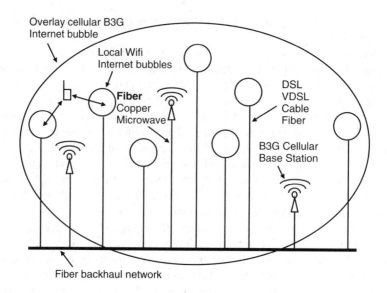

Figure 3.12   Converged cellular B3G and Wi-Fi/fixed-line network infrastructure.

data traffic. It further enables network operators to offer a seamless communication experience to their customers by offering devices that can be used in Wi-Fi networks at home and in cellular B3G networks while on the go. For the mass market it is important to offer and pre-configure services on mobile, nomadic, and stationary devices to work in such a converged network environment. Operators with both assets have a further advantage as they can offer Wi-Fi/DSL, B3G cellular access, devices, and pre-configured services in a single package. Since the concepts of device and application convergence from a network point of view will form an integral part of tomorrow's communication landscape, they are discussed in more detail in the following chapters in this book. Figure 3.12 shows what such a converged network architecture looks like from the user's and the network operator's point of view.

References

1. Prepaid Wireless Internet Access Database (2012) http://prepaid-wireless-internet-access.wetpaint.com (accessed 2012).
2. The European Parliament (2011) http://www.europarl.europa.eu/document/activities/cont/201112 /20111212ATT33766/20111212ATT33766EN.pdf (accessed 2012).
3. Rumney, M. (2007) What next for mobile telephony – examining the trend towards high-data-rate networks. *Agilent Meas. J.*, **3**, 32.
4. Holma, H. and Toskala, A. (2006) *HSDPA/HSUPA for UMTS*, Figure 7.38, John Wiley & Sons, Ltd, Chichester.
5. ECC, CEPT (2011) The European Table of Frequency Allocations and Utilisations in the Frequency Range 9 kHz to 1000 GHz. ERC Report 25, www.erodocdb.dk/Docs/doc98/official/pdf/ERCREP025.PDF (accessed 2012).
6. 3GPP (2008) User Equipment (UE) Radio Transmission and Reception (FDD). TS 25.101, Table 5.0, version 7.8.0. June 6, 2008.

7. Sauter, M. (2007) 3G and 4G Wireless is Private–DSL is for Sharing, http://mobilesociety.typepad.com /mobile_life/2007/06/wireless-is-pri.html (accessed 2012).
8. Douglas, T. (2005) Shaping the Media with Mobiles. BBC News (Aug. 4, 2005), http://news.bbc.co.uk/2/hi/uk_news/4745767.stm (accessed 2012).
9. Holma, H. and Toskala, A. (2006) *HSDPA/HSUPA for UMTS*, Table 8.5, John Wiley & Sons, Ltd, Chichester.
10. Sauter, M. (2011) Sleepless in Cologne but at 16 Mbit/s, http://mobilesociety.typepad.com/mobile_life/2011 /06/sleepless-in-cologne-but-at-16-mbits.html (accessed 2012).
11. Connect (2011) LTE Testing in Cologne, http://www.connect.de/ratgeber/lte-in-koeln-1176249.html (accessed 2012).
12. Dahlman, E., Parkvall, S., Skold, J., and Beming, P. (2007) *3G Evolution; HSPA and LTE for Mobile Broadband*, Chapter 19.2, Elsevier, Oxford.
13. Sauter, M. (2011) Video Stream Data Rates, http://mobilesociety.typepad.com/mobile_life/2011/12/video-stream-data-rates.html (accessed 2012).
14. Sauter, M. (2011) USB Connectivity of 3G Modem, http://mobilesociety.typepad.com/mobile_life/2011/08 /the-usb-cable-ensures-connectivity-again.html (accessed 2012).
15. 3GPP (2008) General Packet Radio Service–Service Description. TS 23.060, stage 2, Release 6, Chapter 9.2.2.1.1, December 11, 2008.
16. 3GPP (2008) Mobile Radio Interface Layer 3 Specification; Core Network Protocols. TS 24.008, stage 3, Release 6, Chapter 10.5.6.12, June 6, 2008.
17. Camarillo, G. and Garcia-Martin, M. (2006) *The 3G IP Multimedia Subsystem*, 2nd edn, Chapter 13.3, John Wiley & Sons, Ltd, Chichester.
18. 3GPP (2007) Multiplexing and Multiple Access on the Radio Path. TS 45.002, Table B.1, version 7.4.0, June 20, 2007.
19. Holma, H. and Toskala, A. (2006) *HSDPA/HSUPA for UMTS*, Chapter 7.5, John Wiley & Sons, Ltd, Chichester.
20. Sauter, M. (2011) *From GSM to LTE–An Introduction to Mobile Networks and Mobile Broadband*, John Wiley & Sons, Ltd, Chichester.
21. Bubley, D. (2011) Has Mobile Data Growth Flattened Off? Are Caps & Tiers Working Too Well? http://disruptivewireless.blogspot.com/2011/11/had-mobile-data-growth-flattened-off.html (accessed 2012).
22. Bundesnetzagentur (2012) EMF Datenbank, http://emf2.bundesnetzagentur.de/karte.html (accessed 2012).
23. Sauter, M. (2009) Satellite Internet on Thalys High Speed Trains–A Report, http://mobilesociety.typepad.com/mobile_life/2009/06/satellite-internet-on-thalys-high-speed-trains-a-report.html (accessed 2012).
24. Sauter, M. (2011) An Update of Thalys on Board Wi-Fi Performance, http://mobilesociety.typepad.com/ mobile_life/2011/02/an-update-of-thalys-on-board-wi-fi-performance.html (accessed 2012).
25. Wireless Intelligence (2011) Germany Rolls Out LTE to Rural Areas, http://www.wirelessintelligence.com/ analysis/2011/06/germany-rolls-out-lte-to-rural-areas/ (accessed 2012).
26. Deutsche Telekom (2012) Mobile Network Coverage Map, http://www.t-mobile.de/funkversorgung/ inland.
27. Dulski, A., Beijner, H., and Herbertsson, H. (2006) Rural WCDMA–Aiming for Nationwide Coverage with One Network, One Technology, and One Service Offering. Ericsson Review.
28. 3GPP (2006) Extended WCDMA Cell Range up to 200 km. RP-060191, http://www.3gpp.org/ftp/tsg_ran/ TSG_RAN/TSGR_31/Docs/RP-060191.zip (accessed 2012).
29. Sauter, M. (2011) UMTS 900 in London, http://mobilesociety.typepad.com/mobile_life/2011/03/umts-900-in-london.html (accessed 2012).
30. HD Voice News (2011) Deutsche Telekom Turns up Mobile HD-Voice in Germany, http://hdvoicenews.com/2011/11/02/deutsche-telekom-turns-up-mobile-hd-voice-in-germany/ (accessed 2012).
31. Sauter, M. (2011) MUROS goes VAMOS and Becomes Interesting, http://mobilesociety.typepad.com/ mobile_life/2011/09/muros-goes-vamos-and-becomes-interesting.html (accessed 2012).
32. Holma, H. and Toskala, A. (2007) *WCDMA for UMTS: HSPA Evolution and LTE*, Chapter 2.8, John Wiley & Sons, Ltd, Chichester.
33. The International Telecommunication Union (2003) One-Way Transmission Time. ITU-TG.114, http://www.itu.int/rec/T-REC-G.114-200305-I/en (accessed 2012).

34. Borman, C. (ed.) (2001) IETF RFC 3095. *Robust Header Compression (ROHC): Framework and Four Profiles: RTP, UDP, ESP, and Uncompressed*, June 2001, http://tools.ietf.org/html/rfc3095 (accessed 2012).
35. International Telecommunication Union (2004) ATM Cell Mapping into PDH. ITU G.804, July 1, 2004.
36. Gast, M. How Many Voice Callers Fit on the Head of an Access Point? http://www.oreilly.com/pub/a/etel/2005/12/13/how-many-voice-callers-fit-on-the-head-of-an-access-point.html (accessed 2012).
37. Free (2012) Accedez au monde de la Convergence Multimedia, http://adsl.free.fr (accessed 2012).
38. Layer 1 ISM Band Tracer (2012) http://www.metageek.com (accessed 2012).
39. IPerf on Sourceforge (2012) Iperf–The TCP/UDP Bandwidth Measurement Tool, http://sourceforge.net/projects/iperf/ (accessed 2012).
40. Sauter, M. (2006) VoIP over Wi-Fi–a Field Report, October 2006, http://mobilesociety.typepad.com/mobile_life/2006/10/voip_over_wifi_.html (accessed 2012).
41. Telegeography (2011) IP Transit Prices Continue to Decline, Geographic Differences Remain, http://www.telegeography.com/products/commsupdate/articles/2011/11/01/ip-transit-prices-continue-to-decline-geographic-differences-remain/ (accessed 2012).
42. Alcatel-Lucent (2012) 7300 Advanced Services Access Manager Outdoor Cabinet, http://www.alcatel-lucent.com/wps/portal/Products (accessed 2012).
43. Wikipedia (2012) E-carrier, http://en.wikipedia.org/wiki/E-carrier (accessed 2012).
44. Dragonwave Microwave Backhaul Solutions (2012) http://www.dragonwaveinc.com/ (accessed 2012).

4

Voice over Wireless

Despite the growing number of Web 2.0 and mobile Web 2.0 applications being used in B3G networks (cf. Chapter 6), voice telephony continues to be one of the most important applications in a mobile network. As mobile operators have a long history of making voice telephony work over wireless networks, and as they control the mobile infrastructure, they are in a good position to secure a sizable market share of tomorrow's wireless telephony business. Up to and including UMTS, voice telephony was tightly integrated into the wireless network infrastructure. As a consequence, mobile operators enjoyed a voice telephony monopoly in their networks. From the user's point of view the situation changed slightly in many countries in recent years due to fierce competition. Mobile Virtual Network Operators (MVNOs) sprang up who bought buckets of voice minutes from mobile network operators for reselling to customers under their own brand. From a technical point of view, however, the network operator remained in charge, as MVNOs were mere resellers of voice minutes.

In Long Term Evolution (LTE) networks, the situation has changed. As already discussed in Chapters 2 and 3, LTE networks no longer have a separate core network for voice telephony and voice optimized protocol stacks in the radio network. As in fixed-line broadband networks, all services and applications are now delivered via the Internet Protocol and a packet-switched connection. This brings both opportunities and challenges. On the positive side, from an innovation point of view, network and services are now decoupled. Thus, mobile services, be they voice or data centric, are no longer solely in the hand of network vendors. Instead, Internet companies are now shaping the service landscape and are working on repeating their success in wireless networks. Since voice telephony has become just one of many services being delivered over IP, fierce competition for next-generation voice services has sprung up between network vendors and operators on the one hand and Internet companies on the other. As far as voice telephony is concerned, no clear winner has yet been determined because for the time being the majority of cellular voice calls are still transported over dedicated and optimized voice channels on the radio interface in 2G and 3G networks. In the radio and core networks, the trend to replace traditional technology with faster IP-based links in effect moves traditional voice service to IP as well. This is transparent to the devices, however, as the channels on the air interface and the higher layer protocols are unchanged.

Both sides have strengths and weaknesses. The following chapter therefore looks at this topic from a number of different angles. First, an introduction is given of how traditional

3G, 4G and Beyond–Bringing Networks, Devices and the Web Together, Second Edition. Martin Sauter.
© 2013 John Wiley & Sons, Ltd. Published 2013 by John Wiley & Sons, Ltd.

voice telephony works in 2G and 3G networks as well as the benefits of migrating this service to the IP world. Afterwards the chapter takes a look at different telephony over IP services from a network operator's point of view and from an Internet company's point of view. Even though there is a general consensus that applications should be access-agnostic, that is that they should not be aware of, or even care, what kind of access network technology (cable, DSL, fiber, wireless, Wi-Fi, etc.) is used, this does not work well in practice for wireless networks. The reason behind this and the implications are also discussed.

4.1   Circuit-Switched Mobile Voice Telephony

The main purpose for which 2G wireless systems such as GSM were designed at the beginning was mobile voice telephony. Since it was the only application, each network element was specifically optimized for it. At the time, the state of the art for voice telephony in fixed-line telephony networks was to establish a transparent full duplex channel between two parties. While the connection is established between two parties in such a system, all data sent by the originator is transparently sent to the other side. The only exception are echo cancellation modules which are put into the transmission chain to improve voice quality.

4.1.1   Circuit Switching

Two parties are connected by a switching center which has a switching matrix to connect any telephone line (circuit) with any other. For long-distance calls, several switching centers are daisy-chained. For 2G and 3G wireless systems, the principle was reused as shown in Figure 4.1. The main difference between a circuit-switched fixed-line telephony system and a wireless circuit-switched telephony system is that a subscriber can no longer be identified by a pair of wires. Instead, a subscriber is now identified with credentials stored on a Subscriber Identity Module (SIM) card. On the network side a database known

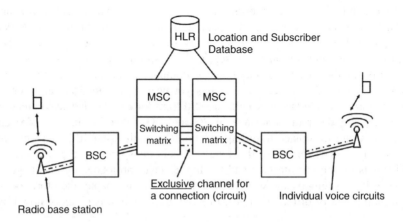

Figure 4.1   High-level GSM network architecture.

as the Home Location Register contains a replica of the user credentials and information on which options and supplementary services (e.g., call forwarding) a user is allowed to use. Except for the additional database in the network, most of the hardware and software of a fixed-line switching center could be reused for the design of a mobile switching center. Since the user is no longer identified by a specific pair of wires, a mobility management software component was added to the switching center software.

4.1.2 A Voice-Optimized Radio Network

Base Stations Controllers (BSCs), as shown in Figure 4.1, are used to decouple the switching center from the tasks required to establish and maintain a radio connection. BSCs are, as the name implies, responsible for communicating with base stations, which were initially designed as modems that convert digital information delivered on fixed-line connections into information sent over the air (OTA). In addition, BSCs are responsible for handing over a voice call to another base station or another radio cell of the same base station in case of deteriorating radio conditions. A more general term for a voice call in this architecture is "circuit-switched call" or "circuit-switched connection," since the switching center uses a switching matrix to connect two transparent circuits together for the duration of the call. On the user plane the Transcoding and Rate Adaptation Unit (TRAU), which is a logical part of the BSC, converts the speech codec used in the radio network to the speech codec used by the mobile switching center. This adaptation is necessary as GSM reused the hardware of fixed-line switching centers until recently, which was based on 64 kbit/s circuit-switched channels, while on the radio network side, it was necessary to compress the voice signal to squeeze as many voice calls as possible through the narrow radio channel. Furthermore, the adaptation is necessary as mobile networks are interconnected with fixed-line networks which also use 64 kbit/s circuit-switched channels for voice telephony.

4.1.3 The Pros of Circuit Switching

As each voice call is transported in a circuit-switched channel, the behavior of the system when setting up a new call is deterministic. If there is a free circuit between the originator of the call, the base station, the base station controller, the mobile switching center, and from there to the terminator of the call, the call is established. If no circuit is available on one of these links, the call request is rejected by the Message Service Center (MSC). During a call, system behavior is also deterministic. As circuits are uniquely assigned to a call, each call is independent and thus they cannot influence each other.

For further details on the design of GSM and UMTS circuit-switched wireless networks, the radio network and call establishment, see Chapter 1 of [1].

4.1.4 The Bearer Independent Core Network Architecture

Over the years, the GSM and UMTS circuit-switched telephony infrastructure, as presented earlier, was enhanced to replace the circuit-switched links between the MSCs with IP-based technology. This way, only a single type of technology is required in the core

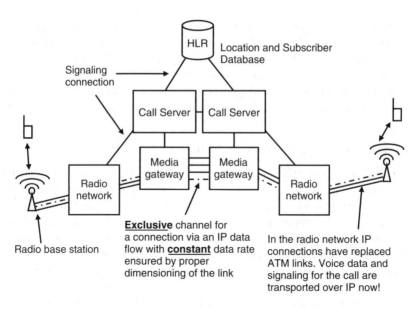

Figure 4.2   Circuit switching with dedicated network components.

network to transport voice and data. IP-based links for voice calls are usually physically or logically separate from links transporting other data and are dimensioned with sufficient capacity for the maximum number of calls to be carried to ensure a quality of service comparable to circuit-switched channels. This approach became known as the Bearer Independent Core Network (BICN) and the (3GPP) Release 4 architecture. In addition, the Release 4 design splits up MSCs into an MSC Call Server component that handles the signaling and a media gateway, which is responsible for forwarding the voice call as shown in Figure 4.2. Instead of fixed connections, media gateways use packet-switched IP connections to forward the call. This removes the necessity to transport the voice data via circuit-switched connections in the core network.

While not part of the Release 4 design, the previously circuit-switched connections based on ATM technology to the UMTS Radio Network Controllers (RNCs) have also been replaced with IP technology. In addition, the links between RNCs and the UMTS radio base stations (NodeBs) have also been migrated to IP. In other words, the previously fully circuit-switched mobile voice telephony service is now based entirely on IP except on the radio interface between a base station and the mobile device. This ensures quality of service for voice calls on the air interface, which is particularly difficult to achieve because of quickly fluctuating capacity and reception conditions. Furthermore, not modifying the air interface had the significant advantage that all changes performed in the network were completely transparent for mobile devices and therefore, no software modifications were necessary.

To use the IP protocol for voice calls on the air interface as well, it was decided in the industry to create a new service architecture from the ground up rather than to further extend the previously circuit-switched Mobile Switching Center-based architecture. The remainder of this chapter therefore now describes these new service architectures.

The classic MSC design and its Release 4 evolution path described in this and the previous section should therefore only be used as a comparison of how voice telephony is handled in wireless networks today and how it is envisaged to be done in the future.

4.2   Packet-Switched Voice Telephony

Designing a network for circuit-switched connections is ideal for voice telephony, fax, and narrowband circuit-switched data calls. Unfortunately, this also limits the use of the network to a narrow set of applications. As networks were designed with these applications in mind, there is no separation between the network and the applications, which ultimately prevents evolution. The SMS service is a good example. Adding SMS to the GSM network meant misusing signaling channels originally designed to carry messages required for voice call establishment. Again the application (SMS) was tightly integrated into the network design and only worked in this specific type of network. As a result, SMS messages could not be sent between wireless networks in the USA that used different kinds of 2G network technology for many years. Thus, tight integration of network and service prevented the take-up of a new service for many years until at last gateways were put in place between networks that convert the SMS signaling on one network to the signaling standard used in another.

Tight integration of applications and networks also prevents the evolution of an application. This is because changing an application also requires changes to the network itself. This is a process which network operators are only doing with great reluctance since changing the network structure is difficult, expensive, and bears the risk of unforeseen side effects on other parts of the network.

4.2.1   Network and Applications are Separate in Packet-Switched Networks

The Internet, on the other hand, follows an entirely different approach. Here, network and applications are independent of each other. This is achieved by creating a neutral transport layer that carries packets. Each packet has a source and destination address, the IP address. Nodes in the network then use the destination IP address to decide on which link to forward a packet to. Packets can be of variable lengths and can be concatenated on higher layers of the protocol stack. Thus any kind of application can efficiently send high and low volumes of data through the network. The network and the applications that run over the network are decoupled since the network does not see applications, just IP packets. At the top of the protocol stack applications do not see IP packets, just streams of data. This separation has worked very well in practice, as can be seen, for example, by the birth of the Hypertext Transport Protocol (HTTP) protocol and Web browsing, which were only invented many years after the invention of the IP protocol and the launch of first networks carrying IP packets. Since then, the Internet, or rather the accumulation of IP networks that form the Internet, has mainly evolved to offer ever higher speeds to the end user and to make access to the network cheaper. Applications have evolved almost independently of the network except for the fact that new applications such as video streaming and Internet Protocol Television (IPTV) require much higher bandwidths.

4.2.2   Wireless Network Architecture for Transporting IP Packets

The support for transporting IP packets in wireless networks was added in several stages. The first stage of the process was to add a packet-switched core network domain alongside the already known circuit-switched part of the network. In addition, the Radio Access Network (RAN) was enhanced to support both circuit-switched and packet-switched services simultaneously. This is shown in Figure 4.3. When a mobile device establishes a connection in a UMTS network, it informs the radio network or the RNC if it wants to establish a circuit-switched connection for a traditional service such as voice telephony or SMS or a packet-switched connection. The RNC then forwards the request to the MSC in case a circuit-switched connection is to be established or to the gateway node of the packet-switched core network (SGSN). Where a packet-switched connection is requested, the mobile device and the SGSN then perform authentication and activate ciphering for the radio link. Afterwards the mobile device requests an IP address in a procedure referred to as the Packet Data Protocol (PDP) context activation. The SGSN then communicates with the gateway to the Internet (Gateway GPRS Support Node, GGSN) which selects an IP address and returns it to the SGSN. The SGSN then returns the IP address to the mobile device and the connection is established. During the PDP context activation the mobile device and the network can also negotiate the quality of service to be used for the connection. An important property of a certain QoS level is, for example, the maximum bandwidth the network will grant to the mobile device and the minimum bandwidth it will try to enforce in case of congestion.

For applications, this process is usually transparent, as the establishment of an Internet connection is usually the task of the device's operating system. On a notebook, for example, the process is usually invoked manually by the user either with a special program delivered together with the wireless network card or by using the dial-up application of the operating system. Applications such as a Web browser, instant messaging (IM) program or VoIP client will just notice that the notebook is connected to an IP network and start using the connection.

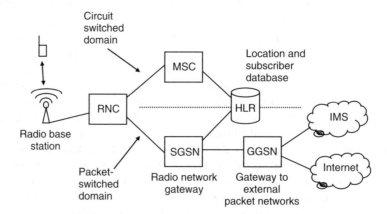

Figure 4.3   The UMTS network architecture with a circuit-switched and a packet-switched domain.

It is important to note at this point that the initially negotiated QoS level of a wireless connection applies to all applications using the connection; that is, it applies for all packets being sent and received. In practice such a QoS negotiation is mainly used to limit the maximum allowed bandwidth per user. Such a general assignment is unsuitable, however, to ensure the priority of IP packets of a voice telephony application over IP packets exchanged between a Web browser and a Web server in case of congestion. Depending on the type of network, it is thus possible to negotiate a separate quality of service level for individual applications or traffic flows. Since this requires the intervention of the application, it breaks the separation of network and application. As a consequence the application must become network-aware. Also, many network operators only allow their own applications to request a certain quality of service for their individual data stream.

While the connection is established, the RNC is responsible on the network side for maintaining the physical connection, enforcing the requested quality of service and handing the connection over to a different cell if the user is mobile. For further details refer to Chapter 2, which discusses these processes for different B3G network types.

4.2.3 Benefits of Migrating Voice Telephony to IP

Given the increase and abundance of bandwidth, it seems like a natural step to also transport voice calls over the Internet instead of over a separate and dedicated network. There are many different angles from which to view this shift. From a network operator's point of view, transporting voice calls over IP reduces cost in the long term since only a single network needs to be maintained. From an enterprise point of view, using the IP network inside the enterprise for voice telephony reduces the number of cables in the building since a desk no longer requires a separate cable for the telephone. From a user's point of view, migrating the circuit-switched voice service to the packet-switched IP domain potentially reduces cost as they no longer have to pay for analog telephony and Internet access separately. Instead of connecting the analog telephone to a traditional telephone jack in the wall, it is now connected to a telephone jack in a DSL router or cable modem. The router contains the required hardware and software to enable call establishment and to digitize the analog signal received from the analog telephone, packetize it and send it over the Internet. If the network is well designed, the change is transparent to the user. Except for the potentially lower price, however, the user has no incentive to migrate to VoIP this way. This applies for fixed-line as well as for wireless networks.

4.2.4 Voice Telephony Evolution and Service Integration

Fortunately there is another good reason for migrating voice telephony to the IP world, both in wireline and wireless networks: service integration. Service integration happens when a communication device is capable of running several services simultaneously and is able to combine them in an intelligent way. Examples of this include telephony plug-ins in Web pages. When a user searches for a suitable hotel she might do this by accessing the Internet and querying a search engine. The search engine will return a number of Web pages of hotels in a city and the user will then go through those pages to see if some of those hotels are suitable. Several hotels are interesting but she wants to be sure that there will be Wi-Fi access and that the costs are reasonable. She therefore decides to call the hotel instead of ordering online. In a circuit-switched telephony world the user now has to pick up the telephone and dial a telephone number. This takes some time and is potentially expensive if this involves making an international call. In a world where voice telephony is just another application running over the same network, the Web page can have an embedded button which will launch a VoIP program on the notebook which automatically calls the hotel. The user is connected instantly, does not have to type in a telephone number and the call is likely to be free no matter what part of the world the hotel is located in.

Another example of service integration is a document with embedded information about those who created it. When downloading an interesting document it is usually difficult to find out who created the document and how to contact them, and then it is necessary to pick up a telephone and dial a telephone number. This means that there is a gap between the desire to communicate and the ability to do so. Once voice telephony has just become another application running over a single network, the information in the document can be used by the voice telephony application to find users and call them. This significantly shortens the gap between the desire and the ability to communicate.

Service Integration also means making the voice telephony service more flexible and grouping new types of communication applications around it. In the IP world, voice telephony can be extended with presence capabilities so a person can see if another person is available before making a call. IM in combination with presence is also an important addition to voice telephony as users often prefer a text-based message to a call.

Voice telephony evolution also means separating the service from the device. Since the network and the application are separate, it is also no longer important which device is used with a certain voice telephony user account. A user can have several devices for voice communication, like a cordless telephone, PCs and notebooks, B3G cellular telephones, a Personal Digital Assistant (PDA), a game console, and so on, that all use the same voice telephony account. Some or even all devices can be attached to the network simultaneously and an incoming call will be delivered to all active devices or a user-defined subset instead of only to a single device, as in the past. The user then accepts the call with the most suitable device in the current situation.

4.2.5 Voice Telephony over IP: The End of the Operator Monopoly

Another important element in the transition of the voice telephony service into the IP world is that it ends the operator monopoly on this service in both fixed-line and wireless networks. This has allowed Internet companies such as Skype, Yahoo, Microsoft, Sipgate, and many others to offer telephony services either with proprietary or standardized protocols. This has sparked competition and innovation.

The remainder of this chapter now looks at a number of different approaches to offering voice telephony in different standardized fashions over wireless B3G networks. Section 4.3 introduces the SIP which is the most popular protocol and service platform for establishing voice calls over IP networks. Since it is completely network-agnostic, it will run over both wireline and wireless IP networks and allows Internet companies to offer voice telephony services.

Section 4.4 takes a closer look at the IP Multimedia Subsystem (IMS), a telephony, and general application framework favored by current fixed-line and mobile operators as their next-generation telephony and services platform. The IMS itself is based on the SIP architecture, which was enhanced with standardized procedures to act as a general multimedia service delivery platform over both fixed-line and wireless networks. The IMS system also contains additions to communicate with the network itself to ensure quality of service, which is something that "naked SIP" (described in Section 4.3) is not capable of doing. Why this might be required in the future and the pros and cons of this are then discussed in Section 4.5.

4.3   SIP Telephony over Fixed and Wireless Networks

The most popular standardized system to establish voice calls over IP networks is Session Initiation Protocol (SIP). It is standardized in RFC 3261 [2] and is the abbreviation for SIP. As the name suggests, the protocol is intended for establishing not only a voice connection between two parties, but also a general messaging protocol to establish a "session" between two or more parties. The term "session" is quite generic and in practice SIP can be used to establish voice and video calls and instant message exchange between two parties; it can carry presence information so users can see when their friends are online and many other things. In practice, however, SIP is mostly used for establishing voice calls today, despite its general nature.

In the wireless world, SIP telephony is used today in various ways. Cable and DSL modem routers often include a SIP application which converts the signals received from an analog voice telephone plugged into a telephone jack at the back of the router. Since from the telephone's point of view the telephone jack behaves just as the standard analog telephone network jack, it can be used with traditional wired and cordless telephones. There are also a number of Wi-Fi telephones available on the market today which use the SIP protocol and a Wi-Fi connection at home or at the office to connect to the Internet. Most importantly, however, smartphones are in addition to 3G or B3G also Wi-Fi capable. For most of them SIP client software is available and they can thus be used as cellular telephones and SIP telephones over either Wi-Fi or the B3G network. Last but not least, there is a great variety of SIP software available for PCs and notebooks.

4.3.1   SIP Registration

Figure 4.4 shows the network components required for voice telephony over IP with SIP. In SIP terminology the VoIP software running on the client device is referred to as the User Agent. This abstraction is a good fit for the B3G world since mobile devices are becoming ever more versatile and, although mobile voice telephony is surely an important application, it is nevertheless just one of several applications on a mobile device.

When the User Agent is started, for example, when a mobile device is switched on, it will register its availability with the SIP registrar in the network in order to be reachable and to allow the user to make outgoing calls. For the following description it is assumed that the mobile is already attached to a wireless network and an IP address has been assigned. This procedure is not part of the SIP specification since SIP voice telephony

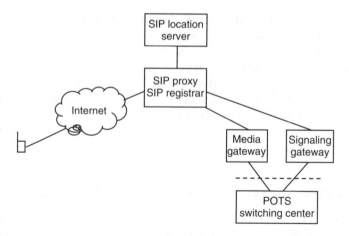

Figure 4.4   The SIP network architecture.

is an application running on an IP network. How a device attaches to the network and how an IP address is requested in Wi-Fi and cellular B3G networks has been shown in Chapter 2.

If an IP address is used to identify the registrar, the User Agent can send a registration message right away. If a domain name is used instead, a Domain Name System (DNS) server has to be queried first to convert the domain name into an IP address.

The most important information elements of the SIP registration message are the IP address and User Datagram Protocol (UDP) port of the User Agent and the SIP identity of the user. The format of the SIP identity, also referred to as the user's SIP Universal Resource Identifier (URI) [3], is similar to that of an e-mail address and contains the user ID and the SIP domain (also known as "realm"). The realm identifies the SIP provider with whom the user has a subscription. Examples of valid SIP addresses are "5415468@sipgate.de" and "martin.sauter@mydomain.com". Both the ID ("5415468" or "martin.sauter") and the realm (sipgate.de or mydomain.com) are required to be reachable by users of a different SIP network provider, as will be shown below.

When the registrar receives a "register" message, it searches the database for the corresponding account information and then attempts to authenticate the user. This is done with a password which is shared between the User Agent and the registrar. Unlike cellular wireless systems such as GSM and UMTS, where the common secret is stored on the SIM card of the mobile device, the password for SIP telephony is usually stored in the mobile device itself and can be changed via a menu in the User Agent software as required. Verifying the password is done by the SIP registrar rejecting the first registration request with an "unauthorized" response message which contains a random (RAND) value for the User Agent, which is referred to as a "nonce." The User Agent then takes the nonce and the password as an input for a shared encryption algorithm to create an authentication response value. This value is then sent back to the registrar in another register message. When receiving the second register message the registrar compares the

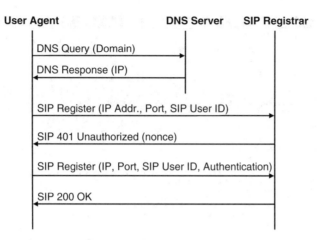

Figure 4.5   SIP registration message flow.

authentication response value to the value it has computed itself. If the values match, the subscriber is authenticated and the registrar answers with an "ok" message. Afterwards the User Agent is available and the user can initiate and receive calls. As a final step the registrar stores the subscriber context (IP address, UDP port number, etc.) in the SIP location server database. Figure 4.5 shows the message flow in a graph.

Associating the subscriber's context with the SIP identity is also referred to as a binding since the registration process binds the user's identity (the SIP URI) to the information on which User Agent on which device the user can be reached. The user can register several User Agents, that is several devices, to their SIP identity. Incoming calls will then be signaled to all User Agents/devices that are bound to a SIP identity. In the circuit-switched telephony world this feature is known as simultaneous ringing or "simring."

Figure 4.6 shows the message content of the second SIP register message. All information in the SIP message is in human readable format. This makes the message quite large (674 bytes in the example), but has the advantage that it allows easier development of new functionality and debugging of a system. The syntax of SIP messages is similar to that of the request and response messaging of the HTTP, which is used by Web browsers to request a Web page. Each SIP request message starts with a request line which announces the type and the intention of message. The message header then contains the information required, depending on the message type. The register message in Figure 2.6, for example, contains the identity of the user and the authorization information. All requests also contain a command sequence number (CSeq) to be able to correlate all messages belonging to the same request/response dialog. This allows simultaneous dialogs for different purposes, like receiving an instant message while at the same time establishing an outgoing call.

Responses to a SIP request contain a status line at the beginning of the message and the message header. The status line contains a numeric status ID and the message name in clear text. The header then contains further response information.

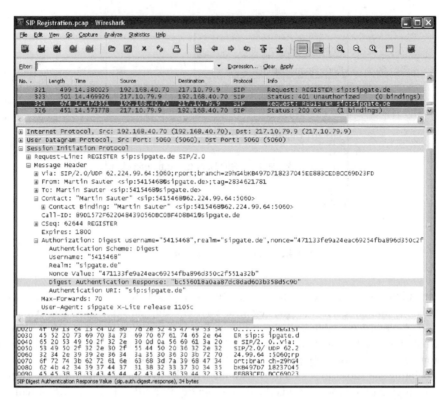

Figure 4.6 A SIP register message with authentication information. (Reproduced from Wireshark, by courtesy of Gerald Combs, USA.)

4.3.2 Establishing a SIP Call between Two SIP Subscribers

Figure 4.7 shows the message flow between two User Agents via two SIP proxies for establishing a voice call. A SIP proxy is usually physically implemented in the same server as the SIP registrar, as shown in Figure 4.4. From a logical point of view, however, the SIP proxy is independent. The proxy gets its name from the fact that the User Agent does usually not know the IP address of the other party the user wants to get in contact with and thus sends the message to the SIP proxy. The SIP proxy then locates the other party and forwards the message on behalf of the user.

The first message sent by the originator to the SIP proxy is a SIP "invite" message. The most important parameter of this message is the destination identity. The identity can have two different formats. SIP subscribers can be identified either with a username (e.g., sip:martin.sauter@sipgate.de) or a standard telephone number to be reachable from external fixed and mobile networks. Where the user dials a standard telephone number the User Agent automatically appends the local realm (e.g., sip:00497544968888@ sipgate.de).

When the SIP proxy receives the message it responds with a "407 Proxy Authentication Required" message to authenticate the User Agent. The User Agent then terminates this invite dialog with an acknowledge message. Afterwards, the User Agent calculates an

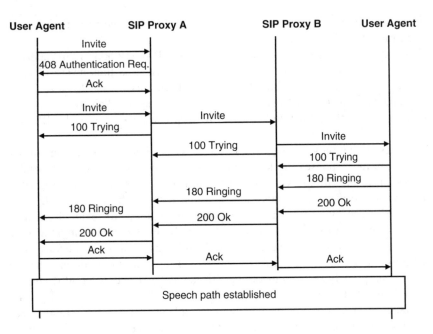

Figure 4.7   SIP session establishment message flow.

answer for the RAND value given back by the SIP proxy and sends another invite message, this time including the security information.

In the next step the SIP proxy verifies the authentication response in the invite message as described in the register dialog above and proceeds with the session establishment by analyzing the destination party identity in the "To:" field of the message header. Depending on the realm part of the identity, the SIP proxy searches either its own subscriber database or a database containing the IP addresses of proxy servers of other SIP networks. In the case shown in Figure 4.7, the realm belongs to a different SIP network and the message is thus forwarded to the SIP proxy server of the other network. To show that the message has been forwarded, the SIP proxy then returns a "100 trying" message back to the User Agent.

When the SIP server of the other SIP network receives the incoming message it also looks at the "To:" field of the "invite" message. Since the realm part of the ID is its own it will query the local user database to retrieve the location (IP address and UDP port number) of the subscriber. It then forwards the "invite" message to the destination subscriber and returns a "100 trying" message back to the SIP server in the other SIP network.

When the destination User Agent receives the "invite" message it also returns a "100 trying" message to the SIP proxy to indicate that it has successfully received the invitation. The User Agent then informs the user of the incoming call and returns a "180 ringing" message to the SIP proxy. This message is then sent back via the SIP proxy in the other SIP network to the originating User Agent. When the user accepts the call, the User Agent sends a "200 ok" message to the originating User Agent via the two SIP proxies. At the same time the audio channel is established between the two User Agents. How this is done is described in the next section. It is important to note that the audio data is exchanged directly between the two User Agents and not via the SIP proxies, as they are only used for establishing a session.

Each proxy that forwards a message to another SIP proxy adds its IP address to the header part of the message. The recipient of the message is thus aware of all the SIP proxies the message has traversed. When returning an answer the User Agent includes these IP addresses in the SIP header of the message again. This way a response can efficiently traverse the SIP network without requiring a decision at each SIP proxy as to where to forward the message.

SIP proxies are not only allowed to forward messages but can also change their content or react to certain responses. If the terminating User Agent, for example, sends back a "busy" response, the SIP proxy can either return this message to the originating User Agent or decide to discard the busy message and forward the call to a SIP voice mail system. Proxies can also fork a message, which is required, for example, if a user has registered several devices to the same SIP identity. In this scenario a single incoming SIP "invite" message is forked and sent to each device registered to the identity. To do this the SIP proxy needs to be "stateful" as it needs to remember how often it has forked a request to appropriately react to incoming SIP messages of the different destinations.

It should be noted at this point that the example in Figure 4.7, which reflects what is done in practice, does not contain messages to authenticate the destination User Agent. This allows a potential attacker who has managed to take over the IP address of the terminating User Agent to accept the call.

4.3.3 Session Description

As SIP is a generic session establishment protocol for many different kinds of media streams, the originator of a session has to explicitly inform the other party what kind of session (voice, voice video, etc.) is to be established. This is done by describing the types of media streams and their properties in the body of the "invite" message. The protocol used for this purpose is the Session Description Protocol (SDP), standardized in RFC 4566 [4].

Figure 4.8 shows the SDP media description in the message body of the SIP "invite" message for a voice call. The most important parameters are:

1. **The connection information parameter** — the originator uses this parameter to inform the other User Agent of the IP address from which the media information for the session will be sent and received.
2. **The media description parameter** — contains information on the type of media stream the originator wants to send:
   (a) The first subparameter specifies the type of media stream (e.g., audio).
   (b) The second subparameter represents the local UDP port number from which the stream will be sent and where the return stream is expected if a full duplex connection is to be established. This will be the case for most types of communication sessions.
   (c) The next subparameter is the description of the transmission protocol to be used for the media stream. Most audio and video applications use the RTP/AVP (Real Time

Transfer protocol/Audio Video Protocol) for this purpose. Further detail on the streaming protocol is given below.

(d) The remaining numeric parameters are RTP payload numbers, which indicate the supported voice and video codecs of the User Agent.

3. **The media attribute parameters** — these follow in subsequent lines and give further media details:

(a) **Rtpmap** — this media attribute gives further details on the payload numbers. Payload number 8, for example, corresponds to the A-Law Pulse Code Modulation (PCMA) codec, which is also used in circuit-switched telephone networks to encode the voice signal for a 64 kbit/s circuit-switched channel. The line also informs the terminating User Agent that the input signal was digitized with a sampling rate of 8000 Hz.

(b) **Sendrecv** — this media attribute tells the remote party that the originator will send a media stream and expects to receive a similar media stream from the remote party as well.

If several media streams are required for a session (e.g., voice video), another media descriptor parameter, and corresponding media attribute parameters are appended to the SDP message.

The remote User Agent sends their session description information as part of the "200 ok" message once the user has accepted the session. The same parameters are used in

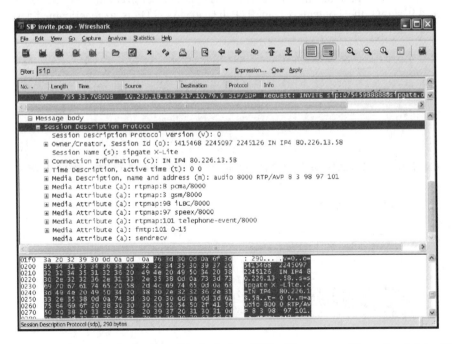

Figure 4.8 Session description in a SIP "invite" or "200 ok" message. (Reproduced from Wireshark, by courtesy of Gerald Combs, USA.)

this SDP message as in the SDP message of the originator. The values transported in the parameters, however, might be different. This is the case, for example, if the remote User Agent supports a different set of media codecs than the originator. If they share at least one codec for a type of media stream the connection can be established. Otherwise the session setup will fail.

Once the "200 ok" message is received by the originator both ends of the connection will start sending their media streams. The media stream of the originator is sent from the UDP port described in the media descriptor parameter of the "invite" message to the UDP port of the remote User Agent given in the "200 ok" message. The media stream of the remote User Agent uses the reverse port combination. The media codec chosen by both ends depends on the local list of supported codecs and the remote list of supported codecs received from the other end.

4.3.4 The Real-Time Transfer Protocol

In circuit-switched networks a media stream can be transparently exchanged between the two parties through a circuit-switched connection. In IP networks, however, transmission is packet-switched and the media stream is encapsulated in IP packets which are then routed through the network. As each IP router between the two parties has to make a routing decision, each packet has to contain the IP addresses of the two parties and the UDP port numbers between which the media is exchanged. On top of IP and UDP, the RTP is used to encapsulate the media stream and to give the destination further information about how to interpret the incoming data. RTP is standardized in RFC 3550 [5] and Figure 4.9 shows the contents of an RTP header. As RTP is a generic media stream

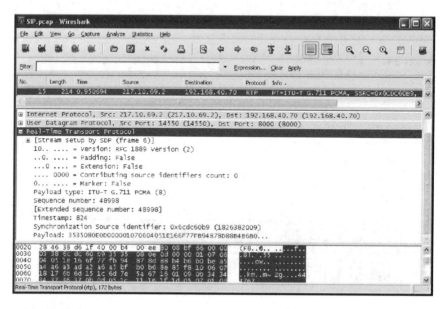

Figure 4.9 Real-time transport protocol header. (Reproduced from Wireshark, by courtesy of Gerald Combs, USA.)

protocol, one of the first parameters of the header is the payload type, which informs the receiver of the protocol and the RTP profile to be used for decoding the payload information. RTP profiles for audio and video transmissions with different codecs are defined in RFC 3550 [6].

In IP networks, the order and timely delivery of UDP packets is not guaranteed. Furthermore, it is possible for UDP packets to be dropped during times of network congestion. For real-time voice and video transmissions this is challenging. To compensate, the receiver uses a jitter buffer in which incoming data packets are stored for a short time (e.g., 50–100 ms) before they are played to the user. If a packet is late, the time it remains in the jitter buffer is reduced. If packets arrive in the wrong order, but still within the limits of jitter buffer, they can be reordered and the event is not noticeable to the user. The longer the jitter buffer stores packets, the higher the chances are that temporary network congestion will have no impact on the voice or video quality. The time for which the jitter can store packets in practice is very limited, however, since this adds to the overall delay of the connection. End-to-end delays including delays caused by the codec, by transmission, and by the jitter buffer exceeding 150 ms are already noticeable to the user.

To control the jitter buffer the RTP header contains two parameters. The first one is the sequence number field which is incremented in each packet. This allows the receiver to detect missing or reordered packets. Constant bitrate codecs can then use the payload size and their knowledge of the sampling rate to determine if a packet is late, early, or on time. Another way to determine the correct arrival time is the timestamp parameter. The use of this parameter depends on the type of audio or video codec used. The Pulse Code Modulated (PCM) codec, for example, samples an audio signal 8000 times a second. With 8 bits per sample the resulting codec rate is 64 kbit/s. The User Agent then typically puts samples of a 20 ms interval into one RTP packet. For each packet the timestamp value is thus incremented by 8000/0.02 160.

Every RTP session is accompanied by a Real Time Control Protocol (RTCP) [5] session that uses the next higher UDP port number. Here, messages are exchanged periodically, in the order of one every 5 s, in which the members of a session inform each other about the number of lost packets and the overall jitter experienced.

4.3.5   Establishing a SIP Call between a SIP and a PSTN Subscriber

Many SIP networks enable their users to call subscribers of a Public (Circuit) Switched Telephone Network (PSTN) or of 2G/3G cellular networks. Since these subscribers can only be reached via a circuit-switched network, a gateway is required as shown in Figure 4.4. The gateway has two logically separate components which can be implemented either in a single device or separately if this is required for scalability reasons. The first logical component is the signaling gateway, which translates the SIP messages shown in Figure 4.8 into Signaling System Number 7 (SS-7) messages, which are used in circuit-switched networks. The second component is the media gateway that takes the digitized voice information from incoming IP packets and puts it into a circuit-switched connection and vice versa. Usually the SIP network and the PSTN network use the same voice codec. In this case the data on the application layer can thus be forwarded transparently. If different codecs are used in the two networks the media gateway additionally transcodes the media stream. Large networks might require several media gateways due to the high number of simultaneous calls or for redundancy reasons. Also, networks usually have at least two signaling gateways to be able to continue service in case one of the gateways malfunctions. The SIP proxy usually contains the functionality to detect that a gateway has failed and automatically redirects new sessions to the alternative gateway.

In the circuit-switched telephony world the terminating switching center is responsible for generating the alerting tone that is returned to the originator of the call until the terminating party has accepted the call. With the message flow as shown in Figure 4.7, the gateway is not able to forward the alerting tone to the SIP originator since the originator first requires the UDP session description from the SIP gateway which contains the UDP port number from which the audio stream originates. The standard therefore already allows the sending of the session description information in the "180 ringing" message or in a "183 session progress" message that is sometimes used as an alternative. This is the preferred solution even for SIP-to-SIP calls as the media streams can be established in the background before the terminator has accepted the call. Thus, no time is lost after the terminator has accepted the call.

Figure 4.10 shows the session establishment message flow from a SIP subscriber to a PSTN subscriber. The message flow up to the Signaling Gateway (SGW) is identical to the message flow for the SIP-to-SIP call shown in Figure 4.7 except for the following: instead of using a "180 ringing" message, the alternative "183 session progress" is used in Figure 4.9, which contains the early session description of the signaling gateway. Also, additional signaling is required to inform the media gateway of the call.

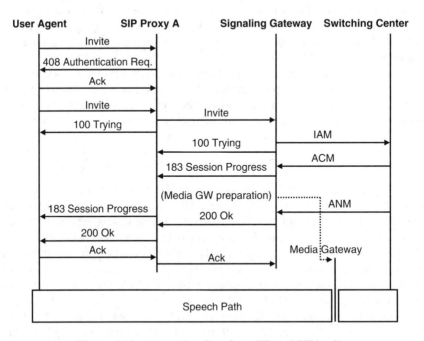

Figure 4.10   Messaging flow for a SIP to PSTN call.

The signaling gateway converts the SIP messages for the Plain old telephone service (POTS), GSM, or UMTS switching center as follows: the parameters of the SIP "invite" message are used to create an SS-7 "Initial Address Message" (IAM), which requests the switching center to establish a circuit-switched connection to a subscriber.

The switching center responds with an Address Complete Message (ACM) which tells the SIP signaling gateway that the call is proceeding and which timeslot on which circuit-switched link between itself and the media gateway will be used for the voice data. This information is then used by the SIP network to configure the media gateway for the session. This is done by informing the media gateway of the link and timeslot number to be used for the session on the circuit-switched side and the IP address and UDP port number of the User Agent on the SIP side. Once the media gateway is ready, a "183 session progress" message is created and sent back to the SIP proxy, which in turn forwards the information to the User Agent. Since the session progress message contains a session description, the User Agent can now already activate the speech path to forward the alerting tone generated by the switching center to the user. Note that in Figure 4.9 the activation of the speech path is shown in the lower part for clarity reasons, even though it can already be active after the session progress message.

When the circuit-switched subscriber accepts the call, the switching center generates an Answer Message (ANM) and sends it to the signaling gateway. The signaling gateway converts this SS-7 message to a "200 ok" SIP message and forwards it to the User Agent via the SIP proxy.

Finally, it should be noted at this point that, from the switching center's point of view, the signaling gateway behaves like another switching center. Thus, no modifications are necessary in the SS-7 messaging. From the SIP proxy's point of view the signaling gateway behaves just like a User Agent. Thus, no changes are required in the SIP messaging either.

4.3.6 Proprietary Components of a SIP System

The functions and entities of a SIP network discussed up to now are only related to establishing a session. In practice, however, a SIP network usually has a number of additional functions:

- Some operators use in-band announcements to inform users about the cost per minute of a call before the connection to the terminating party is established.
- While many operators offer SIP-to-SIP calls for free, calls to PSTN or mobile subscribers are not free. Therefore a billing solution is required. In case of post-paid billing, which means that the subscriber receives a monthly bill, the SIP Proxy or the PSTN gateway needs to forward billing information to a billing system.
- It is also common for network operators to offer prepaid billing where users transfer a certain amount of money to the operator that can then be used for calls. Transferring funds usually requires a Web-based interface. To be able to bill for calls in real time, the SIP proxy or gateway needs to have an interface to a prepaid billing server. Alternatively the prepaid billing solution can also be a part of the proxy or gateway.

- Many SIP networks also offer a Web-based configuration and customer care interfaces with many functionalities. These include user self-configuration of call forwarding, prepaid top-up, review of billing information, and so on.

How and where these functions are implemented is not standardized. From a technical point of view this is not necessary since their implementation has no impact on the SIP signaling between a User Agent and an SIP proxy. Lack of standardization has the advantage that companies can be innovative and offer additional functionality to network operators to make their service more attractive to end users which, in turn, can lead to a competitive advantage. The disadvantage of a lack of standardized interfaces for these functions to SIP components, however, also means that the market is fractured. The availability and evolution therefore depends on the vendor of the SIP equipment. This in turn has the disadvantage for network operators that there is no direct competition in this area, which usually results in higher prices for any additional functionality network operators want to buy once they have bought SIP equipment from a vendor.

4.3.7   Network Address Translation and SIP

Most Wi-Fi home networks connected to the Internet via DSL or cable are only assigned a single IP address by their network operator. To be able to use several devices within the home network the DSL or cable router has to map the local IP addresses to a single external IP address. Since two computers in the private network can use the same UDP or Transmission Control Protocol (TCP) port numbers, these have to be mapped between the private and the public network as well. This process is referred to as Network Address Translation (NAT). Figure 4.11 shows a setup which requires NAT. For applications such as Web browsing this mapping process is completely transparent as they do not use their own IP address and port number on the application layer.

The SIP and SDP protocols, however, use the local IP address and port number on the application layer, as shown in Figures 4.6 and 4.8, to signal to a SIP proxy and to a remote User Agent where to send messages and media streams. As the User Agent on a device is only aware of the private IP address and port number, an additional step is required to determine the external IP address and port used for a request before a SIP operation. This is usually done with "Simple Traversal of User Datagram Protocol Through Network Address Translators," or STUN, which is standardized in RFC 3489 [7].

The principle of STUN is as follows: before a User Agent contacts the SIP registrar to register, it first contacts a STUN server in the public Internet with a "binding request" message. When the "binding request" packet arrives at the DSL or cable router the NAT algorithm changes the IP address of the packet from the private IP address to the public IP address. If the local UDP port number used by the packet is already in use for a conversation of a different client device, it is also changed. The STUN server thus receives the packet with the external identification. The STUN server then sends a "binding response" packet back to the User Agent. At the NAT device the destination IP address (and UDP port) of the packet is changed to the private IP address (and private UDP port). The "binding response" packet, however, also contains the real IP address (and UDP port) in the payload of the packet, which is not changed by the NAT router. This way the SIP User Agent can determine which external IP address was used and if the UDP port

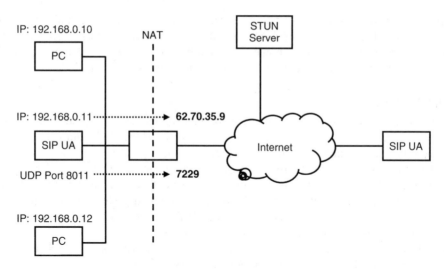

Figure 4.11 Network address translation in home networks.

number was changed as well. As a number of different NAT implementations exist, the User Agent will send additional "binding request" messages to probe the behavior of the NAT device when the STUN server replies to the request from a different IP address and port number. If the NAT device delivers these replies from the STUN server to those requests, the User Agent is aware that it does not need to start sending a media stream to open the NAT port. If no answer is received for such additional binding requests, the User Agent takes this into account later on and does not wait for an incoming media stream from the other end before starting its own transmission.

Figure 4.11 shows one User Agent behind a NAT device and another User Agent directly connected to the Internet. STUN, however, also works for scenarios in which both User Agents are behind network address translators.

It should be noted at this point that there is one NAT scheme which STUN is not able to overcome. Most NAT implementations will always use the same mapping from internal to external UDP port number regardless of the destination IP address. If, however, the NAT implementation uses a new port mapping for each external IP address, it is not predictable which UDP port will be used for the media stream since it will be sent to an IP address that is different from that of the STUN server. In this case, no media connection can be established and session establishment from User Agents behind such a NAT device fail. Fortunately, most routers used in home environments do not use this kind of address translation.

4.4   Voice and Related Applications over IMS

Section 4.3 has shown how SIP is used to establish Voice over IP sessions. SIP, however, is a general protocol to establish any kind of session and is thus capable of much more than just to establish voice sessions. In practice, however, voice telephony is the main application for SIP. As SIP and VoIP are based on the IP protocol, it can also be used in the packet-switched part of 3G networks and of course in B3G networks. From a wireless network operator point of view, however, a number of critical elements are missing in today's SIP specifications to support millions of subscribers. These include:

- Mobility aspects
  - General SIP implementations are network-agnostic and cannot signal their quality of service requirements to a wireless access network. Thus, VoIP data packets cannot be preferred by the system in times of congestion.
  - Handling of transmission errors on the air interface cannot be optimized for SIP calls. While Web browsing and similar applications benefit from automatic retransmissions in case of transmission errors, VoIP connections would prefer erroneous packets to be dropped rather than be repeated at a later time since such packets are likely to come too late (cf. jitter discussed in Section 4.3.4).
  - SIP VoIP calls cannot be handed over to the 2G network such as when the user roams out of the coverage area of B3G networks.
  - VoIP over SIP does not work in 2G networks.
  - Most SIP implementations today use the 64 kbit/s PCM codec for VoIP calls. Compared with optimized GSM and UMTS codecs, which only require about 12 kbit/s, this significantly decreases the number of VoIP calls that can be delivered via a base station. Furthermore, mobile network optimized voice codecs have built in functionality to deal with missing or erroneous data packets. While this is not required for fixed networks due to the lower error rates, it is very beneficial for connections over wireless networks.
  - Emergency calls (112, 911) cannot be routed to the correct emergency center since the subscriber could be anywhere in the world.
- Functionality aspects
  - There is no billing flexibility. Since SIP implementations are mostly used for voice sessions, billing is usually built into the SIP proxy and no standardized interfaces exist to collect billing data for online and offline charging.
  - Additional applications such video calls, presence, IM, and so on, are usually not integrated in SIP clients and networks.
  - It is difficult to add new features and applications since no standardized interfaces exist to add these to a SIP implementation. Thus, adding new features to User Agents and the SIP network such as a video mailbox, picture sharing, adding a video session to an ongoing voice session, push-to-talk functionality, transferring a session to another device with different properties, and so on, is proprietary on both the terminal and the network components. This is costly and the use of these functionalities between subscribers of different SIP networks is not assured.
- Insufficient security
  - Voice is usually sent unencrypted from end to end, which makes it easy to eavesdrop on a connection.
  - Signaling can be intercepted since it is not encrypted. Man-in-the-middle attacks are possible.

– No standards exist on how to securely and confidentially store user data (e.g., user-name/password) on a mobile device.
- Scalability — mobile networks today can easily have 50 million subscribers or more. This is very challenging in terms of scalability since a single SIP proxy in a network cannot handle such a high number of subscribers. A SIP network handling such a high number of subscribers must be distributed over many SIP proxies/registrars.
- There is no standardized way to store user profiles in the network today. Also, no standardized means exist to distribute user data over the several databases which are required in large networks (see scalability above).

To address these missing pieces the wireless industry has launched a number of projects. The most comprehensive is that of the Third Generation Partnership Project, which is also responsible for GSM, UMTS, and LTE standardization. The result of this ongoing activity is the IMS, which is based on the SIP as the core protocol for a next-generation service delivery platform. Additional to the core SIP standards discussed earlier, many new standard documents were created to describe additions required for an IMS system.

Other standards bodies have also started to define next-generation IP-based voice and multimedia architectures. The most noteworthy of those is Telecommunications and Internet converged Services and Protocols for Advanced Networking (TISPAN), which has specified an IMS-like architecture for fixed-line IP networks, and 3GPP2, which has defined an IMS for CDMA-based networks. With Release 7 of the 3GPP standard, 3GPP, and TISPAN have joined forces to create a single next-generation network (NGN) architecture that will work in both fixed and 3G, B3G, and Wi-Fi networks. Furthermore, interworking between the 3GPP IMS and the 3GPP2 IMS is assured. This is good news from a network convergence point of view since operators with fixed and wireless assets are very interested in having a single platform with a single service offering and to allow subscribers to use one subscription with all types of devices and access networks.

In the following sections an overview is given of how the IMS uses the SIP protocol and how it addresses the missing elements described earlier to become a universal session-based communication platform. Because of its centralized design, the IMS is likely to be successful with applications grouped around voice and IM services. The following list shows some applications, which IMS networks could offer to users:

- Voice telephony as the main application — this includes handing over voice calls between networks as the user roams out of coverage of a network. Furthermore, advanced IMS solutions will enable handovers of voice calls to a 2G network when B3G coverage is lost.
- The IMS enables video calls with the advantage over current 3G circuit-switched mobile video calls that the video stream can be added or dropped at any time during the session.
- Presence and IM.
- Voice and video session conferencing with three or more parties.
- Push message and video services such as sending subscribers messages when their favorite football team has scored a goal, when something exciting has happened during a Formula 1 race, and so on.
- Calendar synchronization among all IMS devices.

- Notification of important events (birthdays, etc.).
- Wakeup service with auto answer and the user's preferred music or news.
- Live audio and videocasts of events. The difference between this and current solutions is the integrated adaptation of capabilities on the device.
- Peer-to-peer document push.
- Unified voice and video mail from all devices used by a person that are subscribed to the same IMS account.
- One identity/telephone number for all devices of a user. A session is delivered to all or some devices based on their capabilities. A video call would only be delivered to registered devices capable of receiving video. Sessions can also be automatically modified if devices do not support video.
- A session can be moved from one device to another while it is ongoing. A video call, for example, might be accepted on a mobile device but transferred to the home entertainment system when the user arrives at home. Transferring the session also implies a modification of the session parameters. While a low-resolution video stream is used for a mobile device, the resolution can be increased for the big screen of the home entertainment system if this is supported by the device at the other end.
- Use of several user identities per device — this allows the use of a single device or a single set of devices to be reached by friends and business partners alike. With user profiles in the network, incoming session requests can be managed on a per user identity basis. This way, business calls could be automatically redirected to the voice mail system at certain times, to an announcement or to a colleague while the user is on vacation while private session requests are still connected.

Which of these applications are offered in an IMS network depends on the individual network operator. It can be envisaged, that the services listed above will not remain the only ones for IMS. The Daidalos project [8], which was part of the 6th EU Framework Program, has developed an interesting vision of how these services could be used in practice. A video, available on their Web site, impressively shows the high-level results of their research. In the near-term, it is likely that network operators will first use IMS for a reduced feature set such as the One Voice Profile (voice over Long Term Evolution, VoLTE) and the Rich Communication Service (RCS). Both are described in more detail in the following sections.

IMS systems can also be the basis for many of the Web 2.0 services described in Chapter 6. In practice, however, such services are mostly developed outside the IMS for a number of reasons. Firstly, only a few IMS networks are deployed today. Secondly, Web 2.0 applications are not designed by network operators but by Internet companies. These companies are not keen to integrate their applications with potentially hundreds of different IMS networks since this requires negotiations with potentially hundreds of IMS operators for a global rollout. Finally, it is also quite difficult to integrate IMS applications into IMS clients on mobile devices since there is no universal standard. Web 2.0 applications are thus either Web and AJAX-based and are developed for major mobile device operating systems such as iOS or Android independently from any network operator. Chapter 6 will discuss this topic in more detail.

4.4.1   IMS Basic Architecture

One of the major goals of the IMS was to create a flexible session establishment platform that can be scaled for networks with tens of thousands to tens of millions of subscribers. The IMS standards define logical components, the messaging between them and how external applications can use the IMS to offer services to users. In practice it is then up to infrastructure vendors and network operators to decide which logical components they want to combine into one physical device depending on the size of the network. It is likely that first implementations will co-locate many logical functions in a single device as there will only be a few IMS users at the beginning, which in turn does not require a large distributed system. As the network grows some functions are then over time migrated into standalone physical devices and a single entity might even be distributed over several devices for load sharing and redundancy purposes.

The IMS has been defined by two standards bodies in close co-operation. The main architecture, the logical components and the interworking between them are standardized by the 3GPP. Most of the protocols used between the components, such as SIP, Diameter, Megaco, COPS, and so on, which will be discussed in this section, are standardized by the Internet Society's Internet Engineering Task Force (IETF). This split has been done on purpose to use as many open and freely available Internet protocols for the IMS as possible rather than to use proprietary and closed standards. This allows interworking with other session-based systems using open standards in the future. The split is also beneficial since the IETF has a strong expertise in IP protocols while 3GPP is focused on mobility aspects and network architecture. References to the specifications will be made throughout this section to allow the reader to go into the details if required. All documents of both standards bodies are available online at no cost. To start exploring the standards, 3GPP TS 22.228 [9] and 3GPP TS 23.228 [10] are recommended, which introduce the requirements for IMS and the general system architecture in detail.

Figure 4.12 shows the main components required on the application layer for a basic IMS solution. The IMS standards define several additional functional entities which are only introduced later in this section and are therefore not shown. The figure also does not show the underlying transport functions of fixed and wireless core and access networks. Again, this has been done for clarity reasons and due to the fact that large parts of the IMS are completely access and transport network independent. Only a few parts of the IMS communicate with the transport network to ensure quality of service and to prevent service misuse. These interfaces will also be discussed separately later.

4.4.2   The P-CSCF

From the user's point of view the Proxy Call Session Control Function (P-CSCF) is the entry point into the IMS network. The P-CSCF is a SIP proxy and additionally handles the following IMS specific tasks:

- **User proxy** — the P-CSCF received its name not because it is a SIP proxy but because it represents the user (it acts as a proxy) in the network. All signaling messages to and from the user always traverse the P-CSCF assigned to a user during the registration

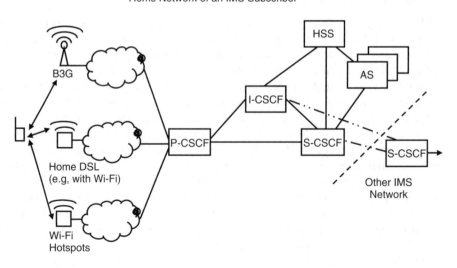

Figure 4.12   The basic components of the IMS framework.

process. In wireless networks such a function is required since a user can suddenly drop out of the network if they roam out of coverage. The P-CSCF is notified of such an event by the access network while a session is ongoing and can thus terminate the session in a graceful way. How this is done depends on the type of access network. In the case of a 3G UMTS/HSDPA (High Speed Data Packet Access) network (cf. Chapter 2) the RNC informs the SGSN that the radio contact with a device was lost. If a streaming or conversational class connection was established the SGSN then informs the gateway node (GGSN) of the radio link failure by setting the maximum bitrate in uplink and downlink to zero [11]. The GGSN will then stop forwarding packets to the user in the case that a remote subscriber involved in a session is still sending its media stream to the subscriber. The GGSN then informs the P-CSCF via the Policy Decision Function (PDF), which is further discussed below, that the subscriber has lost the network connection. The P-CSCF will then act on the IMS application layer on the user's behalf and terminates the ongoing session by sending a SIP "bye" message to the remote user or users in the case of a conference. In LTE networks the process is a bit more efficient since there are fewer nodes between the base station and the P-CSCF. Here, the radio base station signals the radio bearer loss to the Mobility Management Entity (MME) in LTE, which in turn directly forwards the information via the PDF to the P-CSCF.

- **Confidentiality** — SIP transmits all signaling messages in clear text. This is a big security and privacy issue as it allows potential attackers who have gained access to the network at any point between the user and the SIP network to read and even modify the content of messages. The IMS thus requires SIP signaling between the IMS terminal and the P-CSCF to be encrypted. This is done by establishing an encrypted IPSec [12] connection between an IMS terminal and the P-CSCF during the registration process [13]. This encrypted connection will then be used while the user remains registered.

The network behind the P-CSCF is considered to be secure and many network operators will thus not encrypt SIP messages and other signaling traffic between different components of an IMS network.

- **Signaling compression** — as has been shown earlier, SIP signaling messages are quite large. In the IMS the message size grows even more due to additional parameters for functionalities described later on. The larger a message the more time is required for transmitting it. As sessions (e.g., voice calls) should be established as quickly as possible and as there are several SIP messages required before a session is established, it is vital that each message is transferred as quickly as possible. An IMS terminal may therefore request during the registration process to activate Sigcomp [14] compression to reduce the size of the SIP signaling messages.
- **Quality of service and policy control** — the P-CSCF has an interface to the underlying wireless network infrastructure (GGSN, MME, etc.) to ensure a certain quality of service for a media stream (e.g., minimum bandwidth for a voice call to be ensured during a session). Policy control ensures that the link between two subscribers for a session is only used for the types of media negotiated between the subscribers and the network.
- **Billing** — like all IMS components the P-CSCF function can generate billing records for post-paid subscribers which are sent to a billing server for offline processing.

While traditional SIP User Agents are usually configured by the user or out of the box to be aware of the IP address of the first SIP server they need to contact for registration it was felt in 3GPP that this approach is not flexible enough for evolving networks. Here, individual P-CSCFs will be added over time to be able to handle the increasing number of users. Furthermore, self-configuration on startup of the User Agent is highly desired since users should not be required to configure their terminals for IMS should they decide to buy them from a source other than the network operator. The IMS standard therefore defines a number of additional options for a User Agent to obtain the IP address of the P-CSCF at startup. In 3GPP networks (UMTS, HSDPA, and LTE), the P-CSCF IP address can be sent to a device during the establishment of a network connection, which is referred to as the Packet Data Protocol context activation. During this process the device receives amongst other information its own IP address and additionally the IP address of the P-CSCF. Since a PDP context activation is a general process for establishing a packet-switched connection to a 3GPP network, the device stores this information so the IMS User Agent software can request the P-CSCF identity from the device later on. Another possibility to retrieve the P-CSCF's IP address from the network is via a DHCP (Dynamic Host Configuration Protocol) lookup. This protocol is commonly used for another purpose today by Wireless Local Area Network (LAN) and Ethernet devices to request their IP address and the default gateway IP address from the network when they enter the network. Since DHCP is a flexible protocol it was extended for the use in the IMS to send the P-CSCF IP address to a User Agent upon request as well.

4.4.3 The S-CSCF and Application Servers

The central component of the IMS framework is the Serving Call Session Control Function (S-CSCF). It combines the functionality of a SIP registrar and a SIP proxy as defined in

Sections 4.3.1 and 4.3.2. High-capacity networks will require several physical S-CSCFs. One user will be managed by a single S-CSCF while registered to the IMS network and all SIP requests have to be sent to this S-CSCF. Users can be assigned by a load-sharing algorithm to a particular S-CSCF at registration time. It is also possible to assign a particular S-CSCF depending on the services a subscriber is allowed to use since some S-CSCF may only support a subset of services available in the network.

When a user first registers with the IMS network, the S-CSCF requests the profile of the subscriber from a centralized database. The profile contains the user's authentication information and information about which services the user is allowed to invoke. In SIP terminology this database is known as the SIP location server (cf. Figure 4.4) and in the IMS world it is known as the Home Subscriber Server (HSS). This is required as the S-CSCF only stores the subscriber context while he is registered. Downloading user data from a centralized database is also required since the S-CSCF can be a highly distributed system in the case of a large network and it is possible and even likely that a subscriber is assigned a different S-CSCF during a subsequent registration.

After registration the P-CSCF's main task is to forward the SIP messages between the IMS terminal and the S-CSCF. It is the S-CSCF that will then decide how to handle a message and where to forward it. In the case of a SIP "invite" message the S-CSCF has to decide how to proceed with the session setup request. For a simple post-paid voice session establishment for which no additional services are invoked the S-CSCF's main high level tasks are as follows:

- The S-CSCF recognizes from the SDP payload in the invite message that the user requests a voice session and checks the subscriber's service record to confirm they are allowed to originate calls.
- The S-CSCF then analyzes the destination address, which could either be a SIP URI (e.g., sip:martin.sauter@mynetwork.com) or a TEL URL (tel: +49123444456).
- If the user wants to establish a session with a TEL URL the S-CSCF needs to perform a database lookup to determine if the destination is an IMS subscriber, that is if the TEL URL can be converted into a SIP URI. Otherwise the telephone number belongs to a circuit-switched service subscriber and the session has to be forwarded to a circuit switched network. For details on this process see Section 4.4.9 on IMS voice telephony interworking with circuit-switched networks.
- In the case of a SIP URI, the S-CSCF checks if the destination is a subscriber of the local or a remote IMS network by looking at the domain part of the SIP URI (e.g., @mynetwork.com). If the destination is outside the local IMS network the S-CSCF will then look up the IP address of the SIP entry proxy of the remote network (the I-CSCF, see below) and forward the invite message to the foreign network.
- The S-CSCF then stays in the loop for all subsequent SIP messages and creates charging records for the offline billing system so the user can later on be invoiced for the call.

The S-CSCF is also responsible for many supplementary services. If the user, for example, requests that the telephone number or SIP URI is hidden from the destination subscriber, it will again check the service record of the subscriber to confirm the operator allows this operation and then modify the SIP header accordingly. Features such as this are already known in circuit-switched networks and are implemented in a similar fashion in the S-CSCF.

While the S-CSCF is the central part of the IMS its actions are limited to SIP message analysis, modification, routing, and forking to reach all devices registered to the same identity. This already allows basic services and many supplementary services such as identity hiding as described before without any further equipment. More complicated services and applications, however, are not implemented in the S-CSCF but on external Application Servers (ASs). While the application server is responsible for offering the service, the S-CSCF decides which applications are invoked at a session establishment, while a session is ongoing, or when it is terminated. The following example illustrates this approach.

An S-CSCF receives a SIP invite message for a local user. The user however is already engaged in another session and rejects the additional session establishment request. Based on the user's subscription data the S-CSCF can then decide to forward the SIP invite message to a voice mail system. The voice mail system, which is a typical application server, receives the redirected SIP "invite" message and acts as a User Agent to establish the session. After the caller has left a voice message, the Voice Mail system sends an instant message via the S-CSCF to the original destination User Agent to let the user know that a voice mail message was left. In this example the voice mail system acts as both a passive component when receiving the redirected call and an active component when forwarding an IM message. Application servers can therefore not only receive SIP messaging from SCSCFs but can also be the originator of messages based on internal or external events.

Another example of an active application server functionality is an automatic wakeup service. After the user has configured the service the AS providing the wakeup service will automatically establish a SIP session to a user at a predefined time to play a wakeup message, the subscriber's favorite music, the latest news, and so on.

Other functionalities such as presence, IM, push-to-talk, and many other services are also done on an application server rather than in the S-CSCF. More formally application servers can work in the following modes:

- as a User Agent (as described earlier),
- as a SIP proxy by changing the content of a SIP message it receives from an S-CSCF and returning the message back to the S-CSCF,
- as a back-to-back User Agent — this means that it terminates a session as a terminating User Agent toward the S-CSCF and establishes a second SIP session as an originating User Agent.

The S-CSCF's decision when to contact an application server is based on "initial filter criteria" that are part of the user's configuration profile. This topic is discussed in more detail in Section 4.4.6.

4.4.4   The I-CSCF and the HSS

The third logical type of SIP proxy in an IMS network is the Interrogating-Call Session Control Function (I-CSCF). SIP messages traverse the I-CSCF in the following cases.

When a User Agent registers with the IMS, for example, during startup, the first action is to find the P-CSCF as described earlier. Afterwards it will send a SIP registration message to network to announce its availability. The P-CSCF then needs to forward the message to an S-CSCF. Since the IMS can be a distributed system and since the user has not yet been assigned an S-CSCF, the P-CSCF forwards the registration message to an Interrogating-CSCF. The I-CSCF of the user is found using a DNS lookup with the user identity in the SIP registration request. The message is then forwarded to the I-CSCF which will then interrogate the HSS for the S-CSCF name or the S-CSCF capabilities required by the subscriber. When the HSS receives the request, it retrieves the user's subscription record and returns the required information. Based on the S-CSCF name or the services the user has signed up to in combination with the knowledge about the capabilities of individual S-CSCFs, the I-CSCF then selects a suitable S-CSCF and forwards the registration.

The second scenario in which an I-CSCF is required is for terminating a session to a subscriber that is served by a different S-CSCF than the originator of the session. In this scenario the S-CSCF of the originator detects that it does not serve the terminating subscriber and therefore forwards the SIP "invite" message to an I-CSCF. The I-CSCF of the user is found using a DNS lookup with the user identity in the SIP invite request. The I-CSCF in turn interrogates the HSS by forwarding the SIP URI of the destination subscriber and requesting the HSS to return the contact details (e.g., IP address, port) of the responsible S-CSCF. In case the subscriber is currently not registered a suitable terminating S-CSCF is selected since sessions to currently unavailable subscribers might still be terminated, for example, by a voice mail system. The I-CSCF then forwards the SIP invite message to the responsible S-CSCF. As the identity of the S-CSCF is now known further messages need not be sent via the I-CSCF. The I-CSCF therefore does not add its identity to the routing information in the SIP header of further messages.

The HSS is an evolution of the Home Location Register, which was specified back in the early 1990s as the central user database in the first release of GSM standards. In the meantime the database structure has been expanded several times, as shown in Figure 4.13. With the inception of the IMS the Home Location Register (HLR) was renamed to HSS to account for the additional services it offers:

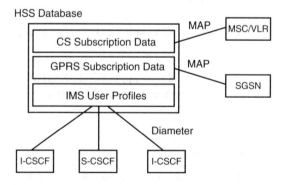

Figure 4.13   The home subscriber server and clients.

For circuit-switched services such as voice calls and SMS the original database entries keep being used. Data stored for a user for the circuit-switched network includes:

- authentication information — a replica of this information is stored on the user's SIM card,
- information about which services the subscriber is allowed to use or is barred from using (e.g., incoming voice calls, outgoing voice calls, incoming/outgoing calls while roaming, presentation, or hiding of telephone numbers of incoming calls, SMS send/ receive capabilities, Intelligent Network service invocations, for example, for prepaid services, and many other things).

Toward the circuit-switched Mobile Switching Center/Visitor Location Register, the traditional Mobile Application Part protocol continues to be used.

When packet-switched services were added to the GSM network with the General Packet Radio Service in the early 2000s, the database was enhanced to store subscriber information for packet-switched services. For UMTS, High-speed Packet Access (HSPA), and LTE the database specification was enhanced several times in a backwards-compatible fashion so the information can be used for 2G, 3G, and 3.5G access networks. Packet-switched subscriber information in the HLR includes:

- the Access Point Names and their properties — APNs are used by the GPRS network to define the quality of service settings of a connection; implicitly APNs are also used to define services (e.g., full Internet access, only Web access, etc.) and billing methods in the network,
- which APNs a subscriber is allowed to use,
- if packet-switched service roaming is allowed.

For the IMS a third block was added to the HSS/HLR where user profiles of IMS subscribers are stored. An IMS user profile includes information such as the user identities connected to a subscription, which services the user is allowed to invoke and information on how incoming requests should be handled when the user is not registered (e.g., forwarding to voicemail). IMS components such as the I-CSCF and the S-CSCF are connected to the HSS via an IP connection. The protocol used for retrieving and updating records in the HSS is Diameter [15]. Diameter is not an abbreviation but a pun on its predecessor the RADIUS (Remote Access Dial In User Server) protocol.

In practice a network usually stores information for a user in all three parts of the HSS. Circuit-switched subscription information is required in case the user wants to use circuit-switched services such as SMS or voice (e.g., when roaming out of the coverage area in which IMS voice calls are supported). The user also needs to be provisioned in the GPRS part of the HSS since using IMS services is only possible from GSM/UMTS/HSPA/LTE networks once a packet-switched connection to the network (via an APN) has been established. Finally, a user also has to have a user profile for IMS services as otherwise no IMS services can be used.

Large networks can have several HSS entities for capacity reasons. In such a case the CSCFs have to query a database which contains the information on which subscriber is administered by which HSS. This task is managed by the Subscription Locator Function

(SLF). The SLF does not sit in the signaling path between the CSCFs and the HSSs but acts as a standalone database that can be queried using the Diameter protocol to retrieve the IP address of the HSS responsible for a subscriber for a subsequent Diameter dialog with the HSS to retrieve subscription information. The SLF is an optional component and thus not shown in Figure 4.13. It is not required if there is only one HSS in the network.

4.4.5   Media Resource Functions

A supplementary service often used in business communication is conferencing. A conference is established, for example, if a third person is to be included in an ongoing conversation. One of the two users will then put the established call on hold to call the third party. If the third party accepts the call, the caller can then conference all three parties together. In circuit-switched networks a conference bridge function is required in the network to mix the speech path of all participants. In VoIP networks two methods can be used to create a conference call. Some networks do not have conference bridge resources in the network and therefore leave it to one of the subscribers to mix the speech signal and distribute the resulting audio stream to the other participants of the conference. While being simple, the downside of this approach is the additional bandwidth required for the device that is mixing the signal. Instead of sending a media stream to a single device, two (or more where there are more than three subscribers in a conference) streams have to be sent, which doubles the required data rate in the uplink. In the IMS, conferencing is therefore done with a conferencing server in the network. The conferencing facility in the network is implemented on a Media Resource Function (MRF). The MRF in turn is split into two logical components as shown in Figure 4.14. Toward the S-CSCF the Media Resource Function Controller (MRFC) acts as an application server and terminates the SIP signaling. The media stream is mixed by the Media Resource Function Processor (MRFP), which is controlled by the MRFC. The protocol between the two entities is not defined. The Media Gateway Control Protocol (Megaco, H.248) which is also used for controlling other media gateways as described further below in Section 4.4.9, could be used for the purpose.

A conference session is established during an ongoing voice session as follows: One of the participants puts the current session on hold and calls the third party. When the third party accepts the call, the first party prepares the conference bridge by establishing a session to the conferencing MRFC and MRFP via the S-CSCF. Once the conferencing facility is ready, the first party sends a SIP message to both other parties to transfer the end points of their voice media streams to the conferencing bridge. For this purpose the MRFC has sent the initiator a unique SIP URI which allows it to identify the incoming SIP requests. Once both remote parties agree to re-route their voice media stream to the conference bridge, the MRFP can start mixing the audio signals of all three participants and the conference is established. Figure 4.14 shows in which steps the voice media streams are transferred to the conference bridge. In the standards, this process is described in Section 5.11.6.2.3 of [10] and details are given in [16].

Optionally IMS clients and the conference bridge can support conference events. When IMS clients register for conference events they are notified by the conference bridge about the identity of the other subscribers in the conference bridge and receive a message when one of the subscribers terminates its session to the bridge.

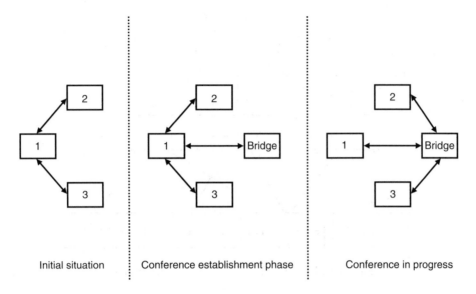

Initial situation   Conference establishment phase   Conference in progress

Figure 4.14   IMS conference session establishment between three parties.

If only two parties remain on the conference bridge the MRFC can then choose to either leave the two parties on the conference bridge or re-transfer the media streams back to a point-to-point connection and remove itself from the session.

An interesting feature which is not possible today with circuit-switched conference bridges is the ability to directly invite a subscriber to join an ongoing conference. This way the originator does not have to put the conference leg on hold to call somebody to join the conference.

4.4.6 User Identities, Subscription Profiles, and Filter Criteria

As it becomes more and more common that subscribers use more than a single device for communication, the IMS specifications have devised a scheme to spread a single subscription over many devices. One private user identity is used per physical device, as shown in Figure 4.15. The private user identity is usually not visible to the subscriber since it is only used during the registration procedure. The private user identity is therefore used for the same purpose on the application layer as the International Mobile Subscriber Identity (IMSI) on the 2G and 3G network layer.

The identification known to a subscriber and under which he can be contacted is the public user identity which can be compared with the Mobile Station ISDN Number (MSISDN) in fixed-line and wireless circuit-switched networks. The public user identity is a SIP URI as described earlier (e.g., sip:martin.sauter@mynetwork.com) or a TEL URL. In the IMS several public user identities can be linked to a single private user identity. This is unlike the 2G and 3G world where a single IMSI is usually tied to a single MSISDN. Due to this 1 : n relationship, the subscriber can be reached via several public user identities on the same device. This allows, for example, a public user identity

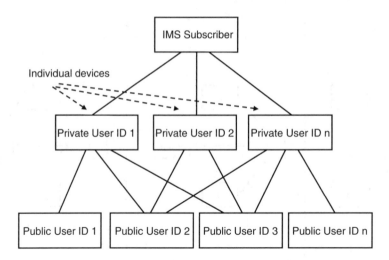

Figure 4.15   IMS subscription information structure.

for business purposes and a separate public user identify for private contacts. Since the user can register to single or multiple public user identities simultaneously, they have the choice of when they can be reached under which identity.

An IMS subscriber can also have several private user identities to which a combination of all public user identities can be mapped to. Annex B of [17] describes this concept in detail. In practice this approach gives network operators and subscribers the flexibility to be reachable via several devices. Additionally, subscribers can choose under which public user identities they can be reached at a certain time on a certain device if the IMS operator chooses to offer such a service.

Network services such as managing when a user will be available via which device where it is registered are performed by application servers. These are put into the communication path by S-CSCFs with the help of what is known as initial filter criteria. These are stored on the HSS alongside the public user identities of a subscriber. Initial filter criteria describe the values that parameters have to have in SIP and SDP messages in order for the message to be forwarded to an application server. The application server may then have additional subscriber information to modify the message content or to act as a User Agent on the subscriber's behalf. An application server may, for example, change the destination public user identification to automatically forward a session initiation request to a voice mail system or another public user identity (e.g., that of a colleague) when it detects that a user is online but does not want to be contacted via this user identity at the time. How the application server obtains the information on which to base this decision is not specified in the IMS standards.

Initial filter criteria can be stacked upon each other so a single SIP message can be sent to several application servers and be modified before the S-CSCF forwards it to a subscriber.

On the mobile device side a subset of the subscriber information is stored on the Universal Integrated Circuit Card (UICC) which is usually referred to as the SIM card.

A SIM card has three functionalities. First, it serves as a secure data storage card in which both open and secret information can be stored. Data is stored in files in a directory structure. Unlike memory cards and hard drives, which use file systems with file names, the SIM card only uses standardized numbers to identify files. This dates back to the days when SIM cards could only store a small amount of data.

Secrets such as passwords and identification material can be used by internal SIM card applications to generate responses for authentication requests and to generate ciphering keys (CKs). These internal applications are invoked from the mobile device via a standardized SIM card interface [18] by requesting authentication and ciphering information, which is then computed by an authentication algorithm based on a RAND number given to the application and the secret key stored on the SIM card.

Only internal applications are permitted read access to the secret data on the SIM card. This is an important security feature as it prevents cloning of SIM cards. Finally, SIM cards can also host JavaCard applications that can be downloaded to the SIM card by the user or via an OTA download by the operator. JavaCard applets are used in practice, for example, to create device-independent applications that can be called via an extended menu structure.

For the IMS the following information and internal applications are required:

- To be able to access a 2G or 3G network in the first place the SIM card contains the IMSI and secret key for 2G and 3G networks. This information is contained in the USIM (UMTS SIM) part of the SIM card. Standards refer to this as the USIM application, which is a little misleading. In general terms the USIM application is the combination of files and internal applications required for UMTS.
- A SIM card can also store authentication information and applications for non-3GPP network access, for example, for public Wi-Fi authentication via EAP-SIM (Extensible Authentication Protocol) (cf. Chapter 2).
- For IMS the SIM card should also contain an ISIM (IMS SIM) application (i.e., a directory for IMS and files for a private user identity, one or more public user identities, the home network domain URI and IMS authentication information. Except for the public user identities, it is possible to derive this information from the UMTS parameters for backwards compatibility. This way it is not necessary to replace SIM card(s) to activate users for IMS services.

It is also possible to store IMS subscriber information directly in the mobile device. This, however, is undoubtedly much less secure and it is much more difficult for subscribers to choose for themselves which devices they want to use with an IMS network and to configure them with the necessary account information.

4.4.7 IMS Registration Process

Now that the basic IMS components and the basic concept of user identities have been introduced, it is time to look at how the IMS uses SIP signaling to offer services to subscribers. Before a user is allowed to establish sessions or, in more general terms, to send SIP messages to network components and other subscribers, it is required that the IMS User Agent registers with the network. The IMS registration process follows the lines of the registration as described in Section 4.3.1 and Figure 4.5. In order to provide security and additional features such as multiple identities, several additional steps are required during this process. This section gives an overview of the different processes during the registration process, the involved components, and the SIP messages used during the registration process.

Before being able to register, a device first needs to connect to the network. In 3GPP wireless networks, connecting to the packet-switched network is done via a PDP context activation as described in Section 4.2. Once the process is completed the device has an IP address and can start communicating with the IMS.

The next step of the process consists of determining the IP address of the P-CSCF because it is the entry point to the IMS network. The device is either informed of the P-CSCF address during the PDP context activation (in case it accesses the network via a 3GPP radio network) or with a DHCP request. Both methods are described in Section 4.4.2.

Once the P-CSCF IP address is known, the IMS User Agent on the subscriber's device assembles a SIP register message. Registrations are performed with the subscriber's private user identity. If the SIM card does not contain an ISIM application, a temporary private user identity is built with the subscribers IMSI. The registration request also contains one of the public user identities stored on the SIM card by which the user wants to be reachable. Since the device does not yet know the S-CSCF's address, it is not included in the registration message. The P-CSCF therefore forwards the register message, as shown in Figure 4.16, to the I-CSCF. The I-CSCF in turn queries the HSS for the user's subscription profile and selects a suitable S-CSCF. This process is described in Section 4.4.4. Afterwards the message is forwarded to the S-CSCF.

When receiving the registration message the S-CSCF detects that the user is not yet known and also requests the user's subscription information from the HSS. In the process the S-CSCF also requests an authentication vector from the HSS, which consists of the following values:

- **RAND** — a random number.
- **XRES** — the RAND number is used by the HSS and by the SIM card together with a secret key and an authentication algorithm to generate a response. During the authentication process the S-CSCF compares the response received from the User Agent with the XRES value sent by the HSS. The values can only match if both entities have used the same secret key and authentication algorithm. As the secret key is never sent through the network and the XRES is kept by the S-CSCF, it is possible to securely authenticate a user if a new RAND number is used for each request.
- **AUTN** — to prevent malicious attackers from acting as the subscriber's IMS network the HSS generates an authentication token (AUTN) with the RAND number and another authentication algorithm. This token is later on used by the device to verify the authenticity of the network.
- **IK** — messages between the User Agent and the P-CSCF are integrity checked by adding a checksum that is calculated with an Integrity Key (IK). The IK is calculated by the HSS with an IK generation algorithm and the RAND number as an input value.

- **CK** — SIP messages are sent through an IPsec encrypted tunnel. The CK is used by the P-CSCF and the User Agent to encrypt and decrypt the messages.

In order to force the user to authenticate, the S-CSCF rejects the User Agent's registration request with a SIP 401 (Unauthorized) message. The body of the message contains, among other values, the five security values listed earlier. When the message arrives at the P-CSCF the IK and CK values are removed from the message since they are secret and must not fall into the hands of eavesdroppers that have gained access to the potentially unsecured communication network that could be between the User Agent and the P-CSCF. Sending these values to the User Agent is not necessary since the SIM card or the User Agent (in case there is no SIM card) calculates the values itself in the same way as the HSS, that is, with the RAND number, the secret key, and the security algorithms.

When the SIP 401 unauthorized message arrives at the IMS User Agent, it will still contain the RAND and the AUTN. Internally the User Agent will then send a request to the SIM card to calculate the AUTN with the RAND. If the result matches the AUTN sent by the S-CSCF, the network is authenticated. In the next step the User Agent then asks the SIM card to calculate the XRES, which the S-CSCF will use later on to authenticate the User Agent. The SIM card also calculates the CK and IK, which will be used to establish a secure tunnel to the P-CSCF. At this point the User Agent can also activate SIP Signaling Compression (SIGCOMP) to speed up signaling procedures.

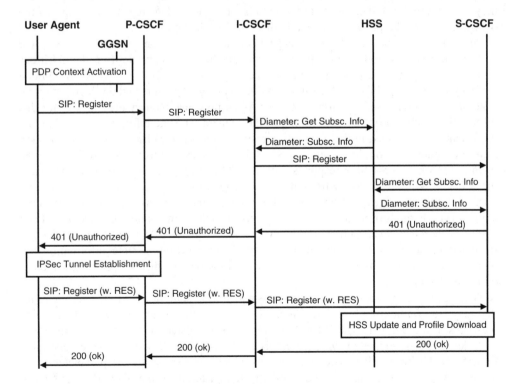

Figure 4.16   IMS registration process — Part 1.

Once the secure tunnel to the P-CSCF is in place the User Agent sends another SIP registration request, this time including the XRES. The message is encrypted and an integrity check code ensures that the message is not tampered with in the very unlikely case the ciphering is broken. Additionally, the message also contains location information in the P-Access-Network-Info parameter. In case of a 3GPP network the parameter contains the Cell Global Identity (CGI) of the cell the device is currently using. The CGI is internationally unique. The P-Access-Network-Info parameter will be included in all future originating messages from the User Agent except for SIP ACK and CANCEL messages. This enables the IMS to offer location-based services via trusted application servers. The S-CSCF also ensures that this parameter is deleted from all messages before they are forwarded to other destinations (e.g., a terminating device).

When the second SIP register message arrives at the S-CSCF the XRES value is compared with the value included in the registration message. If the values match, the subscriber is authenticated and the S-CSCF sets the state of the public user identity to "registered." The user's profile can also indicate to the S-CSCF to implicitly register some (or all) of the other public user identities at this point. These are then also set to "registered." Afterwards, the S-CSCF marks the user as registered in the HSS, downloads the user profile and returns a SIP "200 ok" message to the User Agent to finalize the first part of the registration process. The message also contains all public user identities that are linked to the private user identity but not their current registration state. In a parallel action the S-CSCF also informs the presence server that the user has registered (i.e., that they are now online). This is done by sending an independent register message to the presence server. The presence service is discussed in Section 4.4.10.

The User Agent analyzes the incoming registration response message from the S-CSCF and thus learns which public user identities are attached to the private user identity it has used for the registration process. This is necessary since the SIM card may not contain the complete set of public user identities. As the message does not contain the registration state of these public user identities, the User Agent now has to query the S-CSCF for this information. This is done by subscribing to the registration state information of all public user identities contained in the registration response message. Subscribing to subscription state information is done by sending a SIP subscribe message to the S-CSCF, as shown in Figure 4.17. The S-CSCF responds with a SIP "200 ok" message. Shortly afterwards the S-CSCF will then send a SIP "notify" message with the requested registration state. Further notify messages will be sent to the User Agent if the user registers with other devices (i.e., with a different private user identity) to the same public user identify. Each device is therefore aware which other devices have registered to a public user identity. It should be noted at this point that subscribing to registration state information has nothing to do with subscribing to presence information of other subscribers, which is described later.

The P-CSCF also needs to learn about the registration state of all public user identities that may later on be used by a User Agent as it verifies this parameter before forwarding messages to the S-CSCF. The P-CSCF therefore also registers to the state information.

A device stays registered until it deregisters from the IMS service. A deregistration is performed with a SIP register message in which the expiry (timer) parameter is set to zero. A User Agent can also be implicitly deregistered by the network if it fails to periodically update its registration.

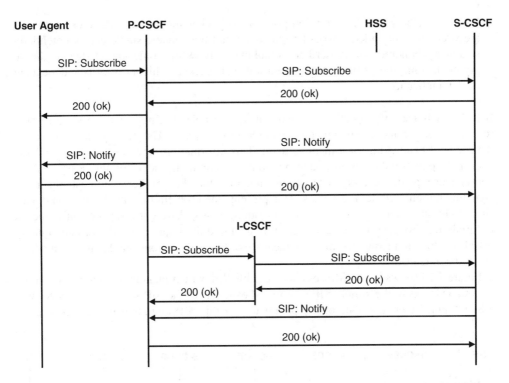

Figure 4.17   IMS registration process — Part 2.

4.4.8   IMS Session Establishment

At the beginning of this chapter, Figure 4.7 showed the basic SIP session establishment messaging. This session establishment takes it for granted that sufficient transmission resources are available and that the media streams in both directions can be sent as soon as required. While this is usually the case in fixed-line IP networks and in Wi-Fi networks, the capacity of cellular wireless base stations in relation to the number of subscribers is much lower (cf. Chapter 3). In addition, packets of real-time media streams should have a higher priority than other packets, like those of a Web browsing session, to reduce jitter. How a certain quality of service for the media stream can be ensured depends on the type of network used:

- **Fixed-line IP networks** — SIP clients can set the DSCP (Differentiated Services Code Point) value to "expedited forwarding." Routers in the path can then forward these packets with a higher priority in case of congestion.
- **802.11/Wi-Fi** — as shown in Chapter 2, the Wi-Fi Multimedia extension of the standard takes the DSCP value to decide which transmission queue a packet should be put into. VoIP packets are classified as media streaming packets and are thus always preferred to other packets.
- **UMTS, HSDPA, and LTE** — due to their architecture, these systems have no visibility of the IP header of a packet and therefore cannot use priority information from this

layer to decide which transmission priority the packet should have. As these networks are wide area networks, it would be unwise to do that since a single cell is potentially shared by hundreds of subscribers simultaneously. With some knowledge of the IP protocol, some subscribers could give all of their packets a higher priority independent of the application.

If the IMS is used via a cellular network it is advisable (but not mandatory) to reserve resources for the media stream. For resource reservation, a User Agent needs to know the required bandwidth for a media stream. This information, however, is only available once both ends of the session have negotiated a common codec for the session. As a consequence, media resource reservation cannot be done before the session establishment signaling but only while in progress. This presents the problem that the destination device should not start alerting the user of an incoming session straight away but only once it is certain that both parties have successfully acquired the required bandwidth and quality of service for the media stream. This means that the SIP session establishment signaling has to be extended.

Figure 4.18 shows how a SIP session is established, which includes resource reservation at both ends. As in the simple SIP example shown earlier the SIP session is initiated with a SIP "invite" message. One of the parameters of this SIP invitation is the precondition

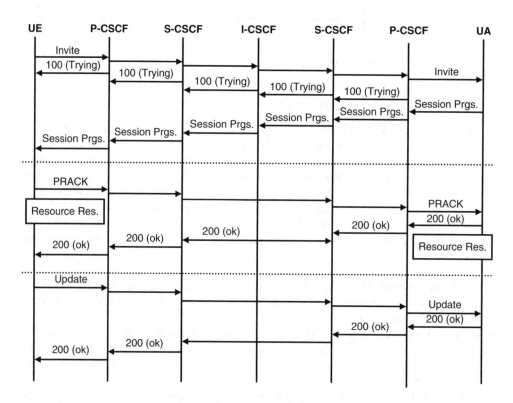

Figure 4.18   IMS session establishment — Part 1.

that the originator needs to reserve resources for the media stream. This tells the called device that it should not start alerting the user until the originator has confirmed that the resources have been assigned. The SIP invitation also contains the supported media codecs of the originator for the types of media streams to be established. The destination then answers with a SIP "183 (Session Progress)" message which contains the list of media codecs supported at this end. As the destination also needs to reserve resources the session progress message conveys this information to the originator as well. At this point the destination is not yet able to reserve resources since it is still not clear which media codec(s) will be used for the session.

When the session progress message arrives at the other end the originating User Agent is then able to select a suitable codec for each media stream. The User Agent then returns a provisional acknowledgement (PRACK) with the information on which codecs were selected back to the terminator and starts the resource reservation process on its side.

Once the PRACK message is received at the destination side the resource reservation process can start there as well, which is confirmed to the originator with a SIP 200 (ok) message.

When the SIP 200 (ok) message arrives at the originator side, a SIP Update message is sent as soon as the resource reservation has been performed successfully. At the terminator side the device starts alerting the user once the SIP "update" message has been received and the required resources for the media stream have been allocated. If resource reservation fails on either side, the session establishment is aborted and the user on the destination side will not be notified.

If resource reservation on the destination side was successful, the device starts alerting the user and a SIP 180 (ringing) message will be sent back to the originator. The reception of the ringing message is acknowledged by the originator and the media stream is fully put into place once the destination user accepts the session in which case the device sends a SIP 200 (ok) message to the originator. The originator returns as a final step a SIP ACK message to confirm the previous message.

In UMTS, HSPA, and LTE networks resources for media streams are requested from the network as follows: as discussed earlier, the aim of the initial PDP context activation is to obtain access to the network and to receive an IP address. This logical connection is then used for SIP signaling. If no resource reservation is required by the IMS, the media stream will simply use this connection as well. Where resource reservation is required, the mobile will request the required bandwidth and quality of service from the network via a secondary PDP context activation, as described in more detail in [19, 20]. Secondary PDP contexts are then used to logically separate the real-time data traffic from background or signaling traffic on the air interface while keeping a single IP address on the mobile device. A secondary PDP context is activated by the User Agent, providing the mobile device, and the GGSN with a Traffic Flow Template that contains the IP address and UDP port of the destination device that will be used for the media streaming. The mobile device and the gateway router (GGSN) in the network will then screen all packets and treat those with the specified IP addresses and port number differently, such as giving them preference over other packets of the same and other users. These mechanisms are applied below the IP layer and thus no changes are required to the IP protocol stack or the way the User Agent handles the media streaming over any kind of bearer. The only mechanism required in the User Agent is the command to activate a secondary PDP context. Afterwards the quality of service handling is completely transparent to the IP stack and the application.

If several different media streams are used for a session (e.g., voice and video), the P-CSCF decides if all streams can use the same secondary PDP context or if individual contexts have to be established. This is done by the P-CSCF by modifying the SIP session establishment messages according to the network's policy.

To prevent misuse of the quality of service feature, the P-CSCF communicates with the PDF to get a media authorization token for the requested media streams. This token is then included in the SIP messages to the mobile device. The mobile device then includes the media authorization token in the secondary PDP activation message to the GGSN. When the GGSN receives this request it will query the PDF to see if the media authorization token is valid and if the amount of requested resources has not been modified by the terminal. In this way the IMS application is not able to request more resources than is required for the media stream. This also prevents other applications from misusing the quality of service features of the network for their own purposes.

The P-CSCFs and S-CSCFs in the network of the calling party and also in the network of the called party can reject a session establishment in the case where a User Agent requests a media codec which is not allowed in the network. This could be the case, for example, if an IMS client suggests using the outdated 64 kbit/s G.711 codec for voice transmission. If the network does not allow this codec to be used, the SIP Invite message is rejected with a SIP "488 (not acceptable here)" message. The User Agent then has to modify the codec list and start another session establishment procedure.

If one of the session participants uses the IMS from a network in which no media authorization is required, some of the SIP messages shown in Figures 4.18 and 4.19 are not required. If the originator, for example, uses a Wi-Fi network for which no media authorization is required, the SIP update message is not required. In this scenario the User Agent on the other side does not expect this message since the originator has indicated in the session invitation that resource reservation is not required on that side. The terminating User Agent can therefore start alerting the user as soon as their own resources have been granted without waiting for the confirmation from the other end.

When a user is registered they can have several simultaneously registered public user identities. When establishing a session the user can decide which of those public user identities shall be shown to the called party. Since the P-CSCF is aware which public user identities are registered, the network is able to verify this choice to ensure a subscriber only uses an identity that is registered and that belongs to them. When the P-CSCF approves the use of a public identity for the session establishment, it modifies the SIP invite message and thus all other SIP routers in the network are aware that the identity has been verified by the network. The originator can also request the network to remove their identity from the SIP invitation before it is sent to the final destination. This way the network ensures anonymity and does not leave this functionality to the terminating User Agent.

Since Release 6 of the 3GPP IMS standards, session establishment messaging flows have been defined to allow the establishment of a session between an IMS terminal and a standard SIP User Agent connected to an external non-IMS SIP network. As was shown at the beginning of this chapter, the signaling flow for a session establishment in a non-IMS network is quite different since no resource reservation is required. As an IMS

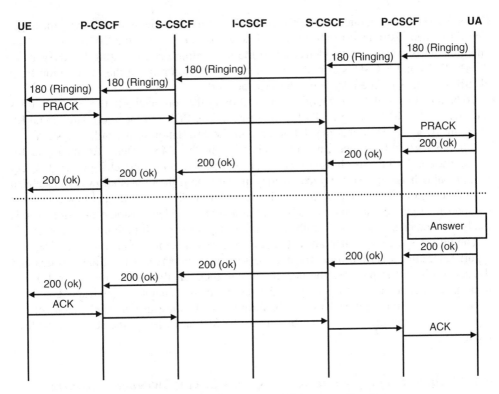

Figure 4.19   IMS session establishment—Part 2.

User Agent cannot know in advance that a called party is a member of a standard SIP network, it attempts to establish a session with preconditions for resource reservation. As the protocol extensions in the SIP invite message required for this are not understood by the terminating User Agent, the invitation is rejected by the SIP client with a SIP "420 (bad extension)" message. The IMS client then assumes that the terminator is a standard SIP client and will construct a SIP invite message without requiring resource reservation preconditions. In most cases not using preconditions for resource reservation should not have an impact since the reservation process is usually very quick and can be finished before the terminating user has a chance to accept the session.

Charging the user for a session is a complicated process since there are many ways to do this. The price for a session could depend, for example, on the type of session (voice, voice video, etc.), the duration of the session, the destination, or the kind of codecs (high bandwidth, low bandwidth, high resolution, etc.) used. For post-paid customers billing information is first collected and then sent to a billing server for analysis. To give the billing system the highest flexibility, each SIP router in the network collects billing information for a session and forwards this information to the billing system. Furthermore, the core network routers also collect billing information (e.g., amount of data transferred) and also forward this information to the billing system. In order to be able to correlate the different messages, they all have to contain the same charging identification. A charging

ID for a session is generated by the P-CSCF when it receives the first SIP invite message. The ID is then distributed to all other IMS network elements via the SIP messages. The charging ID is never directly delivered to the originating or terminating User Agent as the P-CSCFs and S-CSCFs at both ends of the connection remove the information from all SIP messages before they leave the IMS network.

To be able to correlate the charging information of the core network, the GGSN notifies the PDF of the GPRS charging ID for the session. The PDF can then correlate the GPRS charging ID to the IMS charging ID and report the correlation to the billing system.

It is also possible to perform online charging in the IMS system, for example, for prepaid users. In this case billing records are not sent to an offline charging system but to a prepaid charging system which decides in real time if the user is allowed to establish a session or not. There are a number of interfaces from different IMS and transport layer components to the online charging system. For a voice or video session establishment it is likely that the charging request to the prepaid billing system will be initiated by a CSCF.

The billing system can then allow the session establishment and instructs the CSCF to report back after a certain time or once the session is finished. If the timer expires and the session is still in progress the prepaid system then has the possibility to check if the user still has enough credit to continue the session or if it should be interrupted.

Not shown in this section was the forwarding of SIP messages to application servers in case initial filter criteria are met by one of the SIP messages of the session establishment dialog.

4.4.9 Voice Telephony Interworking with Circuit-Switched Networks

Interworking with legacy circuit-switched fixed and mobile telephony networks is an important feature for IMS subscribers since, for the foreseeable future, most wireless subscribers will continue to communicate over circuit-switched networks. A number of logical entities have been defined in the IMS standards for the interworking between the SIP-based IMS networks and external circuit-switched networks. Figure 4.20 shows how these logical entities are connected with each other and how they interact with the S-CSCF when an IMS subscriber wants to establish a call to a circuit-switched network subscriber. For small-scale IMS systems all entities could be implemented in the same physical device while for large-scale systems each logical entity could reside in a separate physical device. For scalability and redundancy each component can also be present several times in a single network.

To reach a circuit-switched user, an IMS user sends a SIP "invite" message with the telephone number of the circuit-switched user to the S-CSCF. The telephone number is formatted as a TEL URL which could either represent a circuit-switched user or the telephone number of another IMS user. IMS users have telephone numbers as well in order to be reachable by circuit-switched users and also in order to make calling IMS users easier from devices that only have a numeric keypad. In a first step the S-CSCF therefore has to discover if the TEL URL sent in the SIP "invite" message belongs to a local IMS user or to a user of an external circuit-switched network. This is done with an ENUM (Electronic Numbering) DNS lookup [21].

Note: The DNS is also used for applications such as Web browsing to convert a domain name (e.g., www.wiley.co.uk) a user has typed into the browser's address line into the

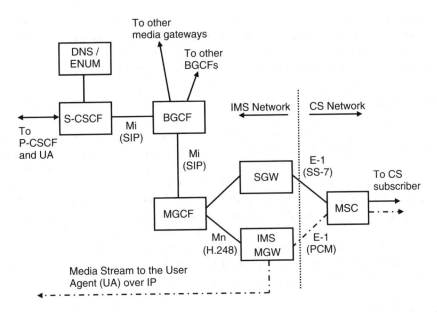

Figure 4.20   IMS circuit-switched interworking with a classical circuit-switched network.

IP address of the Web server. The reuse of DNS to convert a telephone number into a SIP URI fits into the overall concept of SIP and IMS to re-use and extend existing protocols whenever possible rather than to define new ones.

For an ENUM DNS lookup the telephone number (e.g., +443337790) is converted to a domain name which is extended by ".e164.arpa" (e.g., 0.9.7.7.3.3.3.4.4.e164.arpa). The extension signals to the DNS server that the domain name given in a DNS request is not an Internet domain name but a telephone number for which a SIP URI is requested. "E.164" is the numbering plan defined by the ITU for international public telephone numbers. The DNS server then executes a database lookup and if successful returns the corresponding SIP URI to the S-CSCF. In the case of an IMS to circuit-switched call the ENUM lookup is not successful since circuit-switched users do not have a SIP URI. As a consequence the S-CSCF forwards the SIP "invite" request to the IMS circuit-switched gateway functionality. The IMS circuit-switched gateway functionality consists of four logical entities. The first one is the Breakout Gateway Control Function (BGCF), which decides which Media Gateway Control Function (MGCF) to forward the SIP "invite" message to. The BGCF can also decide to forward the "invite" to a BGCF in another network. In this example the BGCF forwards the invite message to a local MGCF. The MGCF acts like a SIP User Agent and terminates the SIP signaling in the IMS network. The components behind the MGCF and the processes invoked there are thus transparent to the S-CSCF.

The MGCF is responsible for controlling both the user plane (the voice stream) and the signaling plane for a connection in a similar way to a User Agent on a mobile device. To be flexible in terms of redundancy and network size, the MGCF does not do these tasks itself. Instead, two logical entities have been defined in the IMS standards which are controlled by the MGCF.

The first logical entity is the SGW. The SGW is, as the name suggests, responsible for converting the SIP messaging used in the IMS network into the signaling protocols used in circuit-switched networks. Two different protocols are currently used in circuit-switched networks, depending on what kind of switching center is used:

- Classic fixed-line mobile switching centers use the SS-7. SS-7 messages are usually carried over one or more 64 kbit/s timeslots in E-1 links (2 Mbit/s). Voice data streams are equally carried in timeslots of E-1 links over short distances or over optical STM-1 connections (155 Mbit/s) over longer distances. This setup is shown in Figure 4.14.
- Newer circuit-switched architectures have separated the circuit-switched mobile switching center functionality into an MSC Call Server and a Media Gateway as described in Section 1.6. For IMS interworking the MSC call server communicates with the IMS signaling gateway with the Bearer Independent Call Control (BICC) protocol, which is usually carried over IP instead of a circuit-switched connection. The IMS media gateway and the circuit-switched media are then connected via an IP connection as the media stream has already been converted from circuit-switched to packet-switched at the circuit-switched media gateway. This setup is shown in Figure 4.21.

The SGW and the MGCF are usually implemented in the same physical device as the protocol between the MGCF and the SGW has not been standardized.

The transcoding of the circuit-switched voice data stream into an IP/RTP data stream is done by the IMS Media Gateway, which is also controlled by the MGCF. The MGCF and

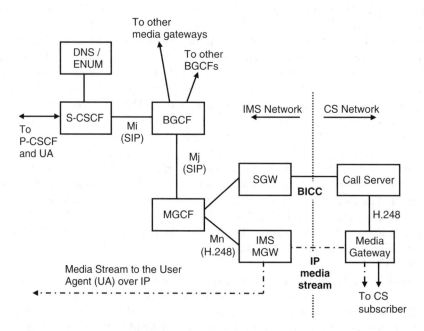

Figure 4.21 IMS circuit-switched networking with a bearer independent core network.

the IMS-MGW (Media Gateway) communicate with the H.248 Media Gateway Control Protocol (MEGACO) [22]. H.248 is used for the following purposes:

- To instruct the media gateway to start/stop converting a voice data stream between a certain timeslot of an E-1 circuit and an IP RTP data stream. The instruction also contains the information on which IP address and UDP port of a User Agent the IP data stream is to be sent to and received from.
- To define which voice codec transcoder is required for a connection. While circuit-switched networks mostly use the 64 kbit/s PCM speech codec, the IMS prefers to use Adaptive Multi Rate with bitrates of 12 kbit/s or less to increase the number of simultaneous voice calls per radio base station.
- To insert tones into the voice stream. For incoming calls this functionality can be used to signal to the originating circuit-switched user that the destination user is currently alerted (alerting tone).
- To insert announcements (e.g., "The number you have dialed is not assigned", etc.) into the voice stream.
- To convert a Dual Tone Multiple Frequency (DTMF) tone message into audible signals. DTMF tones are used, for example, when a user types a code to get access to a voice mail system. A conversion is necessary as some systems carry DTMF tones inside the voice data stream, while other systems such as GSM use DTMF messages and only generate audible signals once the network has been left.
- To insert echo cancellation functionality into the speech path if not already done in the circuit-switched network.

Note that the MSC call server also uses H.248 to communicate with its media gateway. In fact, the IMS has re-used this protocol for its own purposes and has added a number of new functionalities.

Figure 4.22 shows the components involved when a circuit-switched subscriber calls an IMS subscriber by using the IMS subscriber's TEL URL. During the establishment of the call the fixed-line or mobile switching center selects a free timeslot on an E-1 line and sends an SS-7 IAM to the IMS SGW. The message contains the telephone number and the information on which timeslot on which E-1 line the switching center has selected for transmitting the audio stream. The SGW then forwards the information to the MGCF, which starts preparing the IMS-Media Gateway and starts the SIP "invite" dialog. As the MGCF is not aware which S-CSCF is responsible for the subscriber, the "invite" message is first sent to the I-CSCF. The I-CSCF then queries the HSS for this information and forwards the message to the responsible S-CSCF. The S-CSCF then checks the session invitation and if the session is allowed to proceed, it forwards the invitation via the P-CSCF to the IMS device. If the IMS device accepts the session, the MGCF instructs the IMS-MGW to reserve the required resources and to prepare the codec transcoder that the MGCF and the IMS client want to use for the session. When the IMS user answers the call, the IMS terminals sends a SIP "200 ok" message to the MGCF and the MGCF in turn instructs the media gateway to start the media flow. Figure 4.23, which is based on Figure 5.16 in [9], shows a high-level overview of this procedure by combining several messages into a single box.

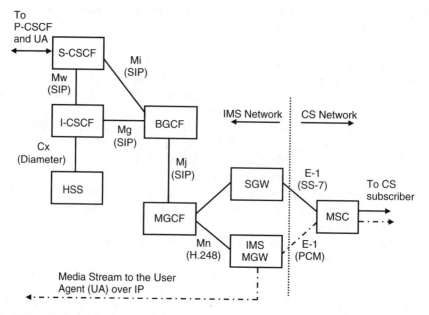

Figure 4.22   IMS components involved in a call setup from a circuit-switched to an IMS subscriber.

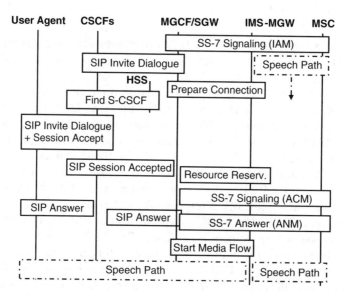

Figure 4.23 Simplified messaging sequence of a voice call establishment from a circuit-switched user to an IMS user.

4.4.10   Push-to-Talk, Presence, and Instant Messaging

In the previous sections the IMS has been presented from a voice session point of view. The IMS, however, is a general session setup platform and can offer, with the help of application servers, a wide variety of session-based communication. The most important session applications that have been specified for the IMS are presence, push-to-talk, and IM.

While the IMS has been standardized by 3GPP, the Push-to-talk over Cellular (PoC) service has been standardized by the Open Mobile Alliance (OMA). The first version of this standard was published in June 2006 and the basic PoC architecture, functionalities, and call flows are described in [23]. IMS interworking aspects, basic PoC session establishment, and timing considerations can be found in [24].

As the PoC service is implemented as an application sever in the IMS environment, the SIP protocol is used for session establishment, talker control, leaving, joining sessions, and so on. The SDP is used for negotiating media parameters such as the voice codec. For transport of the audio stream, the Real-time Transport Protocol is used in combination with the Real-time Transport Control Protocol. It is interesting to note that the PoC service uses the same protocols and standards as other IMS services such as voice and video calls. In PoC sessions, however, speech is exchanged in talk bursts which are originated by the current talker and duplicated and forwarded to one or more listening subscribers by the PoC application server.

The PoC Service has been defined in a very flexible manner and offers the following types of sessions:

- **One-to-one PoC session** — these sessions, which are also referred to as "Personal Instant Alerts" are always established between two subscribers. To establish a 1-1 PoC session, the initiator selects the identity of another PoC subscriber. The identities of other users are usually stored on the terminal and shown to the user in a list. If the requested subscriber is online and attached to the PoC server, the session is established and the two subscribers can start exchanging talk bursts in half duplex mode. In half duplex mode, only one person can talk at a time.
- **One-to-many PoC session** — in this type of PoC session many subscribers can exchange talk bursts with each other in half duplex mode. There can only be one talker at a time. A subscriber can request permission to talk by sending a notification to the PoC server. Once the current talker relinquishes the uplink, the PoC server assigns the uplink to the next person. Details such as priorities, talker pre-emption, and so on, are further described later. A voice channel, that is a data flow, only exists to listeners in the PoC session while there is a talker.

A one-to-many PoC session can be established in one of the following modes:

- **Ad-hoc group session** — for this session mode, the initiator sends a list of subscribers to the PoC server which will form the group. The PoC server then informs all subscribers on the list that they have been invited. The session is running once at least one of the requested participants has joined the session.

- **Pre-arranged group session** — this session mode is initiated by sending a group ID to the PoC server instead of a list of participants. The group ID identifies the responsible PoC server and the group itself. This way it is possible to establish group calls over several PoC servers, which is required in practice as participants might have different national or international home networks.
- **Chat group session** — such sessions are established without notifying possible participants. Subscribers can join and leave chat group sessions at any time and access to a chat session can be restricted to authorized members. More information on group and subscriber management is given below.
- **One-to-many-to-one PoC session** — such talk groups are always pre-arranged. While in other types of sessions all users are equal and all users can hear each other, talk bursts in one-to-many-to-one PoC sessions are only sent to a distinguished participant. Therefore, ordinary subscribers can only hear the distinguished participant while talk bursts sent by other participants are not forwarded to them. Furthermore, only group members that can act as distinguished members are allowed to establish a group session. In practice, such sessions can serve as unidirectional broadcasts from the distinguished participant with an optional uplink.

The PoC standard specifies session establishment as the process and the time it takes from initiation of a new group session until the time the initiator is granted the right to speak. The right to speak is granted once at least one participant has joined the call. Subscribers can join a call by manually accepting it or by setting the device into autoanswer mode.

The right to use the uplink to talk, that is to send a talk burst, is managed by the PoC server. Talk burst control signaling is done via the Talk Burst Control Protocol (TBCP), which defines the messages exchanged between the talker, the PoC server and listeners. In order to receive the right to talk, a subscriber has to send a talk burst request message. The PoC sever acknowledges the request to the subscriber and simultaneously sends out a receiving talk burst indication to all other subscribers to let them know that the uplink is taken and that they shall prepare for the incoming talk burst. The receiving talk burst indication message also contains the identity of the talker unless the talker has chosen to remain anonymous. Once the talker relinquishes the right to speak, a stop talk burst indication is sent to all participants of the session. Furthermore, a no talk burst indication message is sent to inform session participants that the uplink is free.

A subscriber can participate in more than one PoC session at a time. One of the simultaneous PoC sessions is then declared as the primary session. All others are of secondary priority. When a user has joined several PoC sessions at once, the PoC server only forwards media information for one ongoing PoC session at a time. The media stream of the primary session has the highest priority. Other media streams are only forwarded to the user if there is currently no media sent in the primary session.

A user can leave and re-join a group call at any time during an ongoing session. This can be done in two ways. If the user wants to only temporarily leave the call a message is sent to the PoC server to deactivate incoming talk bursts for a specific session until the user reactivates talk burst forwarding again. This can be compared with putting a point-to-point call on hold. The user can also leave an ongoing PoC call. If the user later on wants to participate again, a re-join message is sent to the PoC server. If the session is not ongoing any more at the time, the re-join request will be rejected and the call has to be established again.

A release policy controls the behavior of the PoC server once the originator of the session leaves. Depending on the policy, the PoC server either closes the session or keeps it up until all users have left.

A centralized PoC database referred to as the XDMS (Shared XML Document Management Server) is used to store information about groups, individual PoC parameters per user and other PoC-related parameters. Users can query and modify the database via a standardized XML-based interface over IP. A query or modification request can be sent directly from the handset or by other means, for example, via a Web browser and an AJAX or Java application embedded in the Web page. A set of rules govern which entries of the database can be queried or modified by which users. A subscriber can be the owner of a group in the database and has the right to decide which subscribers are added to the group in the database and what rights they have. Rights management includes assigning the right to initiate and join a group call and if users are allowed to hide their identity on the call (anonymity status).

The XDMS database is also queried by the PoC server, for example, for predefined group session setups. For such a setup a user only sends the group ID, which is then used by the PoC server to retrieve information about all participants from the XDMS database. In addition to the configuration stored in the XDMS database, a number of parameters also have to be configured in the user terminals. While this can be done partly by users, it is also possible to configure the PoC service remotely. This is described in the OMA Device Management [25] and OMA Client Provisioning [26] specifications.

The presence service is another important element in the overall IMS service architecture. In essence it allows subscribers to be informed about the online status of others. Similar services exist in the Internet as part of the Yahoo and Microsoft messengers. In combination with PoC, the presence service is quite valuable for one-to-one and ad-hoc group sessions. Before establishing such sessions it is beneficial to know whether the other parties for the session are online or not. From a technical point of view the presence functionality is independent from the PoC service. Thus, the service requires its own IMS presence application server. Subscribers using the presence service are referred to as presentities. A presentity can, for example, be online or offline. The component on a subscriber terminal that monitors which presentities are online or offline is referred to as the "watcher."

The presence functionality is not only useful in combination with the PoC service but also with IM. In the IMS there are two ways to send messages containing text, speech, and other multimedia information between subscribers. The simple approach is known as immediate messaging which is done by sending a SIP "message" instruction via the IMS to another subscriber. If the subscriber is online, the instruction is delivered right away and the receiving terminal answers with a SIP "200 (ok)" message. If the subscriber is not online, the message could be rerouted by the S-CSCF to an application server which stores the message and forwards it as soon as the subscriber comes back online.

The session-based messaging on the other hand establishes a session between two parties by using the SIP "invite" process. Since no resource reservation is required, the messaging is not as complex as shown earlier for establishing a voice session. Messages containing different kinds of media information are then exchanged with SIP "send" messages. Such a messaging session could, for example, run alongside an established voice session to exchange text messages or pictures while communicating over a voice or video channel. When no more messages are to be sent, the session is terminated with a SIP "bye" message, which is answered by the other party with a SIP "200 (ok)".

4.4.11 Voice Call Continuity, Dual Radio, and Single Radio Approaches

As the IMS supports different types of access networks, the signaling architecture allows to keep a session ongoing when the user moves between different networks. This is especially the case for voice calls when the user leaves or enters his personal Wi-Fi Internet bubble during a conversation. Instead of dropping the session, an intelligent transfer function could quickly move the session to UMTS/HSPA/LTE or even to a circuit-switched 2G network. For this purpose, two IMS extensions that are referred to as Voice Call Continuity (VCC) and Single Radio Voice Call Continuity (SR-VCC) have been standardized in 3GPP. The first VCC functionality that was originally specified requires two active radios [27].

In the network, VCC is implemented as an application server as shown in Figure 4.24, which sits in the signaling path and receives all SIP signaling for voice sessions. Further-more, VCC-capable mobile devices are required, as the request to switch to a different access network must come from the mobile device. This is necessary in this approach as the IMS itself has no information concerning signal levels and cannot therefore trigger a handover of the session to a different access network.

The actions taken by the VCC server in the IMS network for a transfer of a session to another access network depend on the types of networks involved. Roaming into a Wi-Fi

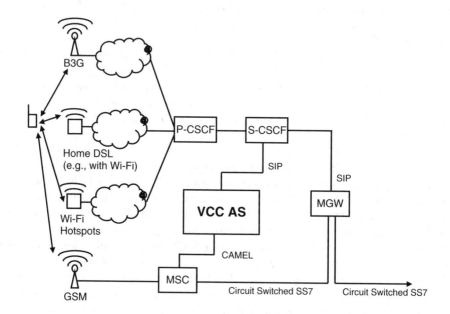

Figure 4.24   VCC architecture.

bubble while having a voice session established via a HSPA or LTE network works as follows: a basic requirement for VCC devices is the ability to be connected (attached) to several radio networks simultaneously. This is usually the case since most 3G/Wi-Fi enabled devices can use both types of networks simultaneously. When the VCC-capable device detects the Wi-Fi network, it establishes an IP connection, registers to the IMS a second time and then sends a SIP Invite request with a special SIP URI. This request is routed to the VCC application server that detects, by analyzing the SIP URI and the identity of the user, that this is a request to hand over the ongoing voice session to, from its point of view, a different IP address. It then instructs both parties to initiate a VCC domain transfer by sending a SIP Update or SIP Re-Invite message with the new IP address and UDP port numbers. It is important to note that this message is originated from the VCC application server and not from the VCC device as the VCC application server acts as an anchor for the SIP communication. This means that the VCC application server terminates the SIP signaling from both parties to allow the VCC subscriber to change IP addresses when transferring from one access network to another. The process in the opposite direction, that is of handing over an ongoing IMS voice session from WiFi to a packet-switched cellular network, is identical.

There are also many cases in which the user roams out of a Wi-Fi area and no high speed packet-switched cellular network is available. In this case the session has to be transferred to a circuit-switched 2G network. For this purpose the VCC application is not only connected to the IMS network but also to the circuit-switched MSC, as shown in Figure 4.24. This signaling link is based on the Customized Applications for Mobile Enhanced Logic (CAMEL) protocol, which is already used by the MSC for other purposes such as the prepaid service for which the MSC has to query the prepaid service to verify if the user has enough credit left to make a call. When the VCC mobile reaches the limit of the Wi-Fi network while a voice session is ongoing, it establishes a circuit-switched voice call to a special VCC roaming number. The MSC then informs the VCC application via the CAMEL interface of the incoming call and the VCC application returns an instruction to the MSC to connect the incoming voice call to the IMS media gateway. The IMS media gateway will then send a SIP invite message to the IMS core, which in turn forwards the message to the VCC application. When the VCC application receives the SIP invite message, it already knows which data stream it should redirect as it has already been informed via the CAMEL interface. It will then immediately send a SIP Update or Re-Invite message that instructs the destination to redirect the media stream away from the current IP address of the transferring subscriber to the IP address and assigned port number of the IMS media gateway.

Transferring a voice call that has initially been established in the circuit-switched cellular network to a Wi-Fi connection has also been specified. For this purpose it is required that outgoing circuit-switched calls of a VCC subscriber are not delivered directly to the destination but instead are first connected to the IMS media gateway. The IMS media gateway in turn communicates with the VCC application via the IMS system since it cannot establish the call to the destination on its own. This way the VCC application is part of the circuit-switched call and can thus later on instruct the media gateway to change the routing of the media stream as required. A 2G to Wi-Fi transfer is initiated by a VCC-capable device that detects during an ongoing circuit-switched voice call that a suitable Wi-Fi network is available. It will then connect to the Wi-Fi network, register to the IMS system and send a SIP Invite message to the VCC application via the S-CSCF, which contains the information on which ongoing call on the media gateway should be redirected. The VCC application will then send the necessary commands to the media gateway to redirect the call away from the circuit-switched bearer to an IP connection and to drop the connection to the MSC. This process is completely transparent for the circuit-switched terminator of the call since their circuit-switched connection to the IMS media gateway remains in place.

The downside of the VCC approach is that it requires two active radios in the mobile device to switch the voice path from one network to another. While this works between cellular and Wi-Fi, a different approach is necessary between B3G networks and 2G networks as LTE, UMTS, and GSM are handled by the same radio chip in a mobile device today and hence, only one radio technology can be used at any one time. As a consequence, 3GPP has specified SR-VCC as part of IMS Centralized Services (ICS) in 3GPP TS 23.216 [28]. Instead of the VCC application server as shown in Figure 4.24, ICS and SR-VCC use a different application server which is referred to as Service Centralization and Continuity Application Server (SCC) as described in 3GPP TS 23.237 [29]. SR-VCC has been designed to hand over the following packet-switched voice bearers:

- From LTE to a circuit-switched UMTS or GSM bearer,
- From UMTS to a circuit-switched GSM bearer,
- From LTE to a circuit-switched CDMA bearer.

The following example describes how an LTE packet-switched voice call is handed over to a circuit-switched GSM channel. To enable communication between the LTE core network and the circuit-switched GSM core network, the Sv interface has been specified between the LTE MME (cf. Chapter 2) and the Mobile Switching Center (MSC). The MSC in turn is connected to the IMS system via a signaling link to the IMS SCC Application server described above.

A mobile device supporting SR-VCC needs to indicate this to the MME during the attach process and later on during tracking area updates. This is done by including a circuit-switched (CS) classmark information element as part of its network capabilities information. The classmark describes the mobile device's capabilities and the supported circuit-switched speech codecs. The MME then answers with a "SR-VCC operation possible" indication if the network supports SR-VCC as well. Once the mobile device is attached to the network, it then contacts the IMS to register as described above. In addition, the HSS database informs the MME during the attach process if the user is allowed to use SR-VCC.

With SR-VCC, it is not the mobile device that initiates the handover to another radio technology but the network. If the network detects that signal conditions deteriorate, it can instruct the mobile device to measure signal conditions on other radio networks. If a 2G or 3G network can be received with a better quality while the user has a voice call established, the network takes steps to prepare the handover. In the first instance, the MME will send a "PS-to CS-request" via the Sv interface to the MSC. If MSC does not control the target radio cell, it contacts the MSC to which the cell is connected to. In the context of this operation, the second MSC is referred to as the relay-MSC. The first MSC will however remain the anchor for all signaling and for the circuit-switched voice bearer. This way, only a single MSC is required in the network that supports the Sv interface. It is then the anchor-MSC or relay-MSC task to prepare a circuit-switched channel in the target cell with the standard handover signaling. This is possible as for the target cell there is no difference between a packet-switched voice call being handed over and a traditional circuit-to-circuit handover.

In the second step, the anchor-MSC contacts the IMS system by sending a "Initiation of Session Transfer" message to the SCC. The SCC then interacts with the IMS to switch the speech path away from the User Equipment (UE) toward the media gateway of the anchor-MSC.

Once the target cell has established a channel for the circuit-switched call, the anchor-MSC responds to the MME's PS-to-CS request. The response contains all necessary information for the mobile device to switch to the right cell and the right channel. This information is then conveyed to the mobile device in an LTE-to-GSM handover command. The mobile device then interrupts the communication with the LTE network and switches to the indicated channel on the GSM network.

Should voice and multimedia sessions over the IMS become popular in the future, it can be envisaged that the VCC and SR-VCC functionality is extended in the long term and combined with intelligent session transfer functions between devices. A call could then, for example, be started at home as a voice video call on a high-resolution device that is connected via Ethernet to the home network and then transferred by the user to a mobile device that uses the personal Wi-Fi network. In the process the resolution of the media stream could be reduced due to the smaller display size. Once the user roams out of their home the call could then automatically be transferred to the 2G network by first removing the video component and then using the VCC functionality to transfer the call to the 2G network. The IMS offers all the tools required for such complicated scenarios which should, however, be as seamless for the user as possible. Before such scenarios become reality, however, more work is required to standardize such inter-device transfers and to develop network and device software.

In the foreseeable future, however, the ambitions of the wireless industry are rather focused on using the IMS with a much narrower feature set. Over the years, a number of different initiatives have been started and at the time of publication of this book, the two most noteworthy are RCS-e and VoLTE. These are described in more detail in the following as examples of how IMS might be put into use at first.

4.4.12 IMS with Wireless LAN Hotspots and Private Wi-Fi Networks

Many mobile network operators are also deploying Wi-Fi hotspots in hotels, airports, and other places where there is a high concentration of people who communicate outside their home and office. These Wi-Fi hotspots complement B3G cellular networks and therefore increase the capacity available at these locations. Current hotspot deployments, however, suffer from a number of disadvantages:

- Wi-Fi hotspots operated by mobile network operators are not connected to their cellular network subscriber database (HLR, HSS). Consequently, a separate authentication and billing system is required. From the user's point of view, this is also less than ideal as they are usually required to pay for access by entering their credit card details.

- Public Wi-Fi hotspots are usually not encrypted on the network layer as the Wi-Fi standards do not include methods to establish an encrypted connection to users who have not previously installed an authentication certificate or password. This makes it very easy for eavesdroppers to intercept data the traffic of others. While the login procedure is usually protected by an encrypted HTTPS (Hypertext Transport Protocol secure) Web session, many other Web services later on use unencrypted connections. The process of eavesdropping in non-encrypted networks has even been automated and tools such as Firesheep [30] and Droidsheep [31] can easily be installed and used with little knowledge. Virtual Private Network (VPN) software can secure the use of a nonencrypted network by establishing a tunnel to an endpoint, usually at a company over which all data traffic is encrypted. Again, this is usually something the user has to start manually as well. Even if such a product is used, most people are not aware of whether all data packets are sent through the encrypted tunnel or if only data packets to IP addresses of the company are protected while traffic to and from the Internet bypasses the tunnel.
- Active attackers have direct access to the device that has attached to the network and can therefore try to exploit operating system weaknesses.
- Current wireless network operator hotspot deployments are also not suitable for connecting to the IMS, which is usually deployed in a secure network environment and cannot be accessed via an unsecured public network.

As shown in Chapter 3, Wi-Fi could not only play an important role in the future to satisfy overall wireless capacity demands by being used as a home network technology but also as a cellular offload technology in public places with a high bandwidth demand. Thus a solution is required for secure, easy, and automated use of public Wi-Fi hotspots. While Wi-Fi networks at home and in the office can be secured using authentication and Wireless Protected Access (WPA) or WPA2 encryption on the radio interface, a different method has to be found for public Wi-Fi hotspots. One of the solutions designed with access to the IMS in mind is 3GPP's Interworking-Wireless Local Area Network (I-WLAN) specification [32].

I-WLAN removes the disadvantages discussed above by standardizing authentication and encryption over public Wi-Fi hotspots as follows. When an I-WLAN-capable device such as a notebook or a B3G/Wi-Fi mobile device detects an I-WLAN capable public Wi-Fi hotspot, it first of all contacts the network as before to establish a nonencrypted connection. In a second step the mobile device then authenticates itself using the EAP-SIM or EAP-AKA protocol [33].

EAP-SIM uses the same authentication framework as described for WPA personal and enterprise authentication as shown in Chapter 2. Figure 4.25 shows the messages exchanged between the mobile station and the authentication server via an EAP-SIM capable access point during authentication. After the Wi-Fi open system authentication and association, the access point starts the EAP procedure by sending an EAP Identity Request, to which the mobile device has to respond to with an EAP Identity Response message. The identity returned to the network in this message is composed of the IMSI, which is taken from the SIM card, and an operator-specific postfix. Alternatively, the mobile device can also send a temporary identity (pseudonym) which has been agreed with the network during a previous authentication procedure. The pseudonym is similar

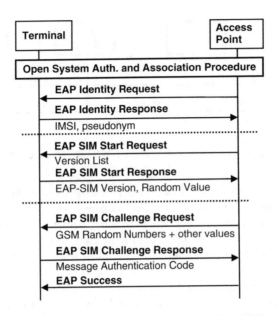

Figure 4.25   EAP dialogue between a mobile device and an I-WLAN access point.

to the TMSI (Temporary Mobile Subscriber Identity) used in GSM networks to hide the subscriber's real identity from eavesdroppers but has a different format.

In the next step, the network sends an EAP SIM start request which contains a list of different versions of supported EAP SIM authentication algorithms. The client device selects one of the algorithms it supports and sends an EAP SIM start response message back to the network. This message also contains a RAND number which is used for a number of subsequent calculations on the network side in combination with a secret (the Kc), which is shared between the mobile device and the network. In this way the network is also able to authenticate itself to the client.

At this point the authentication server in the network uses the subscriber's IMSI to request authentication triplets from the GSM/UMTS Home Location Register/Authentication Center (AuC). Two or three GSM RAND values and GSM ciphering keys returned by the HLR are then used to generate EAP SIM authentication keys, EAP SIM encryption keys, and other values required for the EAP-SIM authentication process. These are sent in encrypted form together with the two or three GSM RAND values in plain text to the client device in an EAP SIM challenge request to the mobile device.

The mobile device then uses the GSM RAND values received in the message and forwards them to the SIM card. The SIM card then generates the GSM signed response and GSM ciphering keys which are used afterwards to decipher the received EAP SIM parameters. If those values are identical to the values used by the network, the mobile device is able to send a correct response message that is then verified on the network side. If verification was successful, an EAP success message is returned and the client is admitted to the network. The Wi-Fi network then also delivers charging related information to the authentication server, such as consumed traffic and online time for online (prepaid) or offline (post-paid) charging.

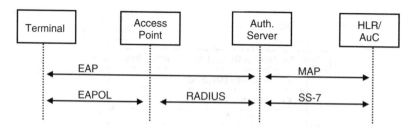

Figure 4.26 Protocol used for I-WLAN authentication.

Figure 4.26 shows the different devices and protocols used during authentication. On the left side the mobile client sends its EAP messages via the Extensible Authentication Protocol (EAPOL) protocol. For the messaging between the access point and the authentication server, RADIUS or DIAMETER is used. The authentication sever communicates with the HLR/AuC via the SS-7 circuit-switched signaling network and the Mobile Application Part protocol.

The authentication and network admission procedure above only allows the device to access the network behind the access point and a direct gateway to the Internet. Access to the IMS network, which is usually located in the secured network of the subscriber's home network operator is not granted. This is because the home network operator has no control or visibility concerning the security of the Wi-Fi network and the interconnection to the home network. For access to the IMS and other services provided by the home network operator it is therefore necessary to establish a secure and encrypted tunnel between the subscriber's device and a gateway located at the border of home network operator. This scenario is shown in Figure 4.27. The secure tunnel can either be established directly after

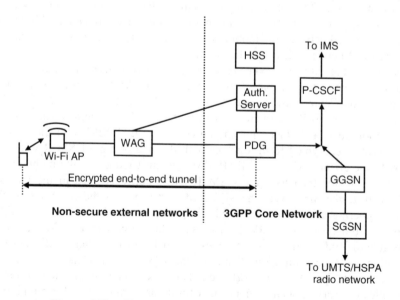

Figure 4.27 IPsec tunnel architecture to the B3G network.

association to the public hotspot or after the EAP-SIM authentication procedure shown earlier. The method used depends on the policy of the home network operator.

There are several possibilities to establish a tunnel with the gateway router, which is referred to as the Packet Data Gateway (PDG). If the mobile device has roamed to a Wi-Fi access point of the home network operator, a DNS query is made to retrieve the IP address of the home network PDG. If the mobile has roamed to a Wi-Fi Access Point of another national or international network operator, the mobile will first try to get access to a PDG in the visited network. The home network PDG is only contacted if this fails. The domain name of the PDG can be built by the mobile device from information stored on the SIM card or in the mobile terminal.

Once the IP address of the PDG is known, the mobile then starts the tunnel establishment procedure with the Internet Key Exchange (IKE) protocol. The PDG in turn retrieves the authentication information required for the user from the authentication server, which acts as a gateway to the HLR/HSS as described earlier. During the key exchange messaging, the PDG and the mobile device authenticate each other. Once the keys for the tunnel are exchanged, IPSec ciphering is activated and the mobile device gains access to the home operator's IP network. As can be seen in Figure 4.27, the P-CSCF is now reachable by a Wi-Fi IMS device. As a final step an IMS Wi-Fi device will now start an IMS registration procedure, which starts with the P-CSCF discovery as described earlier.

Since the establishment of the IPSec tunnel is transparent for the Wi-Fi access point, it can be envisioned that this or a similar process could also be used in the future for IMS access via personal Wi-Fi access points connected to the public Internet via IP. For this purpose the B3G network operator must therefore connect a PDG to the public Internet.

The Wireless Access Gateway (WAG), which is also shown in Figure 4.27, has not been mentioned so far. Its responsibility is routing enforcement; that is, it ensures that the mobile device can only exchange IP packets with the PDG it has an encrypted tunnel with. The WAG is able to enforce such a rule since it can be informed by the PDG during the tunnel establishment process of how to filter packets to and from a certain user device.

Tunnel establishment and the use of IPsec to create an encrypted tunnel between a device and a gateway as shown in this section for B3G core network access is very similar to proprietary security products available for notebooks and mobile devices today.

4.4.13 IMS and TISPAN

In the previous section it has been shown how the 3GPP standard defines how mobile devices can use public and potentially also private Wi-Fi hotspots to connect to a 3GPP IMS network. For private Wi-Fi hotspots, however, the solution offers no quality of service control, since the 3GPP IMS system does not have interfaces to control network elements outside its domain. As such it does not fully meet the requirements of fixed-line network operators who would also like to use IMS as their future service platform. The European Telecommunications Standards Institute (ETSI) has therefore decided to define a broader service architecture in their TISPAN standards project. With Release 8 of the 3GPP standard, both architectures have been merged into a common architecture. In its core, a TISPAN NGN consists of the following entities:

- a subset of the IMS as defined by 3GPP,
- a PSTN/ISDN (Integrated Services Digital Network) PSTN/ISTN Emulation Subsystem (PES),
- other non-SIP subsystems for IPTV, Video on Demand, and other services.

As can be seen in the list, the IMS is only one of several core subsystems of TISPAN. The reason for this is the fact that many services today are not based on SIP and sessions such as IPTV or Video-on-Demand. Many different approaches exist on the market to deliver such services to the user. TISPAN aimed to standardize the way such services are delivered, controlled, and billed, and how such applications can interact with the transport network to request a certain quality of service level for their data packets. For fixed line voice telephony services, however, the IMS is the central instance for both fixed and wireless networks.

It should be noted at this point that there is a fierce discussion ongoing about the advantages and disadvantages of prioritizing some packets over others. While from a technical point of view this has advantages, many fear that network operator-controlled preference of some packets over others has a negative effect on the evolution of the Internet. The proponents of the "all packets are equal" approach do not allow any kind of prioritization in the network. This, they argue, would potentially allow network operators to dictate terms to Internet companies such as Google, Yahoo, and many new startup companies who are offering innovative services over the Internet with high bandwidth requirements. Network operators, on the other hand, argue that quality of service for applications such as VoIP, IPTV, and video on demand can only be achieved during times of network congestion by giving higher priority to data packets of such applications. The debate over this topic is referred to as "network neutrality" and includes a number of other topics, such as whether network operators should be allowed to block or delay packets of certain applications at their borders. Further information can be found in [34].

While in theory the IMS has been defined to be access network agnostic, the 3GPP standards still make a number of assumptions on the kind of access network and subscriber databases to be used. To make IMS usable for DSL and cable operators, it is necessary to fully generalize interaction with the access network and to have a generalized user database in the network. Figure 4.28 shows a simplified model of how the IMS is enhanced by TISPAN for this purpose as defined in [35] and [36].

The first difference between fixed-line and wireless networks is how devices connect to the network. In case of fixed-line access networks, a DSL or cable modem at the customer's premises is usually the gateway device that establishes the connection to the network. One or more devices behind this gateway device will then use the established connection to register with the IMS service. This is quite different to 3GPP, where each device connects both to the transport network (PDP context activation) and to the IMS service (SIP register).

A number of different ways exist today for a DSL or cable modem to attach to the network. TISPAN has made the step to standardize the functionality required in the net-work for user management at the network layer with the Network Attachment Subsystem (NASS). When a DSL or cable modem is powered on, it first communicates with the NASS to authenticate and to obtain an IP address. Protocols used for this purpose are,

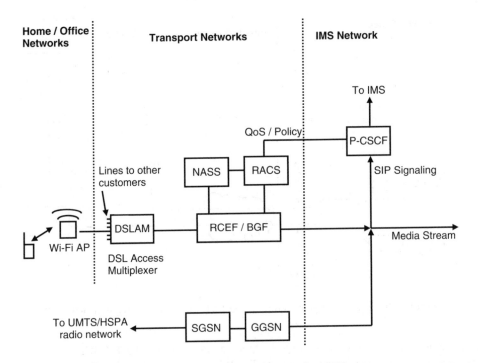

Figure 4.28   TISPAN IMS architecture for xDSL access.

for example, PPPoE (Point-to-Point Tunneling Protocol over Ethernet) and Point-to-Point Tunneling Protocol (PPP) over ATM.

The NASS is reached during the attachment process via the Resource Control Enforcement Function (RCEF)/Border Gateway Function (BGF), which sits between the access network and the core network of the operator. Their tasks are, among others, to route IP traffic between an external network and subscribers and to ensure that quality of service requests coming from the IMS or other core systems listed above are enforced on the transport layer.

The TISPAN Resource and Admission Control Subsystem (RACS) performs a similar task to the IMS PDF. When an IMS session is established by a TISPAN device, the P-CSCF contacts the RACS subsystem to reserve the required resources for the session and to allow IP packets to flow between the participants of the session. The RACS then communicates with the RCEF/BGF to see if enough resources are available for the session and configures them accordingly. While in the 3GPP IMS model resources are reserved by the mobile devices and the P-CSCF, once codecs have been agreed on, the TISPAN P-CSCF contacts the RACS initially when receiving the invite message from the originator. If bandwidth requirements change during the session setup because a different codec has been selected by the devices, the P-CSCF contacts the RACS again to modify the policy. This means that, unlike a 3GPP IMS mobile device, which requests a certain bandwidth and QoS with a secondary PDP context activation, TISPAN

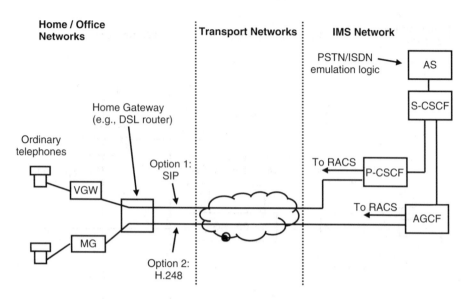

Figure 4.29 High level architecture of the PSTN emulation subsystem with IMS.

IMS devices are not involved in QoS processes at all since the P-CSCF takes care of the whole process. This is necessary as TISPAN devices, unlike 3GPP mobile devices, are pure IP devices and therefore cannot influence the quality of service settings of the network themselves.

Besides the IMS subsystem, the PSTN/ISDNPES is another important element of TISPAN. Its aim is to enable legacy analog and ISDN telephones to be connected to an IP-based next-generation network via a media gateway which is either part of the access modem or a standalone device. An IMS independent implementation for PES is described in [37] while [38] defines how a PES can be implemented with IMS. Figure 4.29 shows how legacy analog (PSTN) or digital (ISDN) devices can be included in the IMS. As legacy devices cannot be modified a gateway has to be deployed on the user's location. On the one side of the gateway the analog or digital telephone is connected to a legacy connector. In case of a standalone gateway an Ethernet connector is used for the connection to the DSL or cable modem. The PES specification in [39] knows two types of devices. Voice Gateways (VGWs) emulate a SIP User Agent on the behalf of the legacy device and communicate with SIP commands with the P-CSCF of the IMS. The second approach is to deploy a media gateway that communicates via the H.248 (MEGACO) to the Access Gateway Control Function (AGCF). In this approach the SIP User Agent functionality is included in the AGCF, that is, not on the customer device but in the network itself. Additionally, the AGCF includes the P-CSCF functionality. During call establishment, the P-CSCF or the AGCF then communicate with the RACS to reserve the required transport resources to ensure the quality of service for a call. In addition, PES requires an IMS application server to emulate the PSTN or ISDN service logic when the PES User Agent sends SIP messages with embedded ISUP messages.

4.4.14   IMS on the Mobile Device

Up to now, this chapter has focused on the network part of the IMS. The IMS implementation on the mobile device, however, is just as important. In practice, deploying IMS applications is done in several ways.

4.4.14.1   Initial IMS Application Deployments

One approach is to create IMS applications, which include the service itself and the necessary IMS software stack to communicate with the network. This approach, however, limits the use of IMS on a device to the applications of a single vendor. In practice, first IMS applications driven by mobile network operators might use this approach, which has the short-term advantage that the complete client application can be developed independently from the operating system and device manufacturer.

Even in this scenario, however, the IMS application has to be tightly connected with other functionality of the phone. For example, if the IMS is used for messaging, tight integration of the IMS application is required with the SMS functionality of the phone as well as the address book to give the user a seamless experience. If the IMS is used for voice telephony, the IMS client must be tightly integrated into the telephony function of the mobile device; so it is transparent to the user if the phone uses a circuit-switched channel, in case no B3G network is currently available and the call is handled a 2G network. Also, address book integration is another must.

In recent years, this has become much easier to accomplish as mobile operating systems such as Android allow software developers to integrate applications very deeply into the overall system. This flexibility is discussed in more detail in Chapter 5.

4.4.14.2   IMS as an API on Mobile Devices

Despite the phenomenal surge of services and applications developed by Internet-based companies in recent years, some network operators might want to deploy IMS-based services beyond those envisaged at first. From a technical point of view, it is then preferable to split IMS applications into the IMS protocol stack, which offers a standardized abstract access to the IMS with a simple to use Application Programming Interface (API) for applications on top. This approach has the following advantages:

- Development of an application that uses the IMS as a service messaging framework does not require detailed IMS knowledge by application programers. This speeds up the development of the application.
- Several IMS applications from different vendors can run concurrently on a mobile device. All applications then use single instance of the IMS environment; that is, only a single copy of the IMS framework is in the memory of the mobile device.
- By giving some degree of control to the IMS environment, it is possible that applications are only started when a SIP message arrives from the network for a specific IMS application.

- A single IMS environment with a specific API can be available for several hardware platforms. All adaptations required for the hardware are encapsulated in the IMS environment. IMS applications can therefore be written in a device-independent way. This would speed up the development and reduces the interoperability issues between the same services on different types of devices.

Today, IMS environments, APIs, and applications are developed by companies such as Ericsson [38], Huawei, Intel (via their purchase of Comneon), Ecrio, and Movial. In other words, these companies are quite different to the big Internet-based companies such as Google on the one hand and small startup companies on the other hand developing IP-based (non-IMS) services today.

Most IMS client environments are offered for a wide variety of different hardware and software platforms. These range from powerful high-end operating systems such as Windows and Linux for notebooks to mobile device operating systems, which are streamlined for much more restricted hardware environments such as smartphones and embedded devices such as set-top boxes.

The IMS specification knows two kinds of applications. Those that are already defined in the 3GPP IMS specifications are likely to be included as part of the IMS environment to a large extent. Applications on the mobile device can then take advantage of these services easily with only a little interaction with the IMS environment. An example of such a service is presence information, which could be used by the mobile device's native address book application to enhance the entry of a user with presence information (online, offline, in meeting, etc.). No IMS knowledge is required as the API function call would reveal presence information to an application while not requiring the application to know where this information came from and how it was obtained.

New proprietary applications require much more interaction with the IMS environment. This includes basic procedures such as registering their service with the IMS framework on the mobile device and interacting with application servers in the network or other service subscribers. All actions are abstracted via the IMS environment's API. This means that application programers do not need to have in depth IMS knowledge such as, for example, how and which SIP messages are used to interact with external elements, how SIP headers of SDP content inside a message are formatted, how quality of service for a media stream is requested (e.g., via secondary PDP context activation), and so on.

For applications written for a specific operating system, IMS frameworks usually offer a C++ IMS or other native programming language API, as shown in Figure 4.30. Native applications then directly interact with the mobile device's operating system for tasks such as file access and the graphical user interface and use the native library from the IMS framework. Many mobile devices also offer a Java virtual machine for platform-independent programing. Such programs, also referred to as Java applets, do not directly interact with the mobile device's operating system, which would make them platform-specific, but do so only via the Java virtual machine. IMS frameworks supporting Java applets have a link into the Java Virtual machine, as also shown in Figure 4.30, and offer a Java IMS API for Java applets. This Java API is specified in JSR-281 [40], which ensures that once a Java applet is developed it will work on any mobile device which includes a Java virtual machine and supports Java JSR-281, again independently from the developer of the IMS framework.

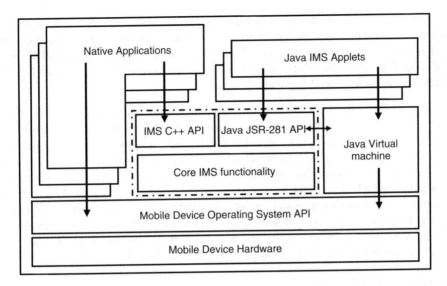

Figure 4.30   IMS framework in the software stack of a mobile device.

Most IMS frameworks are able to interact with many IMS applications running on a mobile device simultaneously where the device supports multitasking of user applications. The IMS framework therefore needs to be able to route an incoming message to the correct application. This is simple for responses to messages originating from a specific application. For unsolicited incoming messages a different method has to be used. Release 7 of [36] introduces the IMS Communication Service ID (ICSI) and the IMS Application Reference ID (IARI) for this purpose. When applications are executed for the first time on an IMS device they register themselves with the IMS framework by specifying their ICSI and IARI. When an unsolicited message arrives at the mobile device that contains these IDs, the IMS framework then routes the message to the corresponding application. If the application is not executed at the time a message arrives, the IMS framework can also launch the application and then hand over the message.

The IMS framework running on a mobile device also takes care of basic IMS procedures on behalf of all applications. For example, the registration process when the mobile device is switched on is such a basic procedure. Depending on the operating system, hardware capabilities, types of currently available networks (HSPA, Wi-Fi, etc.), and user preferences, advanced IMS frameworks can decide if and over which network it should try to register with the IMS network. This process is completely transparent to IMS applications as they just use the API to query the framework for the current registration state or be informed if the registration state changes.

4.4.15 Rich Communication Service (RCS-e)

In 2012, one of the first set of services to be deployed after many years of development is the RCS as part of the GSMA RCS-e specification [41]. Especially network operators in

Europe are working on the deployment [42]. RCS-e is based on the RCS 2.0 specification and provides a simple interoperable extension to voice telephony and text messaging. It is a standalone extension and can run alongside a circuit-switched voice call if used in a UMTS or B3G network, which supports concurrent voice and data use. The following features are supported on RCS-e handsets:

- IM functionality: RCS-e supports direct interactions between two users or with store and forward capabilities in the network similar to how SMS works today. This allows storing instant messages in the network if the destination suddenly becomes unavailable or if a message is sent to an offline subscriber. Once the destination becomes available, the message is automatically forwarded. Furthermore, the IM functionality includes a group chat, which is optional as it requires an IM server in the network, which is not mandatory for network operators to deploy. In addition, RCS-e IM supports notifications that a message is currently composed and that a message has been delivered and is displayed to the user. Such IM features are also included in other IM suits such as, for example, the Yahoo messenger and thus already well known to users,
- File transfers,
- Image sharing during a call,
- Video sharing during a call.

It is interesting to note that presence and status updates (available, idle, away, etc.) are not part of the first version of the RCS-e specification. RCS home services on PCs and fixed line devices are also excluded in version 1.2.1.

If only the mandatory functionality is included in a device, RCS-e is little more than an SMS over IP solution and unlikely to be perceived as different or more attractive. If the optional elements are included in a mobile device and a network, however, RCS-e can be seen as a multimedia extension to the traditional SMS service and might perhaps be able to compete with other IM services offered as part of operating systems such as Apple's iOS, Androids GTalk, and third-party IM client applications.

One of the advantages of RCS-e from a subscriber's point of view is that like SMS, it will work without prior configuration as the service can configure itself. This is accomplished via network-based authentication by the RCS-e server mapping the IP address to an HSS subscriber profile. Also, the address book tie in allows to automatically find other RCS-e users the subscriber knows either at initial configuration or at any time the user searches for contacts in the address book. Also, when a new contact is added to the address book (i.e., a name and a phone number), it is possible to immediately check in the network if that phone number is also tied to an already existing RCS-e account. SIP messages as described above are used for these operations. In case a contact is not subscribed to RCS-e, SMS messages can be used as a fallback option. RCS-e is most useful when several network operators offer the service and the RCS-e systems are interconnected. In Germany, three network operators have committed to launch RCS-e and interconnect their systems [43].

During the initial auto-configuration, a number of parameters such as those shown in the following list are configured on the mobile device:

- The SIP proxy address,
- The XDM server for optional presence,

- The TEL-URI,
- The SIP-URI,
- SIP username and password are used for authentication in case the client is used over Wi-Fi in which the network cannot authenticate the subscriber via the IP address, as it is not assigned by the home network operator but by an external network,
- URI for the IM group chat server,
- A polling period for a periodic capabilities update of the contacts in the user's address book.

The auto-configuration mechanism uses the OMA DM (Device Management) service or a new RCS-e specific mechanism that uses HTTPS and XML to retrieve the settings.

Another aspect of RCS-e is that the service is aware of the capabilities of the type of network that is currently used. If only a 2G network is available, video streaming is automatically deactivated, for example. This is also communicated to a remote user in a conversation when he tries to initiate this service.

For the messaging part of the service, IMS SIP messaging is used to establish the connection between two or more users. Afterward, the Message Session Relay Protocol (MSRP) is used for the message exchange [44]. Depending on whether a direct communication or a store and forward architecture is used, the MSRP media path can be redirected to a server in the network. For the exchange of video streams, the RTP is used.

As RCS-e was designed for both cellular and private Wi-Fi networks, NAT traversal as described above is also part of the specification. The Session Traversal Utilities for NAT (STUN) protocol is used for this purpose as described in TS 24.229 [45] or, alternatively, empty RTP packets with payload type 20.

Specifically not included in the RCS-e standard are voice calls. In many cases, RCS-e services can be used along circuit-switched voice calls or IMS voice calls, like in the case of the Voice over LTE (VoLTE) services, which are described next.

4.4.16 Voice over LTE (VoLTE)

The exact opposite of RCS-e is Voice over LTE (VoLTE), today specified in the GSMA PRD (Permanent Reference Document) IR.92 [46]. As the name implies, this IMS application primarily focuses on voice over IP transmission as part of LTE networks. It is thus regarded as an addition and long-term replacement for the current circuit-switched voice telephony architecture of GSM, UMTS, and CDMA. Verizon Wireless is one of the leading network operators working on the implementation of VoLTE [47]. As IMS is a very flexible signaling infrastructure and offers many ways of implementing a voice service, VoLTE was created for interworking between mobile devices and also for interworking between mobile devices and different networks while roaming. Also, interworking between different networks needs to be assured for calls between networks and also while the user is roaming. As a consequence, VoLTE defines the interworking for the following domains:

- IMS basic capabilities and supplementary services for telephony.
- How the media negotiation takes place at the beginning of a call, how voice calls are transported, and which codecs are required.
- Interaction of the system with the LTE radio and core network.

- Further cross-stack requirements such as international roaming, how the IP protocol is used, and how emergency call services are implemented.
- Interaction with the circuit-switched network for handing over calls to a 2G or 3G network at the border of the B3G coverage area.

A VoLTE compliant device identifies itself using a defined ICSI, which is "urn:urn-7:3gpp-service.ims.icsi.mmtel." To announce its support of SMS over IP, it has to use the feature tag "+g.3gpp.smsip". The type of user identity used by the UE to register to the IMS depends on the SIM card in the device. If the user already has an IMS SIM, which is unlikely today, the IMS public user identity can be used for registration. Otherwise, a temporary public user identity which is derived from the IMSI is used. While authentication is required, the ciphering of signaling messages is only optional as communication is considered secure because of the focus on LTE as a bearer network, which has its own encryption on a per user basis.

As already described in the general IMS introductions section above, an IMS user can be addressed via a tel-URI (e.g., tel: +41134545431) and in addition also via an alphanumeric SIP-URI (e.g., "john.smith@xyz.com"). Like RCS-e, VoLTE makes frequent references to the core IMS specification in 3GPP TS 24.229.

To offer a complete voice service, a VoLTE-compatible device must support a number of supplementary service among which are the display of the calling party phone number, the restriction of forwarding the phone number for privacy reason, all types of call forwarding such as when busy, not available, and after timeout. Also, the barring of calls must be supported as well as call hold and multiparty conference functionality as described in 3GPP TS 24.615 [48]. To configure supplementary services, the XCAP protocol is used as described in 3GPP TS 24.623 [49].

Again, as described in the general section earlier, a VoLTE compliant core network must support proxying of the user so that the Policy and Charging Rules Function (PCRF) can inform the P-CSCF that the user has lost contact to the network. The P-CSCF can then terminate an ongoing voice call with the appropriate SIP messaging on behalf of the user.

On the media codec side, VoLTE requires the mobile device to support all commonly used codecs known from 2G and 3G networks. In addition, wideband Adaptive Multi-Rate (WB-AMR) may be supported by the device and advertized to the other end during the session establishment. This codec is not yet widespread but has been launched by a number of mobile network operators in UTMS networks, for example, in Germany, Austria, France, and the UK. Voice quality is significantly superior compared to the traditional fixed line and AMR codecs and voice and music sounds much more natural. To transport the voice stream, the RTP over UDP is used. The RTCP as described in IETF RFC 3550 [50] is used to control the quality of the speech path. An interval between RTCP messages of 5 s is recommended. For interaction with automated services such as voice mail systems and customer care applications, DTMF tones must be supported.

The VoLTE specification also describes the bearers to be used for communication. On the IP layer, IPv4 and IPv6 must be supported and the VoLTE IMS application on the device has to request an IPv4v6 dual-IP default bearer when initially establishing a connection with the network. If the network is not yet IPv6 capable, it will only return an IPv4 address which is then to be used. For efficient transmission of the small packets containing the voice stream, the support of Robust Header Compression (ROHC) is a mandatory requirement. While signaling data is to be transmitted over an acknowledged mode (AM) RLC (Radio Link Control) bearer, the default for IP-based data transmission today, the voice stream is to be transported over an unacknowledged mode (UM) RLC bearer. This bearer is established by the network initiating a dedicated bearer (also known as secondary PDP context in UMTS) with a traffic flow template that instructs both the network and the mobile device to send IP packets to and from a certain IP address to or from a certain UDP port over this logical bearer. The advantage of this is that voice packets lost on the RLC layer are not repeated OTA interface as their delay would be too great to still be useful. In case voice packets are lost, either silence is inserted in the media stream or the previous voice packet is repeated to mask the missing media. This way, a packet loss of up to 1% can be masked. To reduce the power consumption during an active phone call, the LTE discontinuous reception functionality (DRX), as described in Chapter 2, has to be supported. And finally, the VoLTE specification requires the use of a guaranteed bit rate (GBR) bearer for the IP packets containing the voice data; so the eNodeB scheduler can maintain a minimal data rate for the stream at the expense of lower priority data streams of the same or a different user.

As VoLTE is a long-term replacement for the circuit-switched voice telephony, there is also the regulatory requirement to support emergency calls. This means that the mobile device must be able to recognize dedicated emergency call numbers and initiate a special kind of call for the network to recognize the emergency. The network must then be able to forward the call to an emergency center with additional information. During the introduction period, the network can alternatively instruct the mobile device to use a circuit-switched call on another radio technology for the emergency call.

For operators who's strategy is to roll out VoLTE in LTE networks with inferior coverage to the already existing 2G and 3G networks, the specification describes how the SR-VCC functionality described further above can be used to transition the call from an LTE IP bearer into a circuit-switched 2G or 3G channel.

International roaming behavior is also described in the IR.92 VoLTE specification. If the foreign network also supports VoLTE, it is possible to use a local breakout for the IMS connectivity. This means that unlike today, where the PDN-GW or GGSN of the home network are used when establishing a default bearer (PDP context) for IP connectivity, a local PDN-GW or GGSN and a local P-CSCF can be used. Only from there is the home network IMS contacted. All other non-IMS IP data can use a different default bearer (PDP context), which continues to use the PDN-GW or GGSN in the home network. One of the advantages of this local breakout approach is that no voice data is needlessly sent over an IP tunnel to the home network first, which is especially beneficial if the voice call is to be established to a destination in the visited country.

4.4.17   Challenges for IMS Rollouts

While work on the IMS specification has already started in the year 2000, there are only few IMS systems deployed today. Likewise, there are only a few mobile devices with an IMS stack and IMS applications. This slow uptake is due to a number of reasons that are discussed in the next sections.

4.4.17.1   Circuit-Switched Voice and SMS

The main applications for IMS, namely voice calls and IM, are difficult to introduce due to two in-house competitors, circuit-switched voice telephony, and SMS messaging. These applications work well in wireless networks today and networks have been optimized for these services over many years. Voice and IM over IMS would have to perform at least as well to be accepted by customers. Mobile operators are therefore in no hurry to replace or complement them. In addition, to be fully embraced by customers, IMS voice, and messaging would have to be enriched to a point where users clearly see an advantage over using circuit-switched voice or SMS. This is certainly possible for both services. With IMS, rich media can be added to a voice call while the conversation is ongoing by adding a video stream or by allowing the sending of pictures while a voice session is ongoing. Web integration for IM, presence information, and Web 2.0 community-style extensions for IM (for details see Chapter 6) could similarly make IMS IM preferable to SMS for customers. These enhancements, however, make the service even more complex to develop, test, and deploy to a large customer base.

4.4.17.2   Network Capabilities

Until recently, cellular networks did not have the capability or the capacity to support large-scale migration of voice and video streaming services to IP. With the introduction of B3G networks such has HSPA and LTE, this has certainly improved. The coverage area and in-building penetration of most B3G networks today, however, is far inferior to that of 2G networks. This is due to fewer base stations covering an area and the higher frequency range used for B3G than for 2G networks in many parts of the world (cf. Chapter 3). This means that IMS voice sessions in mobility scenarios will not work as well as today's circuit-switched wireless calls. As shown above, the SR-VCC voice call continuity function could reduce this issue to a certain extent, as it allows an IMS voice session to fall back to a 2G circuit-switched channel.

4.4.17.3   Solution Complexity

The IMS has been designed to be a basic and secure messaging framework for a wide variety of services. As a consequence, development of IMS network components and software stacks for handsets is very complex and time-consuming. While companies are working on IMS developments, standards continue to enhance existing features and specify new features for the framework even before existing features have been tested in networks on a large scale. This makes it difficult for implementers to choose which of the features to implement and from which version of the standard.

Furthermore, interworking with existing core and RAN components to gracefully hand over existing calls to a circuit-switched 2G connection and provide voice and SMS service in general to subscribers outside of the B3G coverage is far from trivial. In effect, this means that the IMS infrastructure needs to communicate with the MSC-based infrastructure. Both services have to be maintained until the B3G coverage area is at least as good as that of the 2G network.

4.4.17.4   Network Interoperability

It is likely that each network operator will want to get its own IMS network to be in control of services and revenue. This means that several IMS networks will be deployed in most countries. Consequently these networks do not only have to allow subscribers to establish sessions to subscribers of other IMS networks but also have to allow them to exchange other information such as presence updates and IM. In today's global environment, many people also want to communicate with people living in other countries and there should be no difference if another person uses the same IMS network, an IMS network of a different national network operator or an IMS network of a network operator on the other side of the world. While the IMS standard specifies how IMS networks and even non-IMS SIP-based networks can exchange SIP messages between each other, an IMS interconnection infrastructure has to be established in practice, since it is impossible to have a dedicated interconnection from each IMS network to all other IMS networks. From a transport layer point of view, this is easy to achieve because the Internet is already in place. From a service point of view, however, this is a complex task because interconnections between IMS networks must be secure. Furthermore, network operators need to reach agreements on how to charge each other for services that are established between subscribers of different IMS networks. If inter-IMS charging models in the future are based on charging models of circuit-switched voice calls, full international IMS interconnection will be a difficult to achieve because operators have to reach agreement on how much to charge for each particular service.

4.4.17.5   Mobile Device Capabilities

Mobile devices for circuit-switched wireless telephony have become very cheap in recent years as the technology has matured, processes are understood, and many of the functions required for telephony are embedded in hardware. For IMS multimedia communication, sophisticated mobile device hardware is required with the following capabilities:

- multiband radio capabilities to support several network types (e.g., 2G, HSPA, LTE, and Wi-Fi),
- a multitasking operating system, ideally, and enough memory to execute several multimedia applications concurrently,
- a sophisticated graphical user interface that supports multimedia applications (video, music, high quality picture capture, etc.),
- high-resolution screens as IMS applications for picture sharing and video streaming only make sense when the multimedia information can be displayed accordingly.

Devices with such potential have only been available since 2005, that is five years after the IMS standardization started. It could therefore be said that IMS standardization was well ahead of its time. In 2012, such devices have become relatively cheap but competition from non-network operator-based voice and non-voice services is strong. Also, there is still a large market for ultracheap mobile devices for basic voice and SMS services, especially in emerging countries and in industrialized parts of the world.

4.4.17.6   Mobile Device and Application Interoperability

Today, there is a rigorous process in place to test a huge number of 2G/3G communication scenarios between a new mobile device and network elements of several network vendors before the mobile device is released to the market. Particular emphasis is put on voice telephony as it is an integral part of the mobile device and is in fact in most cases the main function. IMS services, however, are much more complex than pure voice telephony so interoperability testing between applications on mobile devices and IMS network elements from different vendors is likely to be more difficult. Furthermore, the network will no longer act as a buffer between two mobile devices by rewriting all messages exchanged between devices on the application layer. Therefore, an IMS service does not only have to be tested in combination with network elements but also with IMS devices of other manufacturers.

4.4.17.7   Business Model

There is an established business model in place today for voice and SMS communication based on national networks. As a consequence, mobile operators are likely to require IMS services to generate revenue at the point they are introduced. This is difficult to achieve in practice, which is why most Internet companies first introduce services, see how they develop and evolve and work on a business plan for a particular service only once it has become popular and it has become clear how people use it. Even big mobile network operators only have a small customer base compared with the total number of people using Internet services today and have a strong national focus. As a consequence, development costs for a new service and maintenance once in service have a far greater impact than if a service is available to a global audience. As is shown in more detail in Chapter 6, many communication services only become useful if a critical mass of subscribers is reached. This is much more difficult to achieve in national markets served by mobile operators compared with international markets served by Internet companies. Even if only deployed on a national basis, interoperability agreements for the service need to be put in place between the IMS networks of a particular country to allow subscribers of different networks to use a service to communicate with each other. It is therefore questionable if the current circuit-switched business model will work for IMS or if network operators will at some point have to change their approach.

From a development point of view there are at least two kinds of business models. Some operators may decide to buy services from third-party companies and run them as their own services. For the third-party developer this means that the network operator becomes the customer rather than the end user. This is a difficult constellation for service development as any changes to the service to what was initially agreed require another round of contract negotiations. This makes fast turnaround times very difficult. Therefore this approach is mainly suited to services which, once put into place, are not expected to evolve very much. Another model under which services could be developed is by allowing third parties to develop services for their infrastructure and run them independently with a revenue-sharing contract in place. This gives developers much more freedom on service creation and evolution.

4.4.17.8 Service Development and Processes

As network operators are unlikely to develop services themselves they depend on third-party developers to create services for their IMS. This includes the logic on application servers and the plug-in programs to an IMS framework on mobile devices, as discussed in Section 4.4.14. Access to the IMS network infrastructure for development and later service deployment is difficult, however, as it first requires a business relationship with a network operator. Furthermore, access to users during the development phase is even more difficult since, as discussed above, network operators only have a limited customer base compared with the number of worldwide Internet users. As a consequence most small startup companies or research groups will rather develop their ideas in an Internet environment by either offering their mobile service via downloadable applications or as an HTML5-based application running in a the mobile web browser and release them to a global audience. It is more likely, therefore, that IMS vendors will partner with third-party companies to develop applications for their IMS framework and sell them to network operators. These third-party companies, however, are unlikely to be startup companies but rather established companies and thus strongly revenue-oriented. The types of applications that can be developed in such an environment are very different from those developed in an environment where new ideas can be tried without financial pressure. As a result it is unlikely that IMS services will be developed in an evolutionary way, that is by launching a first version of the service to a global audience, get customers by recommendation and then rapidly evolve the service step by step from feedback and customer behavior.

4.4.18 Opportunities for IMS Rollouts

As shown above, the IMS faces many challenges on its way to become an established platform for services. There are also, however, a number of reasons why it is likely that IMS networks will establish themselves in the future despite the difficulties.

4.4.18.1 B3G Network Design

LTE networks no longer contain a circuit-switched subsystem. Consequently, network operators will have no other choice than to introduce IMS once launching LTE networks if they want to be the provider of voice telephony service instead of leaving this area to a third party.

4.4.18.2 Fixed Line Network Evolution as Role Model and Complement

Another angle to look at the evolution of voice telephony in wireless networks is to analyze the current evolution process in fixed-line networks and to see how this change will in the future also apply to wireless networks. In many countries, fixed-line analog telephony is quickly replaced by DSL lines and voice telephony over IP. The incentive for customers to replace their analog telephone line, which they previously had to have as a precondition to getting DSL service over the same line, is usually a lower price when both services are delivered over IP. Part of such an offer is usually a DSL access device (modem/router/Wi-Fi, etc.) with a built-in SIP/IMS User Agent or media gateway, as discussed in Section 4.4.13 and Figure 4.29. The DSL access device then offers one or more sockets for analog telephones. Since the first edition of this book this trend has accelerated and incumbent network operators such as, for example, Deutsche Telekom in Germany have started the transition process from ISDN-based voice telephony networks to IP-based TISPAN/IMS over DSL solutions.

A similar approach could also be taken by wireless network operators, especially if they also have fixed-line network assets. In the long term, by offering a unified fixed and mobile voice service, they could use their system to let their subscribers register to the IMS system either via a B3G wireless connection or at home via the DSL line, which is also provided by them. The benefit for the user would be a single telephone number through which they can be reached and Internet access wherever they are. At the same time this would also reduce the overall bandwidth required in cellular wireless networks as most people mainly use services requiring a high data rate for a longer amount of time while they are at home or at the office (cf. Section 3.18). By offering such bundles at an acceptable price level users might be less inclined to select a third-party service provider for their voice and multimedia telephony needs that are not as well integrated to the fixed-line and cellular wireless networks they use. With the IMS network, operators also have the ability to offer advanced voice and multimedia services such as call handover from one device to another, which might be more difficult to do for service providers using a non-IMS platform. How well users take up IMS services other than voice and multimedia communication remains to be seen due to interconnection issues between IMS networks and pricing strategies, as discussed in the previous section.

4.4.18.3 Preconfigured Services

As mobile network operators usually offer their services together with a mobile device, they are in the unique position to preconfigure IMS on those devices. This is a huge advantage as third-party services require users to configure their mobile devices or even download and install software. As this is difficult for nontechnical people, this is a huge competitive advantage.

4.4.18.4 National Telecom Infrastructure

From a financial point of view, having a centralized international infrastructure for a service is certainly beneficial. The consequences of a service failure, however, are far more serious as users in many countries will be impacted at once. Furthermore, the risk of a service failure in a single country increases due to the higher number of network components between the user and the service in the network. For critical services such as voice communication, it is therefore advantageous from a security and safety point of view to have several national operators with their own independent IMS networks. These are not impacted when national or international Internet connections are interrupted for whatever reason.

4.4.18.5   Conclusion

When taking the pros and cons of the technology, service development, network deployment, and business cases into account, the IMS shows its strengths for voice, multimedia, and messaging centered services. One main advantage of network-operator-based voice services is the deeper integration into the RAN, which helps to ensure proper quality of service during the call by handling voice packets differently compared to other Internet traffic. Also, calls can be handed over to a circuit-switched channel in a 2G network at the edge of the coverage area. Over the top voice services could in theory also take the advantage of quality of service measures to some extent. However, voice calls of such systems cannot be handed over to a 2G network because of the missing interfaces to the 2G infrastructure, which is a great disadvantage. Also, these voice services will not work at all while the user is in an area that is not covered by a B3G network.

Beyond those basic voice and messaging services, Internet companies are much more flexible in developing new services and in reaching a mass market audience quickly. No contracts between developers and network operators need to be in place to launch a service, and feedback and word of mouth from early adopters help to develop a service. Some of these services might make it into the service portfolio of network operators over time where they do not require an international customer base and a business model can be found that includes at least national interworking between IMS networks of different national operators.

4.5   Voice over DSL and Cable with Femtocells

Many mobile operators are interested in offering voice and data services to users not only while they are on the move but also when they are at home or in the office, without waiting for IMS or mobile devices supporting IMS and Wi-Fi in addition to B3G cellular networks. The easiest way to do this is to offer special pricing while users are in their home zone, which is defined by operators as the cells around the location of the user's home. The problem with this approach is scalability, as even when the user is at home the cellular network continues to be used. To address this disadvantage, several companies are developing femtocells, also referred to as femto base stations. From the mobile device's point of view, femto base stations look and behave like ordinary B3G UMTS/HSPA base stations. In practice, however, they have very limited transmission power and their size is similar to a DSL or cable modem. Instead of being connected via E-1, ATM, or fiber Ethernet, femtocells are connected to the cellular network infrastructure via the Internet and a DSL or cable connection. Femtocells are either integrated into DSL or cable modems or are connected via an Ethernet cable. Once connected to the Internet, a femtocell automatically establishes an encrypted tunnel to a gateway node of the cellular home network and connects to a specialized RNC. This specialized RNC terminates the encrypted tunnel and the femtocells are treated like any other cell of the network. Figure 4.31 shows how such a setup would look in practice. Since until recently there was no common femtocell backhaul standard, the network component that terminates the IP tunnels is usually from the same manufacturer as the femtocells or from a partner

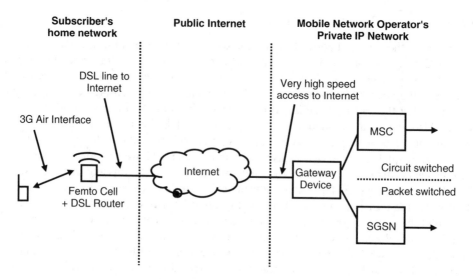

Figure 4.31 Femtocell network scenario.

company. As femtocells behave like standard B3G cells for a mobile device, circuit, and packet-switched voice calls as well as any other kinds of data applications offered by the mobile network operators can be tunneled over a DSL or cable Internet connection. In practice it is important to integrate femtocells with DSL or cable modems for several reasons. First, femtocells are installed by the user and such an approach therefore ensures that the installation is easy and is done properly. Additionally, an integrated device is the only way to ensure quality of service for the femtocell since data traffic generated by B3G voice calls must be prioritized on the fixed-line link over any other traffic. If a femtocell was attached to an existing DSL or cable router which already served other users, the uplink data traffic from these users could severely impact B3G voice calls since ordinary DSL or cable routers do not have quality of service features to ensure that traffic from the femtocell is prioritized. This behavior can already be observed in practice today in other situations. If an ordinary DSL or cable router is used for a VoIP call in addition to a simultaneous file upload, voice quality is usually severely degraded due to the packet delay and insufficient bandwidth availability.

As a consequence, a mobile operator deploying femtocells ideally owns DSL or cable access as well or is at least partnering with a company owning such assets. In this way a single fixed-line gateway could be deployed with Wi-Fi for PCs and other devices and a femto radio module for 3G mobile devices. The single telephone per user idea also benefits from such an approach since owning or partnering for DSL or cable access removes the competition between fixed and wireless voice. This also ensures that a femtocell is only used in locations where the mobile operator has licenses to operate its wireless network since femtocells use licensed frequency bands.

In practice it can be observed today that a number of mobile operators are taking this route already by either buying DSL access provider companies or are at least partnering with them (e.g., Vodafone/Arcor or O2/Telefonica in Germany). This is not done specifically to roll out 3G femtocells at some point, but it seems that such companies have understood that it is vital for the future of a telecommunication company to have both wireless and fixed assets, as the backhaul network can serve both fixed line DSL links as well as macrobase stations. This completely reverses an earlier trend of splitting up fixed and mobile access of a company into separate entities.

Another technical aspect concerning femtocells is interference. In B3G networks, all cells usually transmit on the same frequency and interference is managed by having enough space between them and by adjusting output power and antenna angles. Most 3G operators have at least two frequencies they can use so femtocells could, for example, use the least used second frequency. However, there is still an issue with interference between femtocells of users living in the same apartment building and who have therefore installed their equipment in close proximity. This will result in lower capacity of each cell and might impact quality of service.

At the publication of the second edition, the use of femtocells is not yet wide spread but there are some early adopters such as Vodafone UK, SFR in France, and AT&T in the USA. Vodafone UK and SFR in France, for example, sell femtocells to end customers to improve their reception at home. Access to these femtocells is reserved to those customers and those provisioned by the customers themselves. Devices of other customers are not getting service on such cells and are redirected to the macronetwork. Another way of deploying femtocells is in fully open mode in which they can be used by any customer of that network, including roaming subscribers. From an offloading point of view, such a deployment is preferable and an alternative to the use of indoor repeaters often used today in large indoor areas such as shopping centers and underground car garages. A potential reason why the closed model seems to be more popular for the moment is perhaps the anticipation of network operators that end users might be reluctant to share their private DSL bandwidth with other users. This might change in the future, especially in countries such as France, for example, where DSL network operators have deployed DSL/Wi-Fi gateways for years with an additional and separated Wi-Fi network ID; so the DSL line can be shared with other customers when they are not at home.

The following two sections take a look at femtocells from the operator's point of view and from the user's point of view to analyze in which scenarios femtocells could be successful in the future and how they could fit into an overall B3G network architecture.

4.5.1 Femtocells from the Network Operator's Point of View

In Europe and Asia 3G networks are operated on the 2100 MHz frequency, which is not ideal for in-house coverage. Even in cities it can be observed in practice that dual-mode 2G/3G mobiles frequently fall back to the 2G network. This is because many GSM operators use the 900 MHz band in Europe, which is much better suited for in-house coverage as lower frequencies penetrate walls much better. Femtocells improve in-house B3G coverage, which becomes more and more important with smartphones being often used indoors not only for voice telephony but also for IP-based services, which requires faster speeds than what can be delivered by GSM networks. Especially when users are not in their own home and thus cannot use their Wi-Fi network for fast Internet access for their mobile devices, good B3G in-house coverage is very beneficial.

An improvement could be seen in cases in which the mobile device cannot decide to stick with either the 2G or the 3G network due to changing 3G signal levels. This creates small availability outages while the mobile selects the other network type. During these times, incoming voice calls are either rejected or forwarded to voicemail.

It can also often be observed in practice that a mobile device with weak 3G in-house coverage changes to the 2G network once a IP packets begin to be exchanged (e.g., to retrieve e-mails or to browse the web) and sometimes changes back to the 3G network during the connection. The reason for these ping pong network selections is the changing reception levels due to the mobility of the user and changing environmental conditions. Such network changes result in outage times which the user notices since an e-mail takes longer to be delivered or because it takes a long time for a Web page to be loaded.

Another solution to the issues described above is the use of the 900 MHz frequency band for B3G networks in Europe and Asia and the 850 MHz band in the USA. Many regulators have by now opened the 900 MHz band for B3G technologies in Europe or are in the process of doing so in the near future. Also, the majority of smartphones sold in Europe and elsewhere are UMTS 900 MHz capable by now. Early examples of network operators making good use of this are Orange in France with UMTS 900 MHz deployments in the countryside [51] and O2 in the UK with UMTS 900 MHz coverage in London [52]. While using the lower frequency bands for macronetwork cells helps to improve indoor coverage, the drawback of this solution is that it does not add significant additional capacity.

As the above weaknesses of B3G technologies in higher frequency bands show, it is likely that femtocells can improve customer satisfaction. Putting a femtocell in the user's home would have the additional advantage for network operators of reducing churn, that is customers changing contracts and changing the network operator in the process. Customer retention is all the more reinforced if the femtocell comes in a bundle with DSL access, as further described below since changing wireless contracts also has consequences for the fixed-line Internet access at home.

Another advantage of femtocells is to reduce the gradual load increase on the B3G macronetworks as more people start using B3G terminals for voice and data applications. This could result in a cost benefit since, should the right balance of macro- and femtocells be reached, fewer expensive macrocells would be necessary to handle overall network traffic.

The question is how much these advantages are worth to a network operator since femtocells do not come for free. The options for network operators therefore range from selling femtocells to their customers or over subsidizing them, to giving them away for free as they benefit from a decreased churn or higher monthly usage and revenue.

4.5.2 Femtocells from the User's Point of View

While from the network operator's point of view femtocells have quite a number of advantages, it is far from certain if users will perceive femtocells at home as equally beneficial. While the user shares all operator advantages discussed in the previous section, increasing customer retention and thus churn is not necessarily in the interests of users since it could reduce their choice. Also, it is unlikely that all family members use the same mobile operator and thus could benefit from a single femtocell.

While at the publication of the first edition, most mobile multimedia users were still considered early adopters, this has significantly changed since then. Smartphones and the use of data services are widespread today, with well over half the number of new mobile phones sold in some countries now being in that category [53]. As all smartphones being equipped with Wi-Fi interfaces today, femtocells for multimedia content are not required at home for data services, since Wi-Fi offers at least a similar experience. Data services offered by mobile network operators, however, are usually not available over Wi-Fi. With most services provided by independent Internet companies, however, this is not a big disadvantage in practice.

In public places with insufficient B3G in-house coverage from the macronetwork, femtocells are quite advantageous from a user's point of view as smartphones do not have to fall back to a 2G network. In addition, the user benefits from the fact that femtocells, unlike repeater technology often used today, do not depend on the capacity of the outside macro-network and vice versa.

Monetary incentives could persuade users to install femtocells. Operators could, for example, offer cheaper rates for voice calls via femtocells. Also, the operator could propose to share revenue with femto "owners" if other subscribers use the cell for voice and data communication instead of a macrocell.

Often the argument is brought forward that femtocells allow the marketing of single telephone solutions in which the user no longer has a fixed-line telephone and uses his mobile telephone both at home and on the go. However, such solutions which use the macrolayer instead of femtocells have already been available for several years in countries such as Germany (O2's famous home zone, for example) and are already very popular. Also, it is unlikely that mobile network operators would have competitive prices for all types of calls so many users would still use a SIP telephone or software client on a PC for such calls at home. Calling a mobile number is still more expensive in most parts of the world (excluding the USA) than calling fixed-line telephones, so single telephone offers have to include a fixed-line number. Again, this is already done in practice, for example, by O2 in Germany for a number of years. Femtocells, however, might enable mobile network operators to deliver such services more cheaply than with a macronetwork approach.

4.5.3 Conclusion

When looking at the arguments presented above, femtocells are not likely to be an immediate and outright success. A number of further iterations will probably be needed before the form factor, usability and interoperability with the macronetwork are adequate. Furthermore, mobile operators need to continue their path of buying or partnering with companies owning fixed-line DSL or cable access. However, there is currently enough capacity available in the macrolayer of the network so femtocells are not immediately needed to reduce the load on the network. Therefore, the major immediate benefit of femtocells is improving in-house coverage in cities and rural regions. Femtocells are therefore likely to remain a niche market for now, since 2G, 3G, and home/office Wi-Fi coverage and capacity for urban users are usually sufficient even for in-house use.

4.6   Unlicensed Mobile Access and Generic Access Network

Unlicensed Mobile Access (UMA) is another approach to improving in-house coverage and offloading cellular traffic from the macronetwork to DSL or cable IP connections. The big difference to femtocells, however, is that UMA does not require any special network equipment on the user side. Instead, the mobile device uses its Wi-Fi interface to communicate with a standard Wi-Fi access point. This way, UMA simulates 2.5G GSM/GPRS connections through the Internet while femtocells tunnel B3G UMTS/HSPA traffic through the Internet.

4.6.1   Technical Background

UMA is a 3GPP standard and defined in [54] and is referred to in the standards as Generic Access Network (GAN). The principle of UMA/GAN is simple: it replaces the GSM radio technology on the lower protocol layers with Wireless LAN. A call is then tunneled via a Wi-Fi Access Point connected to a DSL/cable modem via the Internet to a gateway node. The gateway then connects to the mobile switching center for voice calls and SMS and to the Serving GPRS Support Node (SGSN) for packet data. The gateway between the Internet and the network of the mobile operator is called a Generic Access Network Controller (GANC), as shown in Figure 4.32.

In practice, a GAN capable mobile can attach to GSM networks, 3G UMTS/HSPA networks and Wi-Fi networks. To take advantage of Wi-Fi, it is usually configured to prefer using Wi-Fi networks over 2G or 3G networks. Where a GAN mobile detects a Wi-Fi network (e.g., the user's home network) over which it can connect to the 2G/3G core network, it will attach to the Access Point and establish an encrypted connection to the GANC. Moving between a cellular network and a Wi-Fi network is referred to as "roving" in the standards. Once the encrypted IP connection to the GANC is in place, a 2G/3G Location Update Message is sent over the encrypted IP link and the MSC and SGSN will update their subscriber information and the location in the Home Location Register. In practice, the GANC simulates a 2G Base Station Controller for these network elements. Thus, no changes are required on GSM core network elements for GAN.

From the network's point of view, signaling for incoming (or outgoing) calls to the subscriber is the same up to the GANC as for a traditional circuit-switched mobile terminated call. The GANC encapsulates the signaling messages into IP packets and forwards them to the mobile device via the encrypted IP connection. The mobile receives the messages over the IP link and presents them to higher protocol layers in the same way as if they had been sent via the GSM/UMTS network. In the other direction, the higher layers of the GSM protocol stack assemble GSM signaling messages as before. Once these messages reach the lower layers of the GSM protocol stack on the mobile device they are not sent via the GSM protocol stack but via the GAN protocol stack over Wi-Fi via the IP connection. Once the connection is established, the audio stream is also sent over the encrypted connection, encapsulated in RTP, UDP, and IP. Inside RTP, the audio stream is

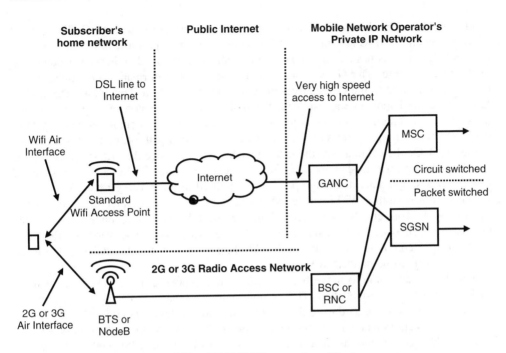

Figure 4.32 UMA/GAN network architecture.

encoded by using the standard GSM Adaptive Multi Rate codec. It is interesting to note at this point that the same protocols are used for audio transmission as in SIP and IMS.

Cell reselection is an action controlled by the mobile and thus does not require any network assistance. Handovers from one cell to another, however, are under the control of the network. In order not to change the software of the existing network infrastructure, a handover between the cellular 2G or 3G network and a Wi-Fi connection works as follows: each cell in a 2G or 3G network has a set of neighboring cells. These are identified with a Location Area Code (LAC), the Cell ID, and on the lower protocol layers with a Base Station Color Code (BCC) and a Network Color Code (NCC). Additionally, each cell operates on a distinct frequency, the ARFCN (Absolute Radio Frequency Number). To enable handovers to Wi-Fi, each cell of the GSM network has to have a neighbor cell in its neighbor list for this purpose. The parameters above are set to predefined dummy values. The same set of dummy values is used for all cells of the 2G/3G network, that is all cells have the same dummy entry in their neighboring cell list. If the mobile decides it wants to handover to a Wi-Fi cell, it first establishes a connection to the GANC over Wi-Fi and creates an encrypted tunnel. Once this is in place it will start sending measurement results to the cellular network that indicate that the dummy cell is received with a much stronger signal then the detected 2G or 3G cells. Based on these measurement reports the 2G or 3G radio network then decides to initiate a handover. Since the dummy cell is under the control of a different BSC (the GANC), the BSC that controls the current 2G cell or the RNC that controls the current 3G cell sends a request to the MSC to initiate an inter-BSC handover, or, in the case of 3G, an inter radio access technology (3G to 2G) handover. The MSC then sends a handover request message to the GANC. The GANC is not yet aware of to which subscriber the handover is to be performed, but nevertheless acknowledges the request. In the next step the current BSC sends a handover command to the GAN mobile. The GAN mobile in turn sends the information contained in the handover command to the GANC over the encrypted connection. This allows the GANC to correlate the handover request from the mobile with the handover request from the network. The handover sequence is finalized with the mobile switching the voice path to the Wi-Fi connection and the GANC connecting the speech path of the GAN connection to the MSC. On the cellular side the previous BSC will then free the timeslot used for the speech path and also finishes the handover procedure. As the GANC acts as a standard BSC, only a simple datafill change per cell is required in the existing network.

The handover from a Wi-Fi cell to the 2G network is a little simpler since the mobile can supply the correct cell parameters of the macrocell to the GANC once it detects that the Wi-Fi signal is getting weaker. The GANC then requests a handover from the MSC which in turn communicates with the BSC/RNC responsible for the 2G or 3G cell the mobile requests to be handed over to.

A mobile is also able to move between the cellular network and a Wi-Fi cell while a packet connection is established. This is done while no data is transferred between the mobile and the network as during those times the mobile device is allowed to select the most suitable cell on its own without the support of the network. Once the access network type has changed, the mobile then sends a location update to the MSC and a routing area update to the SGSN to inform the network of its new location.

4.6.2 Advantages, Disadvantages, and Pricing Strategies

While SIP and IMS are real end-to-end VoIP technologies, UMA could only be considered a semi-VoIP service at its inception, as a call was only transported over IP on the link between the mobile telephone and the UMA Network Controller. On the mobile device and after the gateway, a traditional circuit-switched connection and a Mobile Switching Center are used to connect the call to the destination. With the introduction of 3GPP Release 4 BICN MSCs and the replacement of traditional circuit-switched technologies in the core network with IP connections, GAN has become a fully IP-based voice solution as well as both the voice call signaling as well as the speech path is transported in IP packets end to end. The implications of this are discussed in more detail in Section 4.7.1.

Apart from reducing the traffic in the macronetwork, an additional benefit of UMA for network operators is the fact that a voice call always traverses the core network and an MSC of the mobile operator. This strengthens the relationship between network operator and subscriber and is an interesting way to replace fixed-line telephony by offering a DSL or cable connection and only using the circuit-switched mobile infrastructure. A fixed-line circuit switching center is no longer required. This approach is especially interesting for network operators that can offer DSL and cable access together with a cellular subscription.

As described earlier, UMA replaces one radio technology with another and otherwise leaves the rest of the system unaltered. This makes it difficult to price incoming calls differently for a caller while the called party is at home and using their (cheaper) Wi-Fi/DSL/cable connection compared with calls the called party receives while roaming in the cellular network. This is due to the fact that mobile operators in Europe use special national destination codes in order to be able to charge a caller a different tariff for calls to a mobile telephone user. Therefore, it might make sense for a network operator to also assign fixed-line numbers to their UMA subscribers for incoming calls while they are in their UMA home cell. When roaming in the macronetwork, such calls could then either be automatically rejected, forwarded to a voicemail system or forwarded to the subscriber. In the latter case the terminating subscriber could then be charged for forwarding the call.

It should be noted at this point that in the USA this problem does not exist since both fixed and mobile networks use the same national destination codes. There is no additional charge for the caller as the mobile telephone user gets charged for incoming calls.

As the mobile network is aware that the user is currently in their (cheaper) home Wi-Fi cell, incoming calls could then be charged at a different rate to the terminating subscriber.

Outgoing calls made via the Wi-Fi access point and a DSL or cable connection are also under the control of the mobile operator. It is unlikely that mobile operators will offer outgoing calls for free as is usually the case for connections between two VoIP subscribers, as the call will always be routed through a mobile switching center and a circuit-switched connection instead of being transported via IP end to end.

From an application point of view UMA is transparent to the user on the mobile as the GSM/UMTS telephony application is used for both cellular and Wi-Fi calls. The standard even offers seamless roaming between the two access technologies for ongoing calls, that is, a call is handed over from Wi-Fi to the cellular network and vice versa when a user leaves the coverage area of a Wi-Fi access point or detects the presence of a suitable access point.

UMA also tunnels GPRS services into the core network of the mobile operator. Data speeds are much higher though, which results in a seamless or even better experience for the user while in a UMA Wi-Fi cell, for example, for Web browsing on the telephone, operator portal access, or music downloads.

As UMA/GAN has been present on the market for a few years now, the success of UMA is mixed for now. A number of mobile network operators around the world are offering GAN to their subscribers such as, for example, T-Mobile in the USA and Orange in France and the UK. Others like British Telecom and Telecom Italia have since backed away again and have discontinued the service.

4.7   Network Operator Deployed Voice over IP Alternatives

The Voice over LTE (VoLTE) approach described above is the favorite solution of many mobile network operators for future voice services. The deployment of VoLTE however faces a number of challenges:

- For many years, the coverage of LTE networks is likely to remain inferior to already existing 2G and 3G networks. As a consequence, a tight integration with the MSC-based core network is required to hand over ongoing calls from LTE to a circuit-switched 2G or 3G bearer.
- Interworking between the MSC-based voice network and VoLTE must also be ensured; so a customer is reachable via the same telephone number independently of whether

he is currently being served by a non-IMS-based 2G or 3G network or an IMS-capable network.

- Billing and online charging systems must be enhanced to support IMS-based calling in addition to today's circuit-switched services.
- On mobile devices, the IMS VoLTE client needs to be seamlessly integrated into an overall telephony solution; so it is transparent to the user if a packet-switched LTE or a circuit-switched 2G or 3G bearer is used for a voice call.
- It is uncertain how power efficient early VoLTE implementations will be on mobile devices. The 2G and 3G voice architecture is tightly integrated into the hardware of a mobile device and are thus very power efficient.
- VoLTE implementations should be at least be as good as current circuit-switched technology. This includes voice quality, congestion handling, and handover performance between LTE cells and also 2G and 3G cells as well as call setup times.
- Specified functionalities to ensure proper quality of service in radio networks need to be tested and likely to be optimized.

As a consequence, a number of alternatives to VoLTE have been specified for the short-, medium-, and perhaps even for the long-term by a number of bodies in recent years. These are discussed in the following sections. Each approach addresses one or more of the issues mentioned above. Unfortunately, no approach has a solution for all the issues described.

4.7.1 CS Fallback

One approach to deploying LTE without packet-switched voice call functionality at the beginning is to instruct mobile devices to use 2G and 3G networks when the user makes or receives a voice call and return to LTE afterwards. This solution is referred to as CS fallback and has been specified in 3GPP TS 23.272 [55]. In principle, CS fallback connects the Mobile Switching Center to the LTE MME (cf. Chapter 2) via the new SGs interface as shown in Figure 4.33. The name and functionality of this interface is similar to the optional Gs interface between an MSC and a 2G or 3G SGSN (cf. Chapter 2) for paging and location updating synchronization.

When a mobile device is CS fallback capable, it initially performs a combined CSPS attach to the LTE network. This informs the network that the mobile device wants to use circuit-switched services in addition to IP-based services over the LTE network and is capable of falling back to a 2G or 3G network for incoming and outgoing voice calls. The MME then performs a location update on behalf of the mobile device over the SGs interface with the Mobile Switching Center and the HSS. This MSC is referred to as the SGs MSC in the following to distinguish it from other MSCs that might also become involved during the CS fallback procedure. If successful it signals to the mobile device that it has been registered in the circuit-switched network as well and that incoming calls will be signaled to it.

When an incoming call for the subscriber arrives at the Gateway-MSC, the HSS is interrogated for the location of the subscriber. The HSS then returns the information to the G-MSC that the subscriber is currently served by the SGs MSC. The call is then forwarded to the SGs MSC. The SGs MSC will then send a paging message over the SGs interface to the MME, which will in turn inform the mobile device and require it to

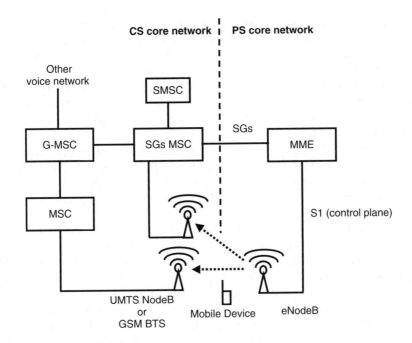

Figure 4.33   CS fallback network architecture.

leave the LTE network to accept the incoming voice call in a 2G or 3G cell. The mobile device will then do as instructed and start communication over a GSM or UMTS cell.

Moving from one radio technology to another can be done in several ways. The basic scenario is a redirect in which the network gives the mobile device an instruction to select a different radio network. The instruction can contain information about the target cells to reduce the time it takes the device to find a suitable cell and establish communication. In a more advanced scenario, a full Inter-radio access technology (IRAT) packed-switched domain handover from LTE to UMTS or GSM is performed which is prepared in the network and thus the interruption time is lower. In this scenario, the network can instruct the mobile device to perform radio measurements. The results of those measurements are then used by the network to select a suitable target cell and give the mobile device precise instructions of how to quickly get to this cell to minimize the handover time.

If the GSM or UMTS cell is in a location area that is different from that in which the mobile device is currently registered, a circuit-switched location update procedure is required before the call can be forwarded. This could be the case, for example, if the SGs MSC connected to the MME is not controlling the GSM or UMTS cell to which the UE is transferred. This could happen in case the mobile device has selected a cell other than the one intended by the network, for example, in areas with a location area border, or in case only a single MSC is SGs capable and hence acts as a mere relay for signaling messages rather than a switching center to which cells are directly connected.

In case of UMTS or GSM Dual Transfer Mode, the packet-switched context can also be moved; so any packet-switched communication can continue while the voice call is ongoing. This is important, for example; so email push and other applications can continue running in the background while the voice call is ongoing. Also, this allows the users to continue using other web-based applications during the phone call, for example, searching for some information on the web during the call, and so on.

In case the SGs MSC does not control the 2G or 3G cell, it is necessary that the SGs MSC redirects the voice path that has been established between the Gateway-MSC and itself to the MSC controlling the cell. This is done with the help of the location update procedure which is invoked as described earlier. Part of the location update procedure is to inform the HSS of the change in location. The HSS in turn informs the SGs MSC that the subscriber has changed the MSC area. The SGs MSC then informs the Gateway MSC of this change with a "roaming retry" message as specified in 3GPP TS 23.018 [56] The Gateway MSC then removes the speech path to the SGs MSC and establishes a new link to the new serving MSC.

If a location update was necessary, the mobile-terminated call is delivered immediately after the procedure has finished. To make the MSC aware that a circuit-switched call is waiting for the mobile device after the location update, it includes a CS fallback (Circuit Switched Mobile Terminated — CSMT) flag in the location update message. This flag allows the MSC to link the location update and the call delivered by the Gateway MSC. In case no location update is required, the mobile sends a paging response message to the MSC, which has the advantage that the call can be established more quickly.

When the user initiates a mobile-originated call, the mobile device contacts the network with an Extended Service Request message, which contains a CS fallback indicator. The network then decides based on its capabilities and that of the mobile device to either perform:

- A packet-switched handover to a GSM or UMTS cell, which is the fastest way to move the mobile device to a radio infrastructure from which the circuit-switched call can be initiated.
- An Radio Resource Control (RRC) release with redirect to GSM or UMTS, optionally with information about possible target cells to decrease the time necessary to find the cell. This is somewhat slower than a handover as the mobile device is required to reestablish contact to the UMTS network on its own without help of the LTE network.
- An IRAT cell change order to GSM. Optionally, the network can include information on potential GSM cells in the area (Network Assisted Cell Change, NACC).

Contacting the network prior to leaving the LTE network is necessary so the mobile device's context in the LTE base station (eNodeB) can be deleted and to get additional information on potential target cells to speed up the process.

The fallback mechanism to GSM or UMTS is also used for supplementary services based on Unstructured Supplementary Service Data (USSD) messages, which are used for modifying the parameters such as call forwarding settings or querying the amount of money left on a prepaid account.

Delivery of SMS messages, however, does not require a fallback to a GSM or UMTS network. This is because SMS messaging is not based on USSD and is not a service implemented in the MSC. Instead, the SGs MSC can forward an SMS message it has received from the Short Message Service Center (SMSC) to the MME via the SGs interface.

The MME will then deliver it to the mobile device via MME to mobile device signaling messages that are transparent to the eNodeB. Mobile-originated SMS messages can be delivered in the same way in the other direction.

As CS fallback is not a Voice over IP technology, it is likely that it will mostly be used in LTE networks before VOLTE becomes available. Furthermore, CS fallback can be used as a backup solution in roaming scenarios in which voice-capable LTE devices are roaming in a foreign LTE network in which VOLTE is not available or in case no roaming agreement is in place for IMS voice services.

The main advantage of CS fallback is that it will enable network operators and device manufacturers to introduce LTE devices with a single cellular radio chip before VOLTE becomes available. The downside of the approach is the increased call setup time required because of the change of the access network and the potential location update procedure that is required in case the new cell is in a different location area before the normal call establishment signaling can proceed. For LTE-to-LTE CS fallback voice calls, the extra call establishment delay is even doubled. The extra call establishment time is likely to be noticed by the customer who expects new technologies to work better than the previous ones and not worse.

4.7.2   Voice over LTE via GAN

Another interesting approach for full VoIP over IP is Voice over LTE via GAN, or VoLGA for short. It was developed by the VoLGA forum [57]. In principle, VoLGA is a modified form of the GAN approach described in Section 4.6. The major difference between GAN and VoLGA is that Wi-Fi as the bearer for the encrypted IP link for voice and data is replaced by the LTE network. From an MSC point of view, this is as transparent as using Wi-Fi and an IP link up to the GANC, which was renamed as VANC (VoLGA Access Network Controller), as the VANC acts as a GSM Base Station Controller or UMTS RNC.

The software of the GANC and a VANC are mostly identical, except for additional software required to hand over an ongoing VoIP call over LTE to a circuit-switched voice call. This is done with the SR-VCC function described earlier for the IMS. To hand over a call, the MME uses the SR-VCC functionality to inform the VANC of the required change of radio access technology. The VANC then signals a handover to the MSC, which then prepares a circuit-switched channel in the target cell. This is transparent for the MSC as the VANC emulates a normal GSM-to-GSM or UMTS handover.

On the mobile device's side, the software written for GAN which is widely implemented, for example, for the Android operating system [58] can be almost entirely reused. The main addition required is support for SR-VCC for the voice call handover to a circuit-switched channel.

On the network side, VoLGA is also easy to implement compared to a full IMS and interworking with the circuit-switched core network via ICS. All that is required is to add a VANC to bridge the gap between the circuit-switched core network and the packet-switched LTE network. For handovers, SR-VCC, originally specified for IMS is reused. With circuit-switched core networks now also transitioning to virtual circuit switching by replacing the Release 99 architecture with a bearer independent approach, which separates the MSC into an MSC server and media gateways, VoLGA is also an end-to-end IP technology. All interfaces of the MSC-Server and the media gateway are based on IP; so legacy interfaces based on E-1 circuit-switched technology are only required for interworking with external legacy networks. In addition, MSC-Servers and media gateways are based on modern off-the-shelf computing equipment in the same way as IMS components. This remarkable transformation from circuit switching to a fully IP-based system is often overlooked in discussions.

Despite its striking simplicity and promise of easy implementation on both mobile devices and networks, VoLGA has so far found little traction with network operators who seem to continue to focus their efforts on a combination of CS-fallback and the introduction of VOLTE at a later stage.

4.7.3 Dual-Radio Devices

Yet another approach to bridge the missing voice over IP solution in current LTE systems is to use mobile devices that are connected to LTE for data and a 2G or 3G radio network for voice services simultaneously. The first network operator to have done so was Verizon Wireless in the USA with the introduction of a dual-radio smartphone in 2011. To support voice and LTE packet data at the same time, the mobile device included two radio chips working independently from each other, except for power output synchronization between the two chips to ensure the device would not exceed the highest output power allowed when voice and data session were ongoing simultaneously [59]. This came at the expense of the device being rather bulky and having had a short overall battery life, caused by the two radio chips operating simultaneously and because of inefficiencies of first-generation LTE radio chips and non-optimized network settings. It showed, however, that devices with two dual-radio chips could be built and it is likely that next-generation radio LTE multimode baseband chips will be less power consuming and allow for smaller device form factors. In terms of size and battery life, early LTE mobile devices are thus similar to early UMTS phones, which were also bulky at the beginning but quickly began to shrink in size and power requirements.

The main advantage over the solutions described earlier is that call setup times equal those of current non-LTE smartphones. Also complex handovers between packet-switched VoIP and circuit-switched voice when running out of LTE coverage are unnecessary. From an evolutionary point of view, however, dual-radio devices are a dead end because the 2G or 3G infrastructure is used for voice as before, in a similar manner as is done with CS fallback.

4.8 Over-the-Top (OTT) Voice over IP Alternatives

While this chapter has so far only discussed future options for network-operator-supplied voice services, it should also be mentioned at this point that a future network that is based on the IP protocol only opens the door for voice services from outside the network operator domain. In fixed networks, this has already happened in recent years and programs such as Skype have become very popular not only for voice but also for video with good quality if a high-bandwidth channel is available. Skype is also available for mobile platforms such as Android, and other companies have innovated in this space as well, such as Apple, for example, with their Facetime video calling solution.

There are however two shortcomings of such solutions, also referred to as Over-The-Top (OTT) offerings as they are not offered by network operators. Firstly, it is not possible to ensure proper quality of service from the Internet in the portion of the network where it matters most, on the air interface. While network-operator-driven voice solutions are integrated into the mobile network and have standardized interfaces to ensure voice packets are preferred over other IP data of all users of a cell, IP packets of Internet-based services are transported on a best effort, first come first served basis. The same interfaces can also be used by network-operator-based voice solutions to ensure that IP packet-carrying voice data can be transported OTA interface without unnecessary repetitions in case of transmission failures and timeslot bundling in the uplink to improve chances of a proper transmission. In some countries, such inequalities might further fuel ongoing debates about whether such a preference should be allowed as in a sense this is against the principle of network neutrality [60].

The second shortcoming of Internet-based voice solutions is that they are unable to hand over an ongoing call to a circuit-switched connection at the border of the LTE or HSPA network. This will become less of a problem as those high-speed networks expand and densify in cities to offer equal coverage and in-house penetration. It is unlikely, however that this will happen in the near future in most cases and will vary from country to country.

4.9 Which Voice Technology will Reign in the Future?

After looking at the different options for IP-based voice services and voice offered in LTE-capable devices by other means (CS fallback, dual radio) and their pros and cons, no clear winner can be declared. While some network operators will start with CS fallback in the hope to bridge the time until VoLTE is ready, others will start with dual-radio devices to achieve the same goal. Internet-based voice services are likely to remain a niche product for some time to come but might have a chance in the midterm if mobile networks are built in a way to offer sufficient quality even in best effort mode, that is, good network coverage and no network overload during busy hour in hotspots. Instead of a single "one size fits all" mobile voice service as in the times of GSM in the 1990s, it is likely that the market will fracture and perhaps recombine around a solution that is able to reduce its disadvantages to become a viable solution for the majority in the future. Which solution this will be, however, is unlikely to emerge anytime in the near future.

References

1. Sauter, M. (2011) *From GSM to LTE*, John Wiley & Sons, Ltd, Chichester.
2. Rosenberg, J. and Schulzrinne, H. (2002) RFC 3261. *SIP: Session Initiation Protocol*, The Internet Society.
3. Berners-Lee, T. (2005) RFC 3986. *Uniform Resource Identifier (URI): Generic Syntax*, The Internet Society.
4. Handley, M., Jacobsen, V., and Perkins, C. (2006) RFC 4566. *SDP: Session Description Protocol*, The Internet Society, July 2006.
5. Schulzrinne, H., Casner, S., Frederic, R., and Jacobsen, V. (2003) RFC 3550. *RTP: A Transport Protocol for Real-Time Applications*, The Internet Society, July 2003.
6. Schulzrinne, H. and Casner, S. (2003) *RTP Profile for Audio and Video Conferences with Minimal Control*, The Internet Society.

7. Rosenberg, J., Weinberger, J., Huitema, C., and Mahy, R. (2003) RFC 3489. *Simple Traversal of User Datagram Protocol (UDP) through Network Address Translators (NATs)*, The Internet Society, July 2003.
8. Daidalos (2008) Designing Advanced Network Interfaces for the Delivery and Administration of Location Independent, Optimised Personal Services, An EU Framework Programme 6 Integrated Project, http://www.ist-daidalos.org (accessed 2012).
9. 3GPP (2008) Service requirements for the Internet Protocol (IP) Multimedia Core Network Subsystem; Stage 1. TS 22.228, June 11, 2008.
10. 3GPP IP Multimedia Subsystem (IMS); Stage 2. TS 23.228, Release 10, 16 December 2011.
11. 3GPP (2008) General Packet Radio Service (GPRS); Service Description; Stage 2, Release 7.5.0. TS 23.060, Chapter 9.2.3.4, October 4, 2008.
12. Kent, S. and Atkinson, R. (1998) RFC 2401. *Security Architecture for the Internet Protocol*, The Internet Society, September 1998.
13. 3GPP (2008) Access Security for IP-Based Services. TS 33.203, June 17, 2008.
14. Price, R., Borman, C., Christoffersson, J. *et al*. (2003) RFC 3320. *Signaling Compression (SigComp)*, The Internet Society, January 2003.
15. Calhoun, P., Loughney, J., and Guttman, E. (2003) RFC 3588. *Diameter Base Protocol*, September 2003.
16. 3GPP (2006) Conferencing Using the IP Multimedia (IM) Core Network (CN) Subsystem; Stage 3. TS 24.147, March 25, 2006.
17. 3GPP (2008) IP Multimedia (IM) Subsystem Cx and Dx Interfaces; Signalling Flows and Message Contents. TS 29.228, June 9, 2008.
18. 3GPP (2008) Subscriber Identity Module–Mobile Equipment (SIM-ME) Interface. TS 11.11, June 12, 2008.
19. 3GPP (2008) General Packet Radio Service (GPRS); Service Description; Stage 2, Release 6. TS 23.060, Chapter 9.2.2.1.1, June 9, 2008.
20. 3GPP (2008) Mobile Radio Interface Layer 3 Specification; Core Network Protocols; Stage 3, Release 6. TS 24.008, Chapter 10.5.6.12, June 6, 2008.
21. Faltstrom, P. and Mealling, M. (2004) RFC 3761. *The E.164 to Uniform Resource Identifiers (URI) Dynamic Delegation Discovery System (DDDS) Application (ENUM)*, April 2004.
22. Cuervo, F., Greene, N., Rayhan, A. *et al*. (2000) RFC 3015. *Megaco Protocol Version 1.0*, September 2000.
23. Open Mobile Alliance (2008) OMA-AD_PoC-V1_020060519-C, Push to Talk over Cellular (PoC) — Architecture.
24. 3GPP Enablers for OMA PoC Services Stage 2. TR 23.979, Release 7, 19 June 2007.
25. Open Mobile Alliance (2007) OMA Device Management, June 19, 2007, http://www.openmobile alliance.org/release_program/dm_v112.aspx (accessed 2012).
26. Open Mobile Alliance (2008) OMA Client Provisioning, http://www.openmobilealliance.org/release _program/cp_v1_1.aspx (accessed 2012).
27. 3GPP (2008) Voice Call Continuity (VCC) Between Circuit Switched (CS) and IP Multimedia Subsystem (IMS), Release 7. TS 23.204, June 9, 2008.
28. 3GPP (2011) Single Radio Voice Call Continuity (SRVCC); Stage 2, Release 9. TS 23.216, June 12, 2011.
29. 3GPP (2011) IP Multimedia Subsystem (IMS) Service Continuity; Stage 2, Release 9. TS 23.237, March 28, 2011.
30. Firesheep on Github (2012) http://codebutler.github.com/firesheep/ (accessed 2012).
31. Koch, A. (2012) Droidsheep, http://www.droidsheep.de/ (accessed 2012).
32. 3GPP (2008) 3GPP system to Wireless Local Area Network (WLAN) Interworking; System Description. TS 23.234, June 9, 2008.
33. 3GPP (2008) 3G Security; Wireless Local Area Network (WLAN) Interworking Security. TS 33.234, March 20, 2008.
34. Wikipedia (2012) Network Neutrality, http://en.wikipedia.org/wiki/Network_neutrality (accessed 2012).
35. ETSI TISPAN (2005) TISPAN NGN Functional Architecture Release 1. ETSI ES 282 001V1.1.1, August 2005.
36. ETSI TISPAN (2006) IP Multimedia Subsystem (IMS); Stage 2 Description. [3GPP TS 23.228 v7.2.0, modified], ETSI ES TS 182 006V1.1.1, March 2006.
37. ETSI TISPAN Protocols for Advanced Networking (TISPAN); PSTN/ISDN Emulation Sub-system (PES); Functional Architecture. ETSI ES 282 002V1.1.1, March 2006.

38. Kessler, P. (2007) *Ericsson IMS Client Platform*. Ericsson Review, 2, pp. 50–59.
39. ETSI TISPAN (2006) Protocols for Advanced Networking (TISPAN); IMS-based PSTN/ISDN Emulation Subsystem; Functional Architecture. ETSI TS 182 012V1.1.1, April 2006.
40. The JSR 281 Expert Group (2008) JSR 281 IMS Services API for JavaTM Micro Edition. Public Draft Version 0.9.
41. GSMA (2011) RCS-e–Advanced Communications: Services and Client Specification, Version 1.2.1, December 16, 2011.
42. Le Maistre, R. (2012) Spain Gives RCS Camp Some Hope, LightReading, January 13, 2012, http://www.lightreading.com/document.asp?doc_id216403 (accessed 2012).
43. Weidner, M. (2012) Telekom, Vodafone und o2 planen Konkurenz zu WhatsApp&Co. Teltarif, January 23, 2012, http://www.teltarif.de/telekom-vodafone-o2-sms-ersatz/news/45397.html (accessed 2012).
44. Campbell, B., Mahy, R., and Jennings, C. (2007) RFC 4975. *The Message Session Relay Protocol (MSRP)*, September 2007, http://tools.ietf.org/html/rfc4975 (accessed 2012).
45. 3GPP (2012) IP Multimedia Call Control Protocol Based on Session Initiation Protocol (SIP) and Session Description Protocol (SDP); Stage 3, Release 9. TS 24.229, January 26, 2012.
46. GSMA (2011) IR.92, IMS Profile for Voice and SMS, Version 4.0, March 22, 2011.
47. Donnegan, M. (2012) VZ Plans Nationwide VoLTE in 2013. Lightreading, January 19, 2012, http://www.lightreading.com/document.asp?doc_id216534 (accessed 2012).
48. 3GPP (2012) Communication Waiting (CW) Using IP Multimedia (IM) Core Network (CN) Subsystem; Protocol Specification, Release 10. TS 24.615 December 20, 2011.
49. 3GPP (2012) Extensible Markup Language (XML) Configuration Access Protocol (XCAP) over the Ut interface for Manipulating Supplementary Services, Release 10. TS 24.623 March 16, 2012.
50. Schulzrinne, H. (2003) RFC 3550. *RTP: A Transport Protocol for Real-Time Applications*, July 2003, http://www.ietf.org/rfc/rfc3550.txt (accessed 2012).
51. Sauter, M. (2010) UMTS 900 Coverage in France, March 14, 2010, http://mobilesociety .typepad.com/mobile_life/2010/03/umts-900-coverage-in-france.html (accessed 2012).
52. Sauter, M. (2011) UMTS 900 in London–A Tough Decision, March 23, 2011, http://mobilesociety .typepad.com/mobile_life/2011/03/umts-900-in-london.html (accessed 2012).
53. Arthur, C. (2011) Half of UK Population Owns a Smartphone. The Guardian, October 31, 2011, http://www.guardian.co.uk/technology/2011/oct/31/half-uk-population-owns-smartphone (accessed 2012).
54. 3GPP (2008) Generic Access to the A/Gb Interface; Stage 2, Release 7. TS 43.318, June 16, 2008.
55. 3GPP (2011) Circuit Switched (CS) Fallback in Evolved Packet System (EPS); Stage 2, Release 9. TS 23.272, December 16, 2011.
56. 3GPP (2011) Basic Call Handling; Technical Realization, Release 9. TS 23.018, December 19, 2011.
57. Wikipedia (2012) VoLGA Forum, February 10, 2012, http://en.wikipedia.org/wiki/VoLGA_Forum (accessed 2012).
58. Sauter, M. (2011) Android and Open Source as A Door Opener for Deep Down Innovation, March 30, 2011, http://mobilesociety.typepad.com/mobile_life/2011/03/android-and-open-source-as-a-door-opener-for-deep-down-innovation.html (accessed 2012).
59. Klug, B. (2011) HTC Thunderbolt Review: The First Verizon 4G LTE Smartphone, April 27, 2011, http://www.anandtech.com/show/4240/htc-thunderbolt-review-first-verizon-4g-lte-smartphone/3 (accessed 2012).
60. Wikipedia (2012) Network Neutrality, February 10, 2012, http://en.wikipedia.org/wiki/Network_neutrality (accessed 2012).

5

Evolution of Mobile Devices and Operating Systems

5.1 Introduction

Mobile devices with wireless network interfaces have gone through a tremendous evolution in recent years. From around 1992–2002, the main development goal was to make these devices smaller. While during that time the form factor of phones shrank considerably, voice telephony and SMS texting remained the main applications and overall functionality changed very little. By around 2002, technology had developed to a point where it became impractical to shrink phones any further from a usability point of view. The Panasonic GD55 is one of the smallest mobile phones ever produced, with a weight of just 65 g, and is smaller than a credit card [1]. To demonstrate the evolution that had taken place in only 10 years, Figure 5.1 shows one of the first GSM phones, the Siemens P1 of 1992.

Once devices could not shrink, any further development has concentrated on adding additional multimedia functionality to mobile devices. At first, black and white displays were replaced by color displays, and display resolutions quickly rose from 100×64 pixels over 640×360 pixels to very high resolutions such as 960×640 pixels on 3.5–4 in. screens. Pixels thus have become so small that individual pixels cannot be seen anymore at a normal viewing distance. High-resolution color displays are a prerequisite for all other functionalities that have been added to mobile phones since. These functionalities include cameras, multimedia mobile e-mail and web browsing, video streaming, and social network interaction, just to name a few.

High-resolution color displays, high processing power with low power consumption and an increase in available memory and storage space have given rise to a number of wireless mobile device categories, whose purpose and range of functionalities has extended and shifted over the years. Today, the most important ones are:

- **Smartphones** — a smartphone can be defined as a combination of a mobile phone, what was formerly referred to as a (non-connected) Personal Digital Assistant (PDA), and an extension of previously desktop-based social media web services to the mobile world. Smartphones now usually include a high-resolution camera for taking pictures and videos, GPS and compass functionality as well as motion sensors, and various

3G, 4G and Beyond–Bringing Networks, Devices and the Web Together, Second Edition. Martin Sauter.
© 2013 John Wiley & Sons, Ltd. Published 2013 by John Wiley & Sons, Ltd.

network interfaces such as high-speed cellular interfaces, Bluetooth, and Wi-Fi. These devices are usually shaped like mobile phones, but are slightly bigger to accommodate a larger screen and additional hardware.

- **Pads/tablets** — over many years, the computing industry has been trying to scale down desktop PCs and make them mobile and portable by adding a touch-sensitive display and adapting the operating system to make its operation touch friendly. This device category did not become successful; however, until screen resolution, power consumption, and battery capacity enhancements were combined with the idea of adapting the user interface (UI) of touch-based smartphones for this type of device rather than using a desktop UI. With sufficient processing power for full-screen web browsers with similar functionality as is found in desktop and notebook PCs and web services adapted for touch-based input, tablets now fulfill many functionalities that were formerly only used with either a smartphone while being underway or with a PC when being at a desk. With cellular and Wi-Fi connectivity, tablets have become an ideal tool for multimedia consumption and for staying connected with friends via email, instant messaging, and web-based social networks without the need to sit at a desk.

- **Netbooks** — formerly also referred to as ultramobile PCs, the idea behind this product category is to reduce the size of a typical notebook while keeping its main characteristics such as the use of a desktop operating system, near full size keyboard, and only a slightly reduced display resolution. Netbooks have a typical screen size of 10–11 in. and a power-efficient CPU, and the mainboard design enables long battery operation times at the expense of processing power. The first models were not very successful as storage and processor capacities were too small for the requirements of Microsoft's Windows operating system. Asus was the first company that developed devices for this category [2]. Instead of using Windows, Asus initially used a Linux-based operating system, which is less resource hungry. Later versions then became powerful enough to host Microsoft's Windows operating system. Even though those devices have been available for several years, the main means for connectivity is still the Wi-Fi interface. Built-in cellular connectivity can only be found in a few models. This is mostly because of the additional price of the cellular network card, which would significantly increase the typical sales price of €250–300 or less. It has thus become quite common to use netbooks and notebooks with a cellular modem via the Universal Serial Bus (USB) port, often referred to as a "3G dongle," or via Wi-Fi tethering to a mobile phone that acts as a Wi-Fi to cellular network bridge to the Internet. Today, netbooks compete with devices from the tablet category, and interest in them has diminished to some extent. While tablets are ideal for information and media consumption away from the desk, the strength of netbooks is their full keyboard integration and desktop operating system, which makes them preferable for many creative tasks that require text input. A significant part of this book, for example, was written on a netbook.

- **Ultrabooks** — devices in this category are usually slightly larger than netbooks with a typical screen size of 13 in. but are significantly thinner and still very light in weight without compromising on battery operation time and processing capacity. This comes at the expense of a significantly higher sales price than netbooks, usually in the order of €800–1200. As with netbooks, USB modems or Wi-Fi tethering is used to connect to the Internet when not at home or at the office.

- **Wireless computing equipment** — a well-established trend for home and office networks is to untether computer equipment such as printers and hard drives (or Network Attached Storage (NAS)) using Wi-Fi. With Wi-Fi chips having become a commodity, the additional price for consumers has dropped significantly and such devices are becoming more and more popular. This device category is different from those previously mentioned because the aim of equipping them with wireless interfaces is not mobility but to reduce the amount of cables in home and office environments. A further advantage of having wireless access to such devices is that it also makes them usable from the mobile devices mentioned above.

Most of the today's connected devices except netbooks and ultrabooks are based on a chip with a processor design from ARM [3]. Although many companies such as Texas Instruments, Marvell, ST-Ericsson, and Qualcomm design and manufacture chips for small devices, most are based on a CPU core licensed from ARM. On the desktop, Intel's x86 design dominates in a similar way. With both architectures now targeting sophisticated mobile devices, these two worlds are about to collide.

5.1.1 The ARM Architecture

The ARM design was initially targeted at ultralow-power embedded devices. As technology evolved so did ARM's processor design and it is estimated that an ARM processor

Figure 5.1 The Siemens P1, one of the first GSM telephones in 1992.

core is used in 95% of mid- to high-end mobile devices today [4]. The current ARM-Cortex A9 and A15 platforms used in high-end smartphones and tablets is the result of a bottom-up approach, as it has evolved from earlier platforms for simpler devices. Today, all mobile devices of mobile giants such as Sony, Nokia, LG, Samsung, and Apple are ARM powered. This shows the flexibility of the ARM architecture since requirements range from voice telephony with very low power requirements to multimedia devices that trade in a higher power consumption for higher processing capabilities.

Today, a lot of operating systems support the ARM architecture. Examples are fully embedded operating systems of low-end to mid-range mobile devices to operating systems for smartphones such as Symbian and Windows Phone. In addition, ARM processors are also used with operating systems that were initially developed for desktop computers such as Linux and Windows 8. Linux is a relatively new operating system for mobile devices as the first mass market device based on Linux was only shipped in 2008 as part of Google's Android operating system [5]. The advantage of using Linux as an operating system for mobile devices is that a significant amount of code can be shared between the desktop and the mobile version of the operating system, as relatively little code directly deals with the differences of the x86 architecture found in the PC world and the ARM architecture found on mobile devices.

It should also be noted at this point that unlike companies such as Intel or AMD, ARM does not produce processor chips themselves. Instead, ARM licenses its processor designs to other companies such as Qualcomm, Samsung, Mediatek, Marvell, Texas Instruments, and many others, which then include the processor designs in their own chip designs. Companies such as Intel, however, go one step further and also design the chips that include their processor designs.

ARM offers several types of licenses. The most basic license only offers the logic of a function block such as the CPU, a bus system, or the graphics chip for use with a chip design program such as Verilog. This is referred to as a soft-macro [6]. The licensees then use those function blocks with self-designed additional circuitry or function blocks bought from other companies and create a physical chip implementation, which is optimized for a certain production process, performance, power consumption, and die size. ARM also offers licenses that already include those steps. Such function blocks are then referred to as hard-macros. And finally, ARM also licenses the ARM architecture itself and allows companies to modify, design, and optimize their own CPU cores and other function blocks. ARM-compatible software still runs without modification on such modified processors but licensees have the opportunity to make enhancements to the architecture independently of ARM. Marvell and Qualcomm are companies that design their own ARM processors instead of buying a finished design.

5.1.2   The x86 Architecture for Mobile Devices

Intel is at the other end of the spectrum and is keen to play a major role in the mobile space with its x86 processor architecture. A few years ago Intel tried to get a foothold in the mobile space by licensing ARM technology and building a product line around that architecture. In the meantime, however, Intel has abandoned this approach and has been refining their x86 architecture for low power consumption and size for several years. In 2012, the size, processing speed, and power consumption of the chipset was for the first time balanced enough for a smartphone-sized device. First prototypes were shown running an x86 version of Android on a form factor smartphone [7], and commercial products based on this design appeared shortly afterward on the market. This rather late competition to ARM's dominance in the mobile space is the result of Intel's approach that is directly the opposite of ARM's as they had to streamline a powerful desktop processor architecture for smaller devices.

Using an x86 platform for mobile devices has the advantage that even fewer adaptations are required for operating systems such as Linux and Windows to use them on mobile devices compared to the ARM approach described above. In the case of Android, most applications are executed in a virtual machine based in Java and only compiled to native code at runtime, so the same executable runs without modification or the need for recompilation on both CPU architectures.

At the time of publication, Intel and ARM have come quite close in terms of performance and power consumption and the two architectures are now competing for use in high-end mobile devices.

5.1.3 Changing Worlds: Android on x86, Windows on ARM

The significant advances in the mobile space in recent years and Intel's inability to establish themselves with their x86 architecture in the mobile domain over many years have also had consequences for traditional alliances formed in the PC space.

Over many years, Microsoft has only developed its Windows desktop operating system for x86-based architectures. With tablet devices having become attractive to end users starting from around 2011 and no company in sight to deliver power-efficient x86-based processors, Microsoft had to choose between extending their relatively novel Windows Phone operating system based on ARM to tablets or to scale down their Windows desktop operating system and adapt it to the ARM architecture. Microsoft chose to do the latter and has developed a new version of the Windows desktop operating system that can also be run on ARM-based devices such as tablets. Such a move would have been impossible only a few years earlier and demonstrates the significant increase in processing power that was achieved on the formerly low-cost low-processing power ARM architecture.

A new UI was developed based on the Windows Phone smartphone UI to complement the existing desktop UI in an attempt to integrate the mobile and desktop computing worlds in a single Windows operating system. This demonstrates how the rise in computing power in mobile devices also has an effect on desktop computing and how the industry is integrating the formerly disparate worlds of low-power mobile devices and high-power desktop computing into a single space.

New alliances are also formed on the x86 side with Android having been ported to the x86 architecture, as described in Section 5.1.2. This in effect enables Intel and other companies producing x86-based processors to move into the high-growth smartphone market, which further blurs the line between mobile and desktop computing.

A further positive effect of the competition between the ARM and the x86 architecture is likely to be further accelerated innovation and falling prices as mobile device manufacturers can now choose between two camps with each one trying to stay ahead of the other with further innovations in the areas of power consumption, processing speed, graphical capabilities, and integration of other components into a single chip.

5.1.4 From Hardware to Software

The following sections now take a look at how mobile device hardware has evolved over recent years and give an introduction to both hardware architectures mentioned above. Different parts of the world use different frequency ranges for wireless communication. This chapter therefore takes a look at the global situation and describes the impact on mobile hardware design and global usability of devices. Adding a Wi-Fi interface to mobile devices has been another important step in the evolution of wireless communication and this chapter will discuss the profound impacts of this step on networks and applications. Finally, this chapter takes a look at the Android operating system for mobile devices.

5.2 The System Architecture for Voice-Optimized Devices

In the entry level segment, mobile phones are sold today both in developed markets and emerging economies that are optimized for voice communication. While the functionality of such phones has not changed much in the past decade, prices have been on a steady decline due to much higher production volumes and reducing the number of required chips and electronic components. This is referred to in the industry as reducing the Bill of Materials (BOMs). Figure 5.2 shows a block diagram of a typical voice-optimized mobile phone computing platform which is offered by many companies. The example in this book is based on Freescale Semiconductor's GSM i.200-22 hardware platform [8], which is optimized for voice communication and even excludes functionalities such as basic General Packet Radio Service (GPRS).

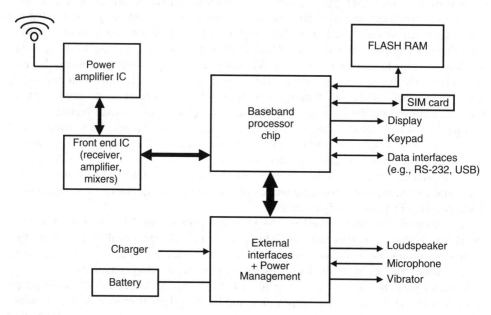

Figure 5.2 Block diagram of a voice-optimized mobile phone hardware platform. (Reproduced from *Communication Systems for the Mobile Information Society*, Martin Sauter, 2006, John Wiley and Sons, Ltd. Ref. [9].)

The core of this chipset is the baseband processor chip. It contains a 32-bit ARM7TDMI-S RISC (Reduced Instruction Set Computer) microprocessor but can be used with a 16- and 32-bit instruction set. While operations that can be performed with the 16-bit instruction set are not as versatile, only half the memory space is required for code compared with 32-bit instructions. Especially in memory-limited devices such as basic mobile phones, this is a big advantage. It is also possible to mix 16- and 32-bit instructions, which enables the software developers to compile their code into 16-bit instructions and profile-specific portions of the software by hand to use 32-bit instructions where more performance is required. The maximum clock speed of the ARM processor used in this chipset is 52 MHz. This is very low compared with processor speeds of 2 GHz and beyond used in desktop systems today, but sufficient for this application. For more sophisticated devices more processing power is required. As will be discussed below, ARM thus offers several processor families and multimedia devices use ARM processor types that offer far better performance at the expense of higher production costs and power consumption. According to [10], power consumption at 52 MHz is between 1.5 and 3 mW. This is at least three orders of magnitude less than the power requirements for notebook processors.

In addition, the baseband chip contains a Digital Signal Processor (DSP) of Motorola's 56x family, which is clocked at 130 MHz. DSP microprocessors are optimized for mathematical operations and run software which is usually designed for specific tasks. Figure 5.3 shows how the RISC CPU and the DSP are used in combination in a mobile phone. The DSP chip is responsible for decoding the received signal from the network and for encoding and decoding the voice signal. There are two main advantages of performing these tasks on the DSP and not on the main processor:

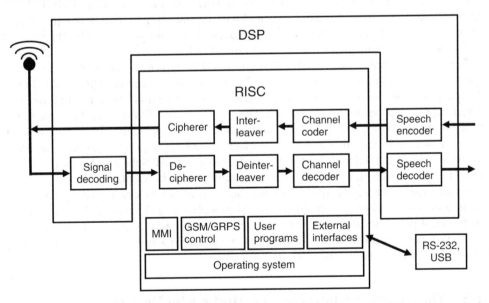

Figure 5.3 Work split for voice telephony in a mobile phone. (Reproduced from *Communication Systems for the Mobile Information Society*, Martin Sauter, 2006, John Wiley and Sons, Ltd. Ref. [9].)

- A DSP has an optimized instruction set for mathematical operations required for dealing with codecs and decoding analog radio signals that have been digitized by an analog-to-digital converter.
- Encoding and decoding external signals is a continuous process and must not be interrupted by other activities such as reacting to user input or updating the display.

A typical voice call is treated by the baseband chip as follows:

- The analog input signal from the microphone is digitized and sent to the DSP chip.
- The DSP applies speech coding and forwards the data to the ARM RISC CPU.
- The ARM processor then packetizes the data stream, adds redundancy to the data (channel coding), changes the order of the bits so block errors can be more easily corrected on the other end (interleaving), encrypts the result and then sends the packet over the air interface.

In the reverse direction, the same actions are performed in the reverse order. In addition, the DSP performs signal decoding. This is a complicated task since the signal sent by the base station is usually distorted by interference. To counter these effects, packets contain training bits (in the case of GSM) that are set to predefined values [9]. These are used by the DSP to build a mathematical model of how the signal was distorted. The mathematical model is then applied to the user data around the training bits to decrease the transmission error rate.

In addition to the tasks above, the ARM CPU is responsible for interaction with the user (keyboard, display), to execute user programs such as Java applications, and to communicate with external devices (e.g., a computer) via interfaces such as USB. As all of these tasks have to run in parallel; a multitasking operating system is required that is able to give precedence to repetitive actions concerning communication with the network and assign the remaining time to less time critical tasks.

For executing programs, about 250 kb of RAM is typically available on the baseband processor chip. In addition, about 1.7 Mb of nonvolatile memory (ROM, Read Only Memory) is available. If more memory is required, the chipset offers an external memory interface that can be used to connect additional RAM and ROM (e.g., flash memory). A 225-pin multiarray ball grid array connects the baseband chip via a 13×13 mm connection field to the other components of the device (cf. Figure 5.2). Other important components of the baseband chip are the module to access the Subscriber Identity Module (SIM) card and a display module for a monochrome or color display.

In addition to the digital processing functionality of the baseband chip, other analog components such as power amplifiers, signal modulators, and functionalities to convert and control power for the device are required. These are implemented in separate chips as analog functionalities require a different manufacturing technology from the purely digital functions of the baseband chip.

5.3   The System Architecture for Multimedia Devices

The design intent for a voice centric mobile device chip set is to strip down the functionality to the bare minimum to reduce the price as much as possible. For high-end wireless mobile multimedia devices, however, the aim is to include as many functions as possible in the chipset. At the same time the device must consume as little power as possible in idle mode in order to achieve acceptable standby times. The chipset has to find a balance between power efficiency and performance while the user interacts with the device.

There are three major building blocks of a high-end mobile chipset today. The first is the application processor unit, which usually consists of one or more CPU cores usually based on a 32-bit ARM architecture. Especially with the ARM Cortex CPU architecture introduced in 2005, ARM has increased performance by introducing a superscalar design that increases the number of execution units a machine instruction passes during its execution. This way, several machine instructions can be processed simultaneously as each can be in a different stage of execution. According to ARM, this increases the performance by a factor of 2–3 compared with the previous ARM processor generation at the same clock frequency. Performance gains for audio and video decoding are achieved with an extension referred to as NEON that allows application of the same operation with a single instruction to several variables simultaneously. This is used, for example, by Android's WebM library to decode this type of video format [11]. Another feature now prevalent in mobile CPUs is a floating point unit to perform non-integer calculations in hardware.

After many years of refinement, Intel presented a design to enter the mobile chipset domain with its x86 architecture as well. Its platform, which is referred to as "Medfield," seems for the first time be able to compete with the ARM design. A comparison between "Medfield" and current high-end ARM designs such as the ARM-Cortex-A9 and A15 can be found in [12]. In addition, Intel has ported the popular Android operating system to its x86 CPUs as well, which significantly helps to make future x86 platforms popular in the mobile domain if power consumption and processing speed develop along similar lines as those of the ARM architecture.

The second major building block of a chipset is the graphics processing unit (GPU). This processing unit is specifically designed to efficiently handle 2D and 3D graphical operations and effects. The calculations required for rendering of web pages and graphical effects such as scrolling a web page and zooming into specific parts, blending screens when changing from one application to another, and rotating the screen when the user changes the orientation of the device are mostly performed in the GPU. Unlike CPUs, which are optimized for sequential program streams, the GPU is optimized to perform many similar operations that are not dependent on each other in parallel. The general functioning of a mobile GPU is the same as that of a GPU in the PC world but its power and processing capabilities are scaled down to adapt to the limited power availability on mobile platforms as well as the limitations imposed by passive (i.e., fan-less) cooling. Today, there are several companies whose GPUs are commonly used in practice. ARM has designed its own GPU, which it has named "Mali." Nvidia, initially a graphics card manufacturer in the PC domain, has also developed a mobile GPU family, which it uses as part of its "Tegra" line of integrated chips for mobile devices. Imagination Technologies "PowerVR" is a division specifically focusing on mobile device GPUs, and its designs are also commonly found in mobile chipsets, with both ARM and Intel CPUs. And finally, Qualcomm also has its own GPU design referred to as "Adreno," which is integrated into their line of "Snapdragon" chipsets. While ARM CPUs from different manufacturers all share the same instruction set, this is not the case for GPUs. To make application programs compatible with different GPUs, a standardized Application Programming Interface (API) is required. This is discussed in more detail in Section 5.4.

The third major building block in a mobile device is the cellular modem, sometimes also referred to as the "baseband" processor. It includes all digital components required to communicate with a cellular network. Analog parts such as power amplifiers, filters, and up- and downlink signal multiplexers are separate components on the motherboard as a different manufacturing process is required for such components. The baseband processor usually consists of an ARM-based processor and additional signal processing components such as a DSP, as described above for voice-optimized devices. While such low-end devices usually only include a GSM modem, baseband processors have significantly grown in complexity and processing power requirements as they now also include software and hardware to process much more complex radio signals than those of GSM, such as, for example, High-speed Packet Access (HSPA) and Long Term Evolution (LTE) (cf. Chapter 2). The baseband processor has its own operating system and communicates via a high-speed (HS) serial connection with the application processor unit on which operating systems such as Android, iOS, Symbian, Windows Phone, and so on are executed. The serial connection over which user data and modem commands are exchanged makes the baseband processor completely independent from the application processor block. When operating systems such as Android are used on the application processor, a modem driver software module simulates several serial connections, typically one for user data and one for modem control commands and feedback messages. Modem commands are, for example, the establishment of an Internet Protocol (IP) connection, and feedback messages contain information about signal strength and other frequently required network parameters. From a mobile operating system point of view, the cellular modem is thus used in a very similar way as a cellular USB dongle connected to a PC and allows the mobile operating system to be easily adapted to different baseband processor implementations as the cellular modem driver module is the only piece of software affected when using different modems. Baseband processors are developed by several companies such as Qualcomm, ST-Ericsson, Marvell, Renesas, and Nvidia.

In the past, baseband processor, application processor, and graphics processor could often be found in dedicated chips. An example is the Nokia N8 that was released in 2010. It was built with a dedicated baseband processor chip of Texas Instruments, a dedicated application processor chip by Samsung based on an ARM11 core, a predecessor of the current ARM-Cortex platform, and a Broadcom graphics 3D processor chip. There is a strong trend, however, to combine all three components on a single System on a Chip (SoC) to save cost, reduce power consumption, and shrink the overall size of the circuit board. Table 5.1 shows examples of fully integrated systems on a chip and their manufacturers.

It is interesting to note that none of the companies listed in the table develop complete mobile devices themselves and also do not develop other components required in a device such as touch screens, displays, batteries, casings, the scratch-resistant glass, camera modules, and so on. In other words, a mobile device contains components from many different manufacturers and the company owning a device and whose logo appears on it is mainly acting as an integrator of the different hardware and software components. When operating systems such as Android and Windows Phone are used, the companies are not even the developer of the operating system software, as that is again done by different companies such as Google and Microsoft.

Table 5.1 Examples of all-in-one system on chip manufacturers

Manufacturer	Platform name	CPU	GPU	Baseband modem
Qualcomm [13]	Snapdragon	ARM, Qualcomm design, "Scorpion," "Krait"	Qualcomm "Adreno"	Qualcomm
ST-Ericsson [14, 15]	NovaThor	ARM Cortex-A9	PowerVR and ARM "Mali"	ST-Ericsson
Renesas [16]	Renesas	ARM Cortex-A9	PowerVR	Renesas
Nvidia [17]	Tegra	ARM Cortex A9	ULP GeForce	Nvidia (Icera)
Mediatek [18]	Mediatek	ARM-Cortex-A9	PowerVR	Mediatek
Intel [12]	Atom	x86	PowerVR	Intel (Infineon)

While CPU, GPU, and baseband modem are the most important components of a mobile chipset, they are by no means the only ones. A modern SoC has a variety of other special hardware units, which are either integrated into the SoC itself or are placed on the circuit board in chips of their own. An example of a possible configuration is shown in Figure 5.4. The example SoC contains two application processor CPUs, typically driven at clock rates of around 1.5 GHz today, the GPU, and the baseband processor on the main chip as discussed earlier.

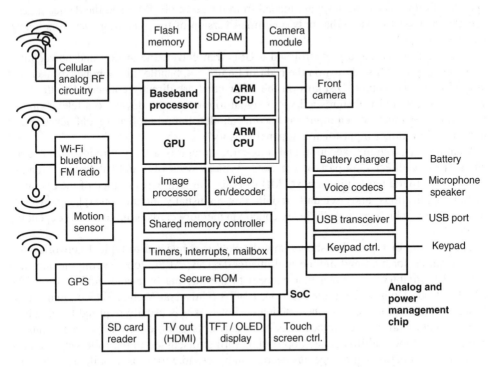

Figure 5.4 Block diagram of a multimedia chipset for a mobile device.

In addition, the SoC in this example contains an image processor unit that is used for processing the input stream delivered from external camera modules. Camera sensors with resolutions of 12–16 MP are supported by high-end chipsets today but not necessarily used to their full capabilities. The image processor converts pictures to compressed jpeg formats on the fly. This significantly reduces the load on the CPU since this computationally intensive task is performed entirely in this dedicated unit.

The video encoder/decoder unit supports the CPU by encoding video streams delivered from the camera to MPEG4 formats. The unit is also used to decode and display videos that have been recorded earlier or downloaded to the device and stored on either internal flash memory or a memory card. Most high-end wireless devices have a main camera on the back of the device to take high-resolution pictures and videos and a small camera on the front for video calls. Consequently the baseband chip has two camera interfaces.

All processing units require access to the main memory to store data and to communicate with each other. A shared memory controller synchronizes the memory requests of the different units and delivers the data over a common bus. The chipset also contains dedicated hardware for mailboxes, which are used for communicating between different tasks, and interrupts, which are used to inform special program handlers of the operating system of external events (e.g., the user has pressed a button).

Finally, the SoC also contains a secure ROM to store confidential information and software, which cannot be accessed from outside the chip. This can be used, for example, to protect the unique equipment identity of the device (IMEI, International Mobile Equipment Identity) and to enforce a SIM lock limitation to bind the device to a specific network operator. Furthermore, the software loaded from the secure ROM when the device is reset can also be used to ensure that only a certified version of the operating system is loaded into memory [19].

In addition to the battery control and a USB transceiver, the analog and power management support chips, shown on the right in Figure 5.4, handle microphone, speaker and keypad input and output. Displays on the other hand are directly connected to the SoC, which in addition also features a TV out signal to connect the device to a television set. In recent years touch screen input has become another important feature and hence, SoCs now provide a direct input for touch panel sensors.

Memory for program execution and long-term storage is usually provided in separate chips and not directly as part of the SoC. Typical sizes of Random Access Memory (RAM) for program execution are 512 MB to 1 GB. For long-term storage of the operating system, applications, and user data, mobile devices usually include several gigabytes of flash memory, which can usually be extended via memory cards of up to 64 GB depending on the device.

The left side of Figure 5.4 shows how the SoC is connected to typical network interfaces required for high-end mobile multimedia devices. At the top, several antennas and analog RF circuitry are shown that connect to the internal baseband modem for GSM, CDMA (Code Division Multiple Access), UMTS (Universal Mobile Telecommunication System), and LTE. More than one antenna is shown as it is usually required to be able to communicate with different types of cellular networks using different frequency ranges. The impact of the multifrequency approach is further discussed in the following sections.

A Bluetooth network interface has become indispensable for most mobile devices today and Wi-Fi interfaces are also becoming more popular. Often, both functionalities are included on the same chip and require only a single antenna as both technologies use the 2.4 GHz ISM (Industrial, Scientific, and Medical) frequency band [20]. In addition, such chips also often contain an FM radio receiver for which a dedicated antenna is required as FM radio is broadcasted between 87 and 108 MHz. As this requires a long antenna due to the longer wavelength of the signals, the headset cable is usually used for the purpose. This means that the FM radio application can only be used when a headset is plugged in.

GPS receivers and 3D motion sensors have also become a standard in today's smartphones. These are either included in the SoC or as a separate chip.

5.4 Mobile Graphics Acceleration

In recent years, the GPU in mobile devices has become at least as important as the CPU itself because of the increasing screen sizes and display resolutions, direct touch screen interaction that requires fast reaction to input gestures, smooth scrolling through web pages, rich UI effects such as blending between screens of different applications, and rotation of the display content when the user turns the device and 3D gaming. Such operations are preferably calculated on a dedicated hardware unit, which is optimized for parallel floating point operations and graphics algorithms already included in the hardware.

As discussed in Section 5.3, there are a number of GPU manufacturers today with different approaches of how the graphical rendering process can be enhanced while at the same time using as little power as possible. As a consequence, each GPU requires the operating system to interact with it differently. A standardized interface is therefore required both for applications and the operating system running on the CPU to manipulate the screen content. As in the PC world, a GPU operating system driver allows to introduce a new GPU by only modifying the graphics driver, while the rest of the operating system remains unaltered.

5.4.1 2D Graphics

Mobile GPUs offer graphics acceleration in 2D and 3D modes. While 3D graphics are mostly used for gaming, 2D graphics is the default way of presenting information to the user in most applications. 2D graphics acceleration works with different 2D layers that are stacked on top of each other. In the web browser, for example, the background layer could contain the web page, while independent layers on top of the background contain other elements such as buttons. The web page on the background layer is usually much larger than the screen size; so, only a portion of it is shown at a time. When the user then scrolls through the page, the CPU simply instructs the GPU to show a different part of the already prepared background layer. This requires little processing power on the CPU and little communication with the GPU as the bitmap for the background layer is already present in the GPU's memory. The GPU can therefore create an appealing scrolling effect with little computational effort involved. The buttons on higher layers remain in their place on the screen even when the web page is scrolled. When the GPU renders a new screen view once every few milliseconds, it uses a different part of the background layer from its memory, which creates the scrolling effect and then just superimposes the graphical elements of higher layers, which do not require a modification of the background layer in memory.

The CPU, on the other hand, is responsible for constructing the full web page once out of the HTML code, JavaScript, and the pictures it receives from the web server. The result is then transferred to the GPU as the background layer. To speed up the process, the CPU usually renders the part of the web page that is visible to the user at first, transfers this part to the GPU so it can be displayed, and then continues to assemble the remaining parts of the page that can only be seen when the user scrolls through the page.

In most cases, the CPU has finished creating all parts of the web page before the user starts scrolling, which makes the delay invisible to the user. Once all parts of a web page are fully rendered and transferred to the GPU memory, the CPU then only reacts to user input by instructing the GPU to show a different part of the page. JavaScript code on the web page remains the responsibility of the CPU, and should the JavaScript code change the web page at runtime the CPU modifies the affected section of the web page and updates the background layer in the GPU's memory.

When displaying a web page that was originally designed for a large PC screen, it is usually presented to the user in a minimized form. The user can then zoom into specific parts of the web page. This part is then enlarged and if the original text would be cut on the right or left because of the magnification level, it is usually reformatted to fit the zoom level. Again, the work is split between CPU and GPU to create an appealing effect during the zooming operation. When the user starts the zoom, the CPU instructs the GPU to zoom into the background layer. The pictures and the text usually blur during this operation as the background layer was not optimized for a higher zoom level. When the user has finished setting the zoom level, the CPU reflows the text in this part of the web page, assembles a higher resolution background layer representing the web page in this zoom factor, and sends the result to the GPU. The GPU then displays the part of the web page the user has zoomed into and the CPU continues calculating the remaining web page; so it is already present in the GPU's memory when the user starts scrolling through the web page.

5.4.2 3D Graphics

For 3D games that project and animate a 3D world on the screen, a different approach is required than that for the 2D layering described above. For such purposes, the OpenGL-ES (Open Graphics Library for Embedded Systems) [21] API has become very popular on mobile operating systems such as Android, Symbian, and iOS. Originally developed and used in the desktop world, OpenGL-ES is a scaled-down version of OpenGL to take the reduced capabilities of mobile GPUs into account. Unlike in the 2D world where the web page is rendered on the CPU and then sent as a background layer to the GPU, 3D worlds are completely rendered on the GPU. The CPU is mainly responsible for defining where objects are located in a 3D space, what their properties are, from which location the objects are viewed by an observer, and how the objects move and change their shape over time. The following paragraphs now give a brief introduction of how this is done in practice. Further details can be found in [22].

To construct a 3D world, OpenGL-ES has standardized commands to define shapes, usually triangles, by defining where their corners (vertices) lie in a 3D x,y,z coordinate system. By combining the triangles, any shape can be constructed. The smaller the triangles, the more are required to form an object and the more realistic it will appear to the user. For defining very complex shapes, such as a racing car in a game, for example, thousands of small triangles are defined in a floating point number array, which is then given via an OpenGL-ES command to the GPU. The GPU is then responsible to render an image based on the coordinates of each triangle's vertices. Triangles can be colorized by assigning an RGB value to the area of each triangle. Bitmaps can be used as textures for triangles. Again, it is the graphics processor that does the image processing. The CPU just transfers the textures and the information of how to apply them to the different triangles to the GPU, which will then do all the computational work required to apply the texture to the shapes defined in the x,y,z coordinate system.

Two more things are required for the GPU to render a 3D scene. First, the GPU needs to know the location of light sources that illuminate the objects, the type of light, its direction, and several other parameters. These are then used to realistically illuminate a scene including the shading created by objects. And finally, the GPU needs to know from which point of view the scene should be shown, in which direction the viewer is looking, and the parameters for his field of view. If, for example, a car is rendered, the viewer could see the car from the front, from the back, from the left, from above, from below, and so on. From each point of view, the rendered image looks completely different. Also, the viewer could be very close to the car and would hence only see a part of it or he could be quite far away. In this case, he would see the car and perhaps also further objects or parts of them beyond or to the sides of the car. Objects more distant appear smaller than objects closer to the viewer. All of these properties can be described by a viewer that is a certain distance away from a rectangle he looks through. The distance to the rectangle determines which part of the 3D world he can see. If he is close to the rectangle, he has a wide angle of view. If he is distant to the rectangle, his angle of view is very narrow. This is shown in Figure 5.5. And finally, the view is limited by a plane in the background. All objects behind the plane are invisible. The viewer, the viewing rectangle, and the limiting plane in the background form a pyramid, with the user being at the peak of the pyramid and the viewing window and the plane in the background forming a trunked pyramid,

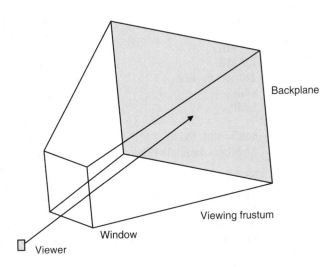

Figure 5.5 The viewing frustum.

which in mathematics is referred to as a frustum. Hence, the definition of the viewing frustum in computer graphics refers to the definition of the view onto a 3D space that the GPU then maps on the 2D display.

As long as the objects in the scene do not change or move, the CPU can instruct the GPU to render the 3D world from a different point of view by only changing the frustum parameters with a simple instruction. The GPU then autonomously recreates the 3D world as seen from the different point. Interesting effects can thus be created, such as flying through the world without the CPU being required to redefine the scene as the user flies through the world. This greatly reduces the work required from the CPU for 3D animations and it can thus concentrate on the calculation of object movements in the 3D worlds and updation of the descriptions of the shapes. In many games, many objects move at the same time. Especially in such situations, several CPU cores are very beneficial, as this means that the movement of several objects can be calculated in parallel instead of sequentially on a single CPU.

5.5   Hardware Evolution

From an end user's point of view, little has changed in the PC world in the past five years. Processors have become faster, graphics capabilities continue to evolve, but usage scenarios and applications are still the same. Mobile computing, however, has evolved significantly since the first edition of the book was published and an inflection point has been reached as mobile Internet access is now widely used. While only few had used Internet-based services on mobile devices only five years ago, over half of the devices sold in 2012 were smartphones in some countries [23]. Table 5.2 compares the typical

Table 5.2   Smartphone hardware comparison over five years

	2007	2012
Typical device	Nokia N95	Samsung Galaxy S-II, iPhone 4G
CPU clock speed	332 MHz	1.2 GHz
RAM	128 MB	512 MB to 1 GB
Screen resolution	240 × 320 pixels	800 × 480, 960 × 640 pixels
Flash	8 GB	32 GB
Camera	5 MP	8–12 MP, 41 MP (Nokia 808)
Video recording	640 × 480 (VGA)	1280 × 720 (Full HD)
Cellular connectivity	3.6 Mbit/s down, 384 kbit/s up	>20 Mbit/s downlink, >5–15 Mbit/s uplink
Infrared data exchange	Common	No longer included
Wi-Fi	New concept, few devices	All smartphones
GPS	New concept, few devices	All smartphones
Motion sensor/compass	Not included	All smartphones
Touchscreen	Few models, slow reaction, single touch	All models, multi-touch
Weight	120 g	120 g

hardware features of a high-end device sold in 2007 when the first edition of the book was written with those of a typical smartphone in 2012. It is likely that this evolution will continue at a similar pace in the years to come and the differences shown in the table give an idea of the potential capabilities devices will have in five years in the future. The following sections now give an overview over the current and future developments of a number of mobile device components.

It should be noted at this point that ever more sophisticated hardware and also software do not only benefit devices in the smartphone category but may also bring about new device types, new applications and usage scenarios, and new types of users. Two recent examples of this are the rising popularity of tablets and e-book readers. While such devices have existed already many years ago, they had only become appealing to a wider public once display resolution, size, and weight reached acceptable values and touch input became refined enough for easy use with little to no perceived delay between an action of the user and a response of the device.

5.5.1 Chipset

In the case of ARM-based chipsets such as those described in Section 5.3, the next few years will see further performance enhancement of all subcomponents. In 2008, most high-end devices using ARM and OMAP products were based on the ARM 11 processor family with processor speeds between 330 and 400 MHz. The modem, the CPU, and the GPU were often implemented in different chips. Current platforms based on the Coretex-A9 architecture exceed the 2008 performance level several times because of the four times higher clock speeds of 1.5 GHz and more, dual- or quadcore technology, and an enhanced processor core. Taken together, it is reasonable to assume a 10 times higher performance compared to 2008. Also, there is a clear trend to integrate the modem, the CPU, and the GPU units in a single chip as described in Section 5.3 and even further components that are now part of every smartphone such as Wi-Fi and Bluetooth support [24].

When looking at the PC sector, it can be noticed that processor clock rates have remained at around 2 GHz for several years now because of the high-power requirements and excessive heat generation at higher clock rates. Instead, processing speeds have increased by enhanced architectures such as hyperthreading, that is, simulating two virtual CPU cores in a single physical CPU and by adding additional CPUs to the die (multi-core operation). Another way of increasing overall processing speed is by using more transistors on the chip to form more complex processing units to reduce the number of clock cycles required to execute a command and by increasing the on-chip memory cache sizes to reduce the occasions the processor has to wait for data to be delivered from external and slow RAM.

While using more than one CPU in a device and increasing the clock rate undoubtedly increase the performance, it also leads to a higher power consumption not only when the CPUs are running at full speed but also when the CPUs are mostly idle. A number of different approaches exist to mitigate this problem. In Nvidia's 2012 Tegra design [25], four HS CPU cores are used to run four independent tasks in parallel. In practice, four cores are only required in few situations and therefore the design allows to deactivate and reactivate individual cores at run time to reduce power consumption. In addition, Nvidia uses a fifth core, which is referred to as a "companion core" that takes over when only little processing power is needed, for example, while the display is off and only low-intensity background tasks have to be served. This is transparent to the operating system; so no software adaptations are required for the switchover. The reason why a fifth low-speed CPU is required instead of just lowering the clock speed of one of the fast CPUs is that the average power consumption is governed by two factors, leakage power and dynamic power. When processors are run at high clock speeds, the chip has to be optimized for low voltage and fast switching operations as the power requirement increases linearly with the clock frequency but in square with the voltage. Unfortunately, optimizing the chip for low voltage and fast switching operation increases the leakage power, that is, the power consumption when voltage is applied to a transistor even when its state is not changed for a long time. It is this leakage power that becomes the dominant power consumption source when the CPU is idle, that is, when the screen switched off, when only background tasks are running, and so on, and hence the processor clock speed is reduced. It is during such times that the main CPUs are switched off and the companion CPU is used. A different manufacturing process is used for the location where the companion CPU is located, which is less optimized for HSs but more optimized for lower leakage power. The companion CPU can thus only be run at clock speeds up to 500 MHz but has the advantage of having a lower leakage power when only slower clock frequencies are required. Switching back and forth between the companion CPU and the four standard cores is seamless to the operating system and can be done in around 2 ms.

Qualcomm has used a different approach in their Krait CPU architecture to conserve power. Instead of running all cores at the same clock speed, which has so far been the default option, each core can be run at a different speed depending on how much workload the operating system assigns to each of the cores. Rather than optimizing one processor for leakage power consumption, this approach conserves power by reducing the clock speed of individual processors when less processing power is required [13].

Other components such as flash memory and RAM for program execution are also likely to continue getting cheaper. In 2007, high-end smartphones featured 128 MB of memory and the design goal was to minimize the RAM in a device for cost and form factor reasons by as much as possible. In 2012, this has changed significantly and high-end smartphones were equipped with 512 MB to 1 GB of RAM. Even netbooks today running Windows 7 do not have more than 1 GB of memory, mostly because of a marketing restriction put in place by Intel to distinguish netbooks from higher priced notebooks. Both smartphones and low-cost netbooks are likely to have their memory "only" doubled in the next few years as memory for program execution is no longer a limiting factor in high-end user devices. In such devices, the limit is rather set by power consumption, battery capacity, weight, size, processing power, and graphics capabilities.

Storage memory in mobile devices has also increased significantly over the past years. While in 2008, 8 GB of built-in flash memory was state of the art in high-end devices, 2012 models include 16 or even 32 GB. In addition, memory card slots allow the extension by another 32 GB. As a result, local storage has almost become limitless for most applications such as the number of applications that can be installed, storage of pictures taken with the camera (1–2 MB per image), and storing data locally such as email file attachments, ebooks, and so on. Only few applications such as storing a complete music library on a device or taking high-definition (HD) videos with the built-in cameras, which require around 100 MB (0.1 GB) of storage space per minute [26], have become applications that can fill up local storage.

After many years of competing formats, microSD has become the de-facto standard for flash memory extension cards for mobile devices. In 2012, the Secure Digital High Capacity (SDHC) version is commonly supported by mobile devices and allows a maximum flash memory size of 32 GB. One of the limiting factors is the FAT16 file system commonly used, which only supports 32 GB volumes. The next version of the microSD standard is referred to as Secure Digital Extended Capacity (SDXC) and enables storage capacities of up to 2 TB. In 2012, storage capacities of up to 64 GB were available on microSD cards. Also, the physical interface itself has been enhanced to allow data transfer rates beyond those of USB 2.0, if supported by the host device and the card.

5.5.2 Process Shrinking

Decreasing the power consumption and increasing the clock speed and the number of components and subsystems on a chip require transistors on the chip to become smaller. Most microprocessors, static RAM (memory), and image sensors are based on Complementary Metal–Oxide Semiconductor (CMOS) technology today. In 2007, chips using ARM11 cores were manufactured with a 90 nm CMOS process. The 90 nm length refers to the average half-pitch size of a memory cell. In 2012, high-end smartphone chipsets are manufactured in 45 nm technology and the next step to 28 nm technology is already in sight for next generation chipsets [27]. Traditionally this value has considerably shrunk over time. In 1972, the Intel 4004 CPU was manufactured in a 10 μm CMOS process, which is 10 000 nm. The CPU speed at the time was 108 kHz or around 2000–3000 times slower than the current clock rates between 2 and 3 GHz for high-end desktop and notebook processors. Since then transistor sizes have been shrinking around 70% every two to three years.

Early CMOS transistors only required power to switch from one state to another while otherwise drawing almost no current. As a consequence the power requirement of a processor could be reduced during idle times by lowering the clock speed. Owing to continuing size reductions, however, the layers between the different parts of transistors have reached a thickness of less than 1.5 nm; that is, only a few atoms separate the different parts of the transistor. This leads to increasing leakage currents while a transistor is idle. This effect is undesired since the leakage is independent of the clock speed. This means that the leakage power, also referred to as static power, remains the same even if the system clock speed is reduced to 1 Hz. In theory, reducing the voltage also reduces the leakage power. In practice, however, chip voltage has already reached a very low level and a further decrease would result in unwanted interference from external components such as conducting paths between different chips. For the moment leakage power can still be controlled by new manufacturing methods which, however, slow down the possible switching speeds of transistors and hence reduce the clock speed improvements that would otherwise result from reducing the size of transistors. Still, Globalfoundries reports that the scaling from a 40 nm process to a 28 nm process results in a 60% higher performance at comparable leakage with a 50% lower energy requirement per switch and 50% lower static power requirement [28]. Nevertheless, industry experts expect that at some point in the future leakage prevention techniques will eat up any performance gains in terms of increased clock speeds that a reduction in transistor size would normally bring about. So while for the moment reducing the size of components still results in faster processors, lower power consumption, and smaller chips, this trend is unlikely to continue with current technologies. The only benefit from reducing transistor sizes in the future is thus a reduction of the size of the chip unless new methods are found to prevent leakage that do not interfere with the speed a transistor can switch from one state to another.

5.5.3 Displays

Despite the limitations imposed by the size of mobile devices, mobile displays have undergone an interesting evolution in recent years. While 2.5–3 in. screen sizes were considered as large, most smartphones in 2012 had screen sizes between 3.5 and 4.5 in. as less space was required for keys because of touch screen-based input. Some smartphones even boasted a screen size of over 5 in. to fill the gap between smartphones and larger tablet devices. Also, display technology has advanced significantly. While a resolution of 320 × 240 pixels was state of the art in 2007, this has increased to 840 × 480 pixels and even further. While LCD displays are still used in many devices, Organic Light-Emitting Diodes (OLEDs) have also become very popular because the power consumption is less, lively colors are available, and there is no need for background illumination. With OLED displays, basic information such as time, date, missed calls, an SMS indicator, and so on can be displayed even when the device is locked without significantly increasing the standby power consumption.

While in 2007, the tablet category did not exist, this device category has become a huge success in 2012 and screen sizes between 7 and 10 in. have become standard. Typical screen resolutions range between 1024 × 600 and in some cases even 2048 × 1536 pixels, a resolution at such a screen size at which the human eye is no longer able to distinguish individual pixels.

Only indirectly related to the display technology is the material used on top of the displays. While plastic was mostly used in 2007 in most phones, it was quickly replaced by glass once display and touch-based input became more popular. At first, the glass was prone to scratches in a similar way as the plastic that was used before. In recent years, however, display glass has become scratch resistant and not even coins, keys, and other objects with sharp edges in pockets are able to damage the surface. As the glass surface adds significantly to the overall weight and dimensions of the device, one area of evolution is to reduce the thickness and thus the weight of the glass without compromising the scratch resistance and touch sensitivity [29].

A different type of display technology that has become very popular especially for e-book reading is electronic paper. Unlike LED (Light-Emitting Diode) or OLED displays, e-paper displays do not emit light themselves but only reflect light from their surface in a similar way as ordinary paper. As a consequence, this display technology is very power efficient and while the content does not change, no power is required to sustain the image. Devices with e-paper displays are thus very power efficient and require only small and lightweight batteries. The overall device can thus be very thin and lightweight, which makes holding such devices, even with large screens and over a longer time, much more enjoyable than tablets with active displays with a high power consumption and a significant battery weight. The downside of e-paper displays is, however, that in 2012, only black and white displays are available and that refreshing the content of the screen is too slow for dynamic content. While this is not an issue for traditional books, which are mostly black and white as well, it limits the use of the device to only a few applications. With the rising popularity of multipurpose tablet devices and multimedia e-book content with animations, and so on, it is difficult to predict if e-paper-display-based devices will remain popular unless their two major limitations, that is, no color and slow display refresh times, are mitigated. It should be noted, however, that e-paper-based e-book readers have become very inexpensive with prices of around €100 in 2012. If prices were to fall further, their popularity and use could remain high as many consumers would not have to choose between an e-paper-based device for e-book consumption or a more expensive multipurpose tablet device on which e-book reading is just one of the many applications.

5.5.4 Batteries

As power requirements of high-end mobile multimedia devices are unlikely to decline in the future, another research focus is better batteries or a different kind of energy storage for mobile devices. Current lithium ion technology is unlikely to get significantly more efficient in the future as physical limits of the technology are almost reached. A possible solution for the future could be fuel cells producing electrical energy from methanol, water, and air [30]. Research has been ongoing for many years now and it is expected that fuel cells can store 10 times the amount of energy compared with current battery technology. However, no major breakthroughs have been reported so far and it seems unlikely that fuel cells will be used in mobile devices during the next few years. It should be noted at this point that new battery or power cell technology would mostly be used to extend the operating time of a mobile device and not to supply more energy to the device itself. Unlike notebooks and PCs, which tend to get warm and require active cooling, fans are not acceptable for mobile devices.

5.5.5 Camera and Optics

In 2007, high-end mobile phones in Europe were equipped with 5 megapixel camera sensors and a typical sensor size of 1/2.5 in [31]. Most mobile devices at that time, however, had a far lower resolution and a smaller sensor. As a consequence, picture quality was significantly inferior to digital compact cameras at the time. In 2012, most smartphones and tablets are equipped with camera components producing at least equal results as those in high-end devices only five years earlier. Typical sensor resolutions in such devices have reached 8 megapixels, and better optics now help to capture images, which come close to what is possible with dedicated entry level point and shoot cameras as long as no optical zoom is involved.

One major shortcoming of many camera phones back in 2007 was the use of fixed-focus lenses. Together with the low-resolution sensors, this often resulted in mediocre pictures. In the meantime, most mid- to high-end smartphones are equipped with auto-focus lenses, which allow wider apertures and thus a higher light sensitivity, or extended depth of field (EDOF) lenses and software for postprocessing to generate a sharp image. While producing similar results in most situations, the major drawback of EDOF is its inability to produce sharp images of objects closer than 1 m, which makes such lenses unsuitable for photographing documents at close distance [32].

Video recording has also improved significantly over the past five years. Owing to dedicated video encoding units in the SoC as described earlier, smartphones are now capable of recording videos in 720p HD resolution at a rate of 30 frames per second. First devices are also capable of 1080p full-HD video recording. Only little compression is applied, and 720p video recording requires about 100 MB of storage space per minute. Uploading a video directly from a mobile device to a video portal such as Youtube is therefore time and data intensive. Once on the server, services such as Youtube then compress the 100 MB file of a 1 min 720p HD stream to about 46 MB while keeping the 720p resolution to 15 MB for the default 480p resolution typically used when videos are viewed with a PC and to less than 6 MB for the 360p version, which is used for streaming to mobile devices [33]. It is interesting to note that this transcoding process is quite time intensive. This might be an indication of the significant processing power required for the transcoding, which is perhaps the reason why this is not already done on the mobile device. In this regard, LTE networks can help as, under good reception conditions, uplink speed transmission is faster compared to that at HSPA (cf. Chapter 2).

Some specialized smartphones trade overall thickness of the device to include enhanced sensors and optics such as, for example, the Nokia N8. A large 1/1.83 in. camera sensor, a resolution of 12 MP, and specially crafted optics produce results with fine grained details without a need for postprocessing such as sharpening, which produces undesired side effects. Even in low light conditions, the sensor produces sharp images and a xenon flash similar to those used in compact cameras, enabling indoor photography with good results.

The largest and highest resolution camera sensor used in a smartphone at the time of publication of the second edition is the 41 MP 1/1.2 in. sensor in the Nokia 808 Pureview. While the size of each pixel on the sensor is the same as in high-end 2012 smartphones such as the iPhone 4S or the Samsung Galaxy S-II, the very high resolution of the sensor is used to combine the information of 8 pixels into one output pixel. This results in a 5 megapixel image of very high quality, especially in low-light situations. Also, the 41 megapixels can be used to act as a zoom without mechanical components, another first in the mobile camera domain. While point and shoot cameras use a mechanical system to move the lenses, which requires significant space and thus makes cameras much thicker than mobile phones, zooming with the 41 megapixel sensor is achieved by only using a part of the sensor and saving the result as a 5 megapixel image. This of course reduces the sensitivity as fewer pixels are used per output pixel, but the size of each physical pixel is still the same as in average smartphones. While being much more compact than mechanical zoom lenses in point and shoot cameras, the camera sensor and optics requires more space and thus the device is significantly thicker than average smartphones with smaller and lower resolution sensors. While this increased size is unlikely acceptable to many users, photography enthusiasts might find such phones appealing. Also, further miniaturization might enable manufacturers to further reduce space requirements for such camera sensors in the future.

With the processing speed available in smartphones in 2012, it has also become possible to analyze the data stream received from the camera in real time, which has enabled functionalities such as face detection for proper focus and smile detection to help the user

Figure 5.6 A 2D barcode on an oversized advertisement poster.

to take the picture at the right moment. Also, smartphones have become powerful enough to assemble a panorama from several pictures in near real time.

Another interesting use of mobile phones and their built-in cameras is barcode scanning. While 2D bar codes have become popular in Japan many years ago, a similar trend took much longer to emerge in other parts of the world. Different bar code formats were developed and deployed, which made it difficult for the user in practice as applications could often only decode some of the different formats. In the meantime, however, a number of bar code uses have become quite popular, for example, on advertisement posters to get more information as shown in Figure 5.6 and also when shopping to compare prices of a product found in one shop to see if the product is cheaper elsewhere.

On the basis of the evolution of camera module hardware in the past, it is likely that there will be a further reduction in size and cost of camera modules while keeping camera sensors big enough for high-quality imaging. Image quality is likely to further improve, partly also because of intelligent hardware and software combinations that have enabled cheap EDOF lenses to produce good results except in the very near field.

5.5.6 Global Positioning, Compass, 3D Orientation

In 2007, one of the major new features in high-end mobile phones was the inclusion of a GPS receiver. The Nokia N95 released in early 2007, for example, was Nokia's first GPS-enabled smartphone [34]. By 2012, GPS has become a standard feature in all smartphones, and first chipsets such as the Qualcomm Snapdragon platform have included a GPS receiver in the main chip and thus a specialized GPS chip is no longer necessary.

Advances have also been made in obtaining the first position after activating the GPS component in a device (Time To First Fix, TTFF). Calculating a location requires the reception of the signals of at least four GPS satellites. In addition, the receiver must know the exact orbit of each of these satellites, which is referred to as the "ephemeris" data.

The ephemeris data is transmitted by the GPS satellites themselves. At a transmission rate of 50 bits per second, however, reception of the complete information can take several minutes, especially when several iterations of the data have to be received because of reception errors due to a weak signal or user movement, which can easily interrupt the data stream. Mobile devices thus usually get the ephemeris data from a network-based server over a cellular or Wi-Fi connection. Initial signal acquisition is also sped up by knowing the approximate location of the device. In combination with the ephemeris data, the device can quickly calculate which satellites should be received at the approximate location. The GPS receiver can then try to receive those signals first. Getting the approximate location is usually done by collecting information about nearby Wi-fi networks and cellular network base stations (Wi-Fi SSIDs (Service Set IDs) and network/base station identifications). These are also sent to a network-based server with a database of network identifications and their locations. Using the approximate location and ephemeris data received via the Internet, the TTFF is usually below 5–10 s. This is the case even if the GPS receiver is activated at a totally new location, for example, after an intercontinental flight. In this case, the previous location data still stored in the device that may otherwise also be used to improve the acquisition speed is invalid.

It is interesting to note at this point that the location information obtained by querying a database in the network for a coarse location is also useful to load initial maps data and show an approximate location to the user before a more precise location is available via GPS. As the range of Wi-Fi networks is quite short, location information obtained this way is usually accurate within a few tens of meters, which is sufficient for many applications such as street navigation. While specialized companies were needed initially to obtain the location of Wi-Fi access points by driving through countries with measurement equipment picking up Wi-Fi signals and storing this data with location information, it is the navigation and mapping applications on the user devices today that provide this information to databases in the network by reporting Wi-FI SSIDs and MAC addresses together with GPS information. This helps companies such as Google, Nokia, Microsoft, and others to keep their location databases up to date.

In addition to GPS, other navigation systems have also become operational. The Russian GLONASS system, for example, is now also available worldwide at the time of writing, and the Chinese Beidu navigation system covers parts of Asia. And finally, the European Galileo positioning system is also set to become operational in this decade. Some GPS chips now also support GLONASS and use both systems to obtain location information [35]. This increases the number of visible satellites in challenging situations, for example, when the user is located between high rising buildings or in narrow streets where only a few degrees open sky is visible. Having more satellites thus reduces the times in which no valid positioning data is available and thus increases the usability of street-side navigation.

When location information is used for applications such as route navigation in cars, the direction the user is moving toward can be easily obtained from the directional movement. The map displayed on the screen can then be rotated accordingly. For street-side navigation, however, this is often not possible as the user is often immobile while deciding the direction in which to walk. In 2012, electronic compass components have also become a default feature in smartphones and offer an easy way to rotate maps according to the direction the user is looking. This significantly helps street-side navigation as the user is instantly aware of the direction in which he is looking. Knowing not only the location of the user but also the direction he is facing has enabled other applications that use the camera, location information, and the direction from the compass to overlay a video stream with additional information, for example, Wikipedia entries for locations and objects in the area a user is currently located in. Examples of such applications are Wikitude and Layar [36].

Motion sensors, also referred to as accelerometers, are also common in mobile devices now. These sensors can detect 3D motions, which can be used, for example, by astronomy programs to show stars and constellations the user points the device at. For this application, it is not only necessary to know the approximate location of the user and his viewing direction but also the angle above the ground the device is held at to display stars and other objects at the right location. Examples of astronomy applications making use of motion sensors are Google Sky Maps and Astroller. Other uses of accelerometers include pedometer apps to count steps, sleep phase alarm clock apps that detect motions while the user is sleeping and adapt the wake up call accordingly, and also motion detection in games for controlling 2D and 3D movements (e.g., a virtual steering wheel). These are just a few examples that demonstrate the many uses for location, direction, and motion information.

5.5.7 Wi-Fi

It was only in 2006 that first mobile phones included Wi-Fi functionality. Early examples are the Nokia N80 [37] and N95 [38]. At that time, it was by far not certain if this would be a continuing trend as not all market participants were happy that Wi-Fi was included in a mobile device as a means to access the Internet. In the meantime, however, Wi-Fi has become a standard feature even in cheap smartphones and other mobile devices, and some versions of mobile devices such as tablets and ebook readers are available with Wi-Fi connectivity but without a cellular interface. Mobile network operators also benefit from the Wi-Fi interface as many users prefer to connect to their fixed line Internet connection at home via Wi-Fi because of mobile data costs and therefore reduce the amount of traffic transmitted via the cellular network. Some mobile network operators have also deployed Wi-Fi hotspots in hotels, train stations, and airports and offer combined Wi-Fi and cellular plans to their customers.

As described in Chapter 2, most mobile devices now include a Wi-Fi interface according to the 802.11n standard. While the standard offers many options to increase the data rates, most mobile devices only use a single antenna, that is, no Multiple Input Multiple Output (MIMO) operation, and are designed for single 20 MHz channel operation. Also, most devices only use the 2.4 GHz band and do not operate in the 5 GHz band. In 2011, highly integrated Wi-Fi front-end-modules optimized for smartphones have become available that can operate in the 2.4 and 5 GHz bands. Together with the emerging 802.11ac standard (cf. Chapter 2), the support of the 5 GHz band might thus become more common in mobile devices in the next few years.

As the Wi-Fi interface of a device does not only offer Internet connectivity but also enables local communication in the home network to other connected devices, device manufacturers now include software to interact with connected TVs to show pictures and videos stored on the device or to connect to Hi-Fi equipment to stream music stored on the device. In the reverse direction, it has also become possible to use a mobile device to listen to music stored on a server in the home network via streaming over

Wi-Fi. Such functionality is, for example, enabled by the Digital Living Network Alliance (DLNA) standard and while still not a mass market application in 2012, there is a rising number of devices supporting it. For exchanging any kind of files in the home network between PCs, NAS drives, and mobile devices, protocols such as Microsoft's Server Message Block (SMB) can be used, as this protocol is supported by all desktop operating systems and also by mobile operating systems such as Android in combination with third-party file manager applications.

As Wi-Fi chips can be used in client mode and access point mode, many smartphone operating systems now include this functionality and a smartphone thus becomes a bridge to the Internet for other Wi-Fi devices such as notebooks. Some network operators restrict the use of this functionality in devices sold over their sales channel but most smartphones and tablets sold via independent sales channels (e.g., via Amazon) have this functionality activated.

In an ordinary Wi-Fi network, a data packet from one device to another must always pass through an access point. This in effect cuts the available data rate into half as each packet has to be transmitted twice. As the use of an access point complicates many usage scenarios such as the ad hoc transmission of a picture or file to another device not connected to the same network, the Wi-Fi Alliance has created the "Wi-Fi direct" specification [39]. As the name of the specification implies, Wi-Fi direct allows ad hoc communication between two Wi-Fi devices for any kind of data transfer. This is achieved by one device becoming a Wi-Fi direct access point for the time of the transfer to which the other device can connect to. The Wi-Fi Protected Setup (WPS) procedure (also specified by the Wi-Fi Alliance) is used to establish a secure and encrypted connection with minimal user interaction. In essence, Wi-Fi direct thus works in a similar way as file transfers or data exchanges over Bluetooth. In 2012, only few devices included this functionality but since Google's Android operating system has started to support the functionality in version 4 for use cases such as file transfers and the exchange of address book and calendar entries, the number of supporting devices is growing.

Some devices have also started to support Wi-Fi as a bearer for Bluetooth connections. Some Symbian-based devices make use of this for establishing a Bluetooth connection to transfer calendar and address book entries. Once the devices detect that they not only support Bluetooth but also Wi-Fi, the bulk of the data is transferred over the much faster Wi-Fi channel. In effect, this functionality is similar to that offered by the Wi-Fi direct specification and might thus be superseded by it.

Another interesting aspect of including a Wi-Fi interface is to use it for positioning purposes. As Wi-Fi access points are usually always at the same location and emit a unique name (SSID) and MAC hardware address, this information can be used in combination with a location database located in the network as an additional means to quickly pinpoint the user's approximate location indoors when no GPS signal is available or while the GPS receiver is still decoding the information for an initial set of coordinates.

One important topic that is often overlooked when discussing Wi-Fi on mobile devices is security. While home and office Wi-Fi networks that are using WPA (Wireless Protected Access) and WPA2 authentication and encryption are well secured against external attacks, there are significant security issues when using public Wi-Fi hotspots. Here, encryption is usually not used, which means that applications themselves have to encrypt the data they transmit. This, however, is still often not the case and tools such as Firesheep and Droidsheep make it easy for attackers to spy on the data traffic [40]. Companies therefore often protect their employees with Virtual Private Network (VPN) products that offer strong authentication and encrypt all incoming and outgoing data. Private users, however, usually do not use VPN products because of the overhead, complexity, and cost involved. If the public Wi-Fi hotspot is managed by a mobile network operator, another possible mitigation would be to use authentication and encryption based on the Extensible Authentication Protocol — Subscriber Identity Module (EAP-SIM) protocol, which uses the credentials stored on the SIM card. While this standard has existed for years, there has been little take up so far and it is not certain if there will be more interest from the operator community in the future. As a consequence, care should be taken when public Wi-Fi hotspots are used. Applications transmitting private data should be disabled unless it is certain that the application offers adequate application-level authentication and encryption. As a general rule, 3G or 4G cellular connectivity should be preferred in public places whenever possible.

5.5.8 Bluetooth

A wireless technology that has been part of mobile devices for many years is Bluetooth. Over the years, the specification has been continuously enhanced and evolved. Most applications that Bluetooth is used for today are still the same as when they were first developed:

- Wireless headsets for bidirectional voice telephony,
- Wireless headsets for high-quality music streaming,
- Exchange of data between devices such as files, pictures, calendar, and address book entries,
- Wireless keyboards and mice for notebooks, smartphones, and tablets (note that wireless mice and headsets in the PC sector are often not using Bluetooth but a proprietary protocol).

Some applications where Bluetooth was used in early years such as offering Internet connectivity to PCs and notebooks via a mobile device have mostly disappeared in the meantime because of the integration of Wi-Fi access point functionality in smartphones and data speeds in mobile networks exceeding the maximum Bluetooth data rates of about 2–3 Mbit/s. Also, ideas such as Bluetooth marketing, that is, sending messages to mobile phones of customers in shopping malls and other places have not proved to be very successful. On the other hand, there are a number of applications that make use of Bluetooth, which have become quite successful such as, for example, wirelessly connecting heart rate sensors for sports purposes [41].

Most of the Bluetooth functionality used today is part of the Bluetooth 2.1 specification, which was released in 2007. Many devices today are declared as Bluetooth 3.0 compliant, as they are using a number of evolutionary enhancements to existing functionalities contained in this version. In addition to those enhancements, version 3.0 of the specification offers an alternate MAC/PHY layer specification, which enables devices to use the Bluetooth protocol stack for establishing an initial connection and then using a Wi-Fi bearer for transmitting the actual user data. To inform users that this optional feature is supported by a device, the +HS designation was created by the Bluetooth Special Interest Group

(SIG). Devices supporting combined Bluetooth and HS Wi-Fi transmissions are thus to be labeled as Bluetooth 3.0 + HS compliant. In practice, however, there are currently only few Bluetooth 3.0 devices supporting this feature.

The latest Bluetooth version at the time of publication of the second edition is Bluetooth 4.0, incorporating a very low energy physical layer implementation, which it has inherited from the WiBree project. This very low energy physical layer has little in common with the traditional Bluetooth standard except for the use of the license exempt 2.4 GHz band that it also shares with Wi-Fi. Its very low power requirements make it ideal for very small sensors with a very limited power supply, for example, via coin cells. Mobile devices could act as UI for such sensors but so far, only few implementations have appeared on the market. With competing technologies such as Wi-Fi and the emerging Near Field Communication (NFC) standard now more and more to be found in mobile devices, it is likely that Bluetooth will mainly be refined in the coming years for the purposes it is already used today rather than being used for revolutionary new applications.

5.5.9 NFC, RFID, and Mobile Payment

Another area of ongoing evolution is embedding NFC chips in mobile devices for local exchange of information and mobile payment functionality. In Japan, Sony's FeliCa NFC system is already widely used as a payment system for public transportation, convenience stores, and vending machines. Embedding the chip in a mobile phone and making it accessible to applications running on the mobile phone extends the use of the payment system for online ticketing for flights and sporting events via a mobile device. Tickets can then be printed at the airport, at the stadium, or the embedded FeliCa chip used directly to gain entry.

As the FeliCa system is designed as a micropayment system, linking the system to the Internet via a mobile device extends the payment system beyond direct interaction with an NFC reader and it is likely that many online services beyond the two examples given above will make use of such a system in the future.

In addition, FeliCa-equipped mobile phones can be used for non-payment purposes as well, for example, as door keys [42]. If the door lock is connected to the Internet, users can even check remotely with the mobile phone if the door is locked.

In other parts of the world, it was and still is more difficult to introduce a standardized NFC-based local information exchange and mobile payment system because of the number of different market players and the uncountable number of public transport companies, each using their own ticketing system. Thus, it will take much longer for a single system to reach a critical mass. Having a universally adopted standard, however, is critical for mobile device manufacturers to consider including NFC technology in their mobile devices, which are often produced for a global market without national hardware variants. Some progress has been made in this regard with companies such as Google, Nokia, and RIM backing the NFC-Forum [43] based standard for local information exchange as a basis for mobile payment services. In 2011, first Android- and Symbian-based mobile devices with NFC chips appeared on the market and many others have followed since.

Payment and non-payment NFC services rely on the same NFC protocol stack and only differ in how this stack is made use of by the applications mentioned here. The following paragraphs therefore first describe the basic operation of the NFC protocol stack and how it is used for different purposes.

On the physical layer, NFC uses a carrier frequency of 13.56 MHz to transmit data between two NFC devices via inductive coupling as specified in ISO/IEC 14443 [44]. Two transmission modes are specified. The passive mode is used in combination with NFC RFID (Radio Frequency ID) tags that do not have their own power source. In this mode, the originator's RF field is used as an energy source by the passive RFID tag to return a message. The message is usually short in nature and can, for example, be an address of a web page (URL) or a business card entry in vCard [45] format.

It is also possible for two active NFC devices to communicate with each other such as two mobile devices or a mobile device and a point of sales (POS) terminal. In active mode, both devices are equal and transmit and receive alternatively using their own power source. The practical working distance between the two NFC devices is 4 cm, but much longer distances can be achieved with high-gain antennas. As a consequence, sensitive data exchanges must be encrypted and applications using the communication channel for such information need to authenticate each other to prevent over-the-air attacks such as eavesdropping, information theft, and man-in-the-middle attacks. Supported data rates over the NFC air interface are 106, 202, and 424 kbit/s.

The next higher level of the NFC protocol stack defines different message formats. A popular format is the NFC Data Exchange Format (NDEF) [46] defined by the NFC forum. The format defines a general message structure and allows devices to identify what kind of information is contained in the message it has received. The following list gives some examples of message types that can be encoded in an NDEF message:

- Universal Resource Identifier (URI) (web address and email address);
- Plain text;
- Smart poster = text + uri;
- Bluetooth and Wi-Fi parameters;
- Business card (vCard format);
- Signature.

Once an NDEF message has been received, a content-dispatching system forwards the content of the message to an application that has registered itself for a certain content type. This way, different applications can register themselves for different message types and the dispatching system automatically forwards the message to the application that can handle the particular message type. When several applications register themselves for the same message type, a dialog box is usually presented to the user to choose which application should receive the message. When a web address has been received, for example, it could be sent to the web browser while an NDEF message containing a vCard would trigger the address book application that has previously registered itself as a receiver for this kind of content.

Applications on a mobile device can not only receive NFC messages but also transmit them. The address book application, for example, can be extended to send NFC messages, for example, an address book entry to the next NFC-enabled device that comes in range.

Third-party applications can also register themselves as recipients of NDEF messages with a certain content type and can also trigger the transmission of messages themselves.

An example of such an application is the Foursquare mobile application [47]. The aim of this application is to let users "check-in" at certain locations and share their current location with their friends by various means. The traditional "check-in" procedure uses the GPS receiver in a mobile device to pinpoint the user's location. The location of the user is then sent to a database in the network to retrieve all places near this location that are registered with Foursquare. The user can then select the place he is visiting. With NFC tags, this process is significantly simplified as the "check-in" is now performed by the user holding his device close to a Foursquare RFID tag. The type of content of the tag is then forwarded to the Foursquare application on the device, which then registers the user in the network at the current location. NFC functionality can also be used to exchange information between two Foursquare apps on devices of different users to make friend requests and exchange location lists. This is done by launching the Foursquare applications on both devices and then holding the two devices close together.

For mobile payment purposes, the same NFC hardware and software is used as described above with one major addition. As the payment software (e.g., from Visa, Mastercard, American Express, a local transportation company, etc.) and the data identifying the user for the payment process are highly sensitive, it must not be accessible from outside to prevent unauthorized monetary transactions or theft of personal information. Consequently, the payment software and the user data have to be stored in a secured area in the mobile device, the "secure element." One approach is to use the SIM card for this purpose as it is already used as a secured storage element for the user's mobile network subscriber data. Another approach is to include an independent secure element chip in the mobile phone. The banking industry, for example, requires a secure element that is compliant to the Global platform specifications [48].

When the user places his mobile device on a POS terminal, the terminal then sends a message identifying the payment applications it is compatible with. The mobile device or the user may then select one of the payment applications stored in the secure element to perform the transaction. For performing a payment, the secure element requires two interfaces, one to the NFC chip to be able to exchange messages with the POS terminal and another to interact with the user. In other words, there must be an application on the mobile device that can be reached by the secure-element-based payment software; so information such as, for example, the amount to be paid or a PIN input screen can be shown to the user.

Several companies are involved in this value chain. On the one end, there is the bank that issues the payment software and the user identification that is to be stored in the secure element. On the other end is the issuer of the secure element, which is the mobile network operator in case the SIM is used as the secure element, or the device manufacturer or the user himself in case a dedicated chip is used. A tricky aspect in this regard is how the software and information can be securely transferred between the issuer (e.g., the bank) and the secure element, which again depends on who is in control of the secure element. Some types of payments such as, for example, NFC-based entry systems to public transportation systems do not require an interaction with the user at the time the device is brought into contact with the NFC-based entry system. This is the case, for example, when the user buys a weekly or monthly ticket in advance or pre-pays a certain amount of money that is then automatically deducted when the device is held over an NFC reader at an entry or exit gate. In such cases, however, an application might be supplied by the public transportation provider that interacts with the software application on the secure element so the user can later on access details of the purchases made, that is, the trips he has made and the amount of money that was deducted from his account.

It should be noted at this point that there is no standardized process for mobile payments; so different banks/card issuers will have their own software stored in the secure element. Also, the payment procedures may be different in different parts of the world; so it is not certain that the NFC payment processes used in Europe will be compatible with those used in North America. Further details on NFC and mobile payment processes can be found in [49].

5.5.10   Physical Keyboards

One of the main limitations of mobile devices because of their size and primary use as an information consumption device rather than an information creation device is the missing or miniaturized keyboard. This makes text input difficult, slow, and error-prone, which significantly inconveniences activities beyond information consumption such as writing e-mails or longer documents. A solution is external keyboards that are connected to devices via wireless technologies such as Bluetooth. Even though the keys are sometimes slightly smaller than on a full notebook or desktop keyboard, typing with 10 fingers is nevertheless possible. This makes text input for e-mail, mobile blogging, and many other activities very simple and convenient while preserving mobility. Alternatives to foldable or fixed add-on keyboards presented over the years are rollable keyboards and laser-projected keyboards. Over the years, however, such products did not have much success.

With the increasing processing capabilities especially of tablet devices, add-on keyboards further blur the line to traditional computing devices such as notebooks and netbooks with full keyboards that are part of the overall product design. In 2012, processing power and battery capacities already came close to the performance of early single CPU-based netbooks introduced only a few years earlier. The biggest difference between the two categories is thus the operating system and its optimization. Tablet operating systems are tailored for touch input, the use of only a single application at a time, and the functionality of applications is often limited to one purpose without many options. In contrast, netbooks, notebooks, and other "traditional" devices use full desktop operating systems that are optimized for keyboard and mouse input to facilitate fast switching between applications and focus on information creation rather than consumption. Also, programs tend to be more complex to be more versatile.

A trend can be observed to make desktop operating systems more touch friendly for use in the tablet space as well, and it remains to be seen how far such an adaptation can go before usability on traditional devices is negatively affected. Physical keyboards being used as an extension to tablets and other mobile devices will surely influence the changes in mobile operating systems as they extend tablet usage scenarios while at the same time offering a traditional and effective way of interacting with a computing device. This in turn reduces the number of changes required for operating systems to be usable on tablets and also on more desktop-oriented devices.

5.5.11   TV Receivers

Many new hardware functionalities were added to mobile devices over time but not all of them have been successful. One interesting example is mobile TV receivers. In the first edition of this book, DMB, DVB-H, MediaFlo, and MBMS have been looked at from a critical point of view because of the strong competition from Internet-based audio and video streaming applications and downloading content for later use. These services are in many cases better suited for mobile use as content can be consumed at exactly the time the user has time to spend (e.g., while waiting for the bus) rather than only at predefined times. In addition, content that has previously been retrieved can be consumed at places where no or only sporadic network coverage does not allow receiving a continuous data stream. The main advantage of mobile TV was thus broadcasting of live events such as football games, Formula 1 races, and so on. Over the years, this prediction has held and only few devices appeared in Europe and the USA including a digital television receiver. Falling prices for mobile Internet use and Wi-Fi having become a standard feature, mobile devices have made streaming and downloading of content easy and affordable. For the remaining real time services where mobile TV broadcasting has an advantage, pricing and marketing strategies for mobile TV have not found widespread user acceptance. As a consequence, mobile TV services launched in a number of countries were closed again after a few years despite significant investments in standardization, development, and build-out of broadcasting infrastructure.

A few mobile devices appeared on the market that included a standard DVB-T receiver [50] had the advantage of being able to receive the free-of-charge standard terrestrial digital television signal and not the handheld adaptation DVB-H, for which a monthly service fee had to be paid. Nevertheless, these devices had only limited success and no successor models were produced. Since then no further large-scale attempts have been made in practice to include life television broadcast reception in connected devices such as smartphones and tablets.

It should be noted at this point that a number of mobile network operators offer TV streaming services of select content such as soccer games. These offers, however, are based on individual user data streams (IP unicast) rather than broadcasting a single signal to all users. This has the advantage that no special hardware is required in the device for the reception of the stream. Instead, an application on the device receives the TV stream over a mobile Internet connection. The downside of this approach is the amount of data that has to be streamed from a central place, which increases with the number of users and the limit of the number of concurrent user per cell.

5.5.12   TV-Out, Mobile Projectors, and DLNA

With increasing multimedia capabilities and storage capacities, mobile devices are commonly used today for storing pictures and videos and for presenting them to other people. The biggest disadvantage of smartphones compared with prints or to using a tablet or notebook for this purpose is the size of the screen, which is limited by the portability and mobility of the device. One current solution is to include a TV out port so the device can be connected to a standard television set (cf. Figure 5.4). While early devices had an analog TV-out port, current devices feature a mini-HDMI output port and can thus be connected to computer LCD screens and modern television sets. The disadvantage of this solution is that the smartphone needs to be close to the TV set or computer screen as the HDMI signal requires short cables. Also, carrying an HDMI cable and a mini-HDMI converter is often inconvenient.

A potential alternative are miniaturized projectors, which are small enough to fit into ultramobile devices such as smartphones and tablets. First working samples have been shown on occasion [51], and standalone devices are available on the market today. Smartphones equipped with small projectors have, however, so far not gone into mass production. With devices becoming slimmer and slimmer, it is not certain if combined smartphone/projector products would become a success, especially since in the meantime software alternatives to show pictures and videos on other devices such as, for example, the DLNA [52] protocol, which allows streaming data over Wi-Fi to TVs and other connected devices, have become available.

5.6   Multimode, Multifrequency Terminals

While in the past, only a few frequency bands were used for cellular wireless systems, their number has risen rapidly in recent years. Table 5.3 shows some of the bands currently assigned and used in different parts of the world for 3GPP GSM/UMTS/HSPA [53] and LTE [54] networks.

While it is unlikely that further bands will be actively used for UMTS and LTE in Europe in the next few years beyond those listed in the table, it is likely that new bands will be added in the North American market for the use of LTE. This is due to the different ways the bands are defined. While in Europe, large spectrum allocations are put into a single band, North American regulators have split up the available frequency bands

Table 5.3 Popular frequency bands for 3GPP UMTS and LTE

Band	Uplink bands (MHz)	Associated downlink bands (MHz)	Locations used (not exhaustive)
UMTS			
I	1920–1980	2110–2170	Europe, Asia, Australia
VIII	880–915	925–960	Europe, Asia, Australia (GSM refarming)
II	1850–1910	1930–1990	North America
IV	1710–1755	2110–2155	North America
V	824–849	869–894	North America, Australia
LTE			
1	1920–1980	2110–2170	Europe, Asia, Australia (UMTS refarming)
3	1710–1785	1805–1880	Europe, Asia, Australia (GSM refarming)
7	2500–2570	2620–2690	Europe, Asia, Australia
20	832–862	791–821	Europe, digital dividend
4	1710–1755	2110–2155	North America
13	776–787	746–757	North America
17	704–716	734–746	North America

especially in the 700 MHz band in a way to only allow relatively small pieces of spectrum (e.g., 2 × 10 MHz) to form a single band. For spectrum in the 2600 MHz band, the use of Frequency Division Duplexing (FDD) or Time Division Duplexing (TDD) has been left open, which will further add to the number of used bands being put into service over time.

In practice, the growing number of bands has a number of undesired implications for both users and network operators. In 2007, typical mobile devices supported only one or two 3G frequency bands like, for example, the 2100 MHz band for European models or 850/1900 MHz for US models. In addition, such high-end phones usually also supported four bands for 2.5G GSM/GPRS, that is, 900 and 1800 MHz for Europe and 850 and 1900 MHz for the USA. In 2012, band support has risen to typically four UMTS bands, for example, 2100 and 900 MHz for Europe and 850 and 1900 MHz for North America. In some devices, penta-band radios are used that also include the 1700/2100 MHz band combination, which is the third band used in North America for UMTS services. The Nokia N8 was one of the first smartphones that supported five UMTS frequency bands [55].

While connectivity with such devices is currently guaranteed in most places in the world, the support of 5 UMTS bands is far from the 16 bands currently defined in 3GPP for UMTS (excluding the Chinese TDD bands) and 37 bands currently defined for LTE (including the TDD bands). In the future, mobile phones thus have to support additional bands in order to ensure that travelers can use their wireless equipment globally.

For device manufacturers and network operators, the increase in the number of bands is equally disadvantageous. T-Mobile, for example, who have bought spectrum in the USA in band IV (1710–1755 MHz and 2110–2155) [56], have struggled over the years to get a wide variety of devices for its UMTS services in this band as it is specific to the USA and Canada. This means that the volume of devices using this band is very limited compared with global sales of billions of devices for more popular frequency bands. Furthermore, incentives are small to include this local band, only used by a few operators, in mainstream devices since the support of each new frequency band adds to the production cost per device. Since the market grows only insignificantly for a device, if this band is included this reduces sales margins and will thus inhibit production of such mobile devices for global sales. In addition, operators using local bands cannot profit from users roaming into the country from abroad who have at best a quad-band 2.5G or quad-band UMTS device. Already today this is felt by network operators not using GSM or UMTS, since they cannot profit from visiting GSM and UMTS roamers [57].

Adding support for an additional frequency band in a mobile device has no impact on the digital components of the phone since processing of signals once they are digitized are independent of the band. The software for the DSP and the radio protocol stack software of the baseband chip, however, have to be adapted to the additional frequency bands. Examples of such changes are the support of the additional channel numbers of a new frequency band, scanning of all available frequency bands for networks at power-up or while in idle mode and changing messages, and parameters to inform the network of the additional capabilities of the device.

The analog part of the mobile consists of antennas, front-end filters, and RF chips. Here, each additional frequency requires additional hardware components. To limit the number of additional components, hardware designs are usually multiplexing certain components for use with more than one frequency band. As multiplexing and switching components reduce sensitivity, each additionally supported frequency band further decreases the reception sensitivity of the device. As technology improves, this is usually compensated for by improved hardware designs.

It is estimated in [58] that the analog part of a mobile device is responsible for 7–10% of the cost of a mobile device, independent of whether it is a low-, mid-, or high-end handset. This is due to the fact that, while low-end devices are cheaper, they do not support as many radio interfaces and frequency bands as high-end devices. The report also estimates that the hardware cost per supported frequency band is around US$2. For GSM devices this includes the frond-end switching and routing functionality and an additional duplex filter. For an additional UMTS or LTE frequency, additional duplex filtering is required. In addition, a separate antenna is required if the band is too far away from other bands. Furthermore, high-end devices require a special chip and board design to prevent unwanted inter-modulation effects between the cellular, Wi-Fi, Bluetooth, and GPS radio units. The white paper further estimates that, in addition to the hardware costs, the engineering cost for a new band is in the order of US$ 6 million. This includes development costs, type approval, and testing. With annual global GSM 900 MHz phone sales of several hundred million devices, the development costs per device are only a few cents. For national frequencies for which only a few million devices are sold per year, the development costs for supporting the frequency band per device can easily exceed the price of the hardware.

From a development point of view, many manufacturers thus prefer to focus their design resources to decrease the price of their next-generation designs and to improve sensitivity rather than to add exotic frequency bands. In other words mainstream bands attract much more engineering effort resulting in less expensive hardware with higher sensitivity.

In the future, analog hardware in mobile devices is likely to get more expensive since technologies such as HSPA+ and LTE require at least two antennas and receiver chains for the MIMO transmission (cf. Chapter 2). Standards even include data transmission modes for 4×4 MIMO, which requires four separate antennas and receiver chains in mobile devices. As the hardware components cannot be shared between receiver chains, this increases the number of required components and thus increases costs. If in addition to Wi-Fi, Bluetooth, and GPS, several frequency bands for cellular B3G networks are supported, physical limits will limit either the number of MIMO antennas per band or the number of supported bands per device.

With the launch of LTE networks, two issues are further complicating a generic cellular modem design for mobile devices that can be sold globally. The first is the different evolution paths to LTE, which are incompatible to each other. On the one hand, the 3GPP GSM and UMTS standards have evolved to LTE. Devices sold globally should thus incorporate these three radio technologies. On the other hand, the CDMA standard, mostly popular in North America because of its use by Verizon, Sprint, and a few others, also uses LTE as an evolution path. As a consequence, there must be devices for these network operators that support CDMA and LTE. While in the past, LTE devices either contained GSM/UMTS or CDMA for backwards compatibility, first chipset designs now incorporate all radio technologies [59]. This makes it easier in practice to reuse most of the design of mobile devices for different parts of the world. This might thus perhaps lead to devices that are capable of operating in GSM/UMTS, CDMA, and LTE networks. At the time of publication, however, no such products were yet announced.

The second issue that complicates a generic cellular modem design is the intentionally missing circuit-switched voice capability of LTE (cf Chapter 2). While this is not a problem for data centric devices such as tablets and netbooks, smartphones and other devices which are also used for voice calling require a telephony integration that not only works over an Internet connection when an LTE network is available but also over circuit-switched channels provided by CDMA, GSM, or UMTS networks. As discussed in Chapter 4, there are different approaches to this, which are summarized again from a cellular modem point of view, as hardware support is required for each:

- Concurrent use of CDMA and LTE: Early mobile devices required two wireless radio chips; so CDMA voice and LTE data could be used simultaneously [59]. Newer chipset designs now include CDMA circuit-switched and LTE data in a single modem chip. This is also referred to as "Simultaneous Voice and LTE Data" (SVLTE).
- GSM/UMTS network operators have chosen to use a mechanism referred to as CS-fallback. Incoming voice calls are signaled over LTE to the mobile device, which then requests from the network to transfer to a GSM or UMTS overlay network to receive the call over a circuit-switched bearer. The advantage to the dual-mode approach mentioned above is that the device needs to observe only a single network at a time. The downside is that the fallback to another radio technology requires several seconds, which thus significantly increases call setup times.
- Single Radio Voice Call Continuity (SR-VCC): This method is foreseen to be used with IP Multimedia Subsystem (IMS) based IP telephony in LTE networks (cf. Chapter 4). When the user leaves the LTE coverage area, an ongoing voice call can be handed over to a GSM or UMTS circuit-switched channel. At the time of publication, no network operator has deployed this method in practice.

All methods are thought to be temporary in the standardization bodies as the final goal is to have nationwide LTE networks with a coverage area that is equal to that of already installed networks. When such a level might be achieved is difficult to predict. The majority of mobile network operators are unlikely to make this step in the next five years. Until then, companies offering chipsets and devices that have integrated as many solutions as discussed above will have a significant advantage in terms of volume shipments of a single hardware version of a device.

5.7 Wireless Notebook Connectivity

One of the most important factors of the success of Wi-Fi has been Intel's push for an embedded Wi-Fi chip in all notebooks with its mobile Centrino chipset. For some time, Intel was considering repeating this approach with a combined Wi-Fi/HSPA wireless chipset, but they gave up on these plans when their alternative WiMAX strategy emerged. This approach has not been successful, however, and Intel has since given up support on integrated WiMAX solutions.

Instead, several other possibilities have emerged to connect notebooks to cellular networks as shown in the list below. These have gained significant popularity among travelers and users who are using a wireless broadband Internet connection as an alternative to fixed line connectivity via Digital Subscriber Line (DSL) or cable:

- A built-in wireless broadband network card via a mini PCI expansion slot. Such notebooks are sold, for example, by mobile network operators in combination with a 12 or 24 month contract and a monthly fee. While an interesting concept, it has remained a niche market as the majority of people have continued buying their notebooks from other sources.
- A USB stick. This sort of wireless broadband network adapter is completely external and has the advantage that it can be placed in a convenient position and orientation, which is especially important when the signal strength to the wireless broadband network is weak (cf. Chapter 3). Sizes of external USB sticks have shrunk considerably in recent years. Also, prices have reduced significantly and entry-level devices are available in many countries with prepaid SIM cards for Internet access for less than €30–40.
- In some countries, the concept of a wireless broadband router has also had reasonable success. The dedicated router device would connect to notebooks and other devices over Wi-Fi by acting as an access point, and the Internet connection is established over a wireless broadband network such as UMTS or LTE.
- Another connectivity option for netbooks, notebooks, and other devices that enjoys rising popularity is the use of a smartphone or tablet as a Wi-Fi access point in a similar way as the router device above while outside the home or office. This is also referred to as "wireless tethering." While already envisaged in 2006 [60], it took until 2010 when Google introduced this feature in version 2.2 of their Android operating system [61]. The downside of this approach is the limited operation time of a smartphone as the constant operation of cellular radio and Wi-Fi quickly depletes the battery. The use of a power adapter is therefore required for longer sessions. Another unintended consequence is that while routers and wireless broadband USB sticks usually have an advanced antenna design (e.g., a diversity antenna) to improve the performance, this is usually not the case in smartphones. Here, the space for an antenna is much more limited, which results in a performance degradation.

In recent years being able to access the Internet with productivity devices such as notebooks while on the go has become almost ubiquitous in many countries. This has been made possible by falling prices and cheap USB wireless broadband dongles. With monthly charges even on prepaid SIM cards between €10 and 20 for moderate use, this has triggered a significant adoption of cellular network connectivity with notebooks. In Austria, for example, about 30% of broadband subscriptions were already wireless (mostly over HSPA networks) in 2008 instead of DSL and cable [62].

5.8 Impact of Hardware Evolution on Future Data Traffic

In 2003, one of the most sophisticated connected mobile devices available in Europe was the Siemens S55 GSM/GPRS phone. It was one of the first mass market devices with a Bluetooth interface and a stable GPRS protocol stack already in the first version of the device's software. The S55 was also the first device in this product line with a real color display with a resolution of 101×180 pixels and 256 colors. Memory card slots in mobile phones were not yet available, so file storage space in the built-in 1 Mb flash memory was limited to a few hundred kilobytes. Even if the device would have had a multimegabit

B3G network interface, it would not have been possible to take full advantage of such capabilities. The simple Web browser was only suitable to load and display small Web pages designed for mobile use with a size of only a few kilobytes.

Five years later in 2008, the mobile device landscape had changed completely. Cellular data speeds had evolved from 45 kbit/s to several megabits per second, onboard memory has grown from a few hundred kilobytes to several gigabytes and memory slots allowed an expansion of the storage capacity to tens of gigabytes. Device sizes on the other hand have only increased slightly, mostly to accommodate bigger and higher-resolution displays with resolutions exceeding 320×480 pixels and 16 million colors. At the same time, the sales price had remained the same. Web browsers had evolved to handle standard Web pages composed of hundreds of kilobytes of information. This is more than the total available flash memory storage of the S55 from 2003.

From around 2008, Internet applications have started to enjoy mainstream success and required a high amount of data to be transferred over the network. Such applications were and still are today podcasts, videocasts and music downloads. Until a few years ago, users have mostly used cable and DSL access networks to first download such files to the computer and from there transfer them to a mobile device. This process is also referred to as sideloading. With devices now incorporating Wi-Fi and broadband cellular network interfaces, it is no longer required from a technical standpoint to place a computer between the source in the Web and the mobile device. Widespread direct downloading of music files, podcasts, and video via Wi-Fi and cellular networks while having only been envisaged back in 2008 is now a reality and has reached the mass market.

This trend is further strengthened by players such as Google and Intel entering the market who do not see voice telephony as the primary application for a mobile device and a cellular wireless network but rather Internet connectivity itself. With iOS- and Android-based smartphones and tablets having become main stream devices in 2012, over half the mobile devices sold in 2012 in countries such as Germany are Internet-connected devices and hence, are usually connected to the Internet around the clock [63]. Without connectivity, these devices become almost useless. The ideas behind such devices are very different from those of previous mobile phone design as they no longer focus on local applications such as calendars and address books that run well even without network connectivity. On mobile Internet devices, applications such as Web browsing, feed reading, Web radio, on-demand videos, Voice over Internet Protocol (VoIP) telephony, video chatting, messaging, and gaming are prevalent. Calendars and address books are only playing a secondary role and even these applications are tightly integrated with an Internet-based backend; so users synchronize them between different mobile devices and PCs. Over the past five years, this trend has seen data volumes in wireless networks increase every year between 50% and 100% as more and more people have bought a connected mobile device and because of the rising number of connected applications. It should be noted however, that it has been reported by the end of 2011 that this year on year growth has slowed to some extent in some networks. Potential reasons cited for this are more stringent data caps and take-up of mobile Internet-based devices by users who, unlike those who were taking-up these devices earlier, are not using their mobile devices as often or for as many services [64].

Applications that require a significant amount of bandwidth are music and video streaming. Streaming a 30 s video from platforms such as Youtube generates around 2.7 MB of data, that is, around 5.4 MB per minute [33]. Another data-intensive application is radio streaming. Streaming a radio station 8 h a day with a data rate of 128 kbit/s, for example, results in a data volume of over 460 MB per day or 13 GB per month. As users mostly watch video clips and listen to radio at home or in the office, the Wi-Fi interface in mobile devices is an ideal way to offload this type of traffic from the cellular network to DSL and cable connections.

With the examples above it becomes clear that mobile devices today and in the future, especially those dedicated for being used mainly with Internet-based applications will continue to drive the amount of data in mobile networks. Also, first attempts are made to build devices where all applications are executed in a web browser such as, for example, the Google Chrome book [65]. The applications and the data they use are downloaded from a server in the network and only cached locally. While still in its infancy in 2012, this approach might become more interesting to users with broadband networks becoming more and more ubiquitous. It is thus likely that network operators will continue to increase the overall capacity and reach of their cellular networks. This will likely change the overall network design over time as discussed in the next section.

Another future challenge will be the management of several devices per user. Requiring a separate contract or prepaid SIM subscription for each device will become impractical once users are connected with their notebook, their private phone, their business phone, their dedicated MP3 player, an Internet tablet, their car, and so on.

5.9 Power Consumption and User Interface as the Dividing Line in Mobile Device Evolution

Tablets, also referred to as pads, are an interesting new device category since Apple has launched the first iPad back in April 2010. Many manufacturers have followed since with their own devices, mostly based on the Android operating system or Amazon, with their pads primarily designed for reading e-books. Microsoft had been a bit behind other players in this area and has thus decided to follow a different approach than that of Apple and Google.

The software used by Apple and Google for pads initially came from the low end of computing, that is, from smartphones. Smartphones are optimized for low power consumption, they have a low-power CPU, low-power components, and the operating systems were designed around low power consumption as well. Previous mobile operating systems might have been more power optimized but iOS and Android were specifically built from scratch to adapt to this environment. No legacy requirements and application had to be catered for, even though these operating systems use kernels once designed for the desktop PC, for example, Linux in the case of Android. But those kernels were shrunken, unnecessary parts were removed, and the graphical UI was designed from scratch. It was these operating systems that were then subsequently used as the basis for the new tablet device category.

From a user's point of view, the UI on smartphones and pads running the same operating system look very similar, and most applications running on smartphones can adapt to the higher screen resolution of tablets without modification. Pads might have the screen size of netbooks or even small notebooks but they have to be light, which limits battery capacity.

Consequently, the processors used in tablets are the same as those in smartphones. This can be noticed when the processor is asked to do complex tasks such as rendering graphics-intensive web pages including flash content. Such web pages are rendered more slowly than on a PC and scrolling is not as smooth. It would of course be possible to design a pad with a faster processor but this would come at the expense of how long the device will run on a single battery charge. And, when power consumption rises so does the heat generated, which is immediately noticed by the user. The speed compromise works well in most cases but power consumption can be seen as the dividing line to netbooks and notebooks and their operating systems.

Microsoft, however, is approaching tablets from a different point of view in an attempt to do things in a different way compared to their already established competitors. Microsoft has decided to scale down their Windows desktop operating system to run on the ARM platform. In addition, they have also ported their office suite and other desktop programs to run on this platform as well. In theory, only an additional Bluetooth keyboard and perhaps a mouse are required and the tablet might become a replacement for a notebook. Whether this will work in practice remains to be seen for a number of reasons: First, there is the power divide described above. While complex office suites runs smoothly on high-power Intel platforms, it is not yet certain how they will perform on a platform that has only a fraction of the processing power by design to conserve energy and limit the operating temperature. Second, success will also depend on the UI: On a tablet, big buttons and other UI elements are required that can be touched reliably with a finger. Also, current tablet devices usually use the complete display for a single application, which, despite multitasking capabilities, significantly limits the use of a tablet for complex tasks that require the use of several applications simultaneously. Simultaneous use of multiple applications can be made much simpler with a UI that allows instantaneous task switching with the help of a keyboard and a mouse. In other words, the tablet UI today is much more suited for information consumption than for interactive creation of content. A taskbar, for example, is ideal to switch between many applications instantly. A taskbar, however, is usually not part of a tablet UI, where holding a menu button for a second is the typical way to open the task manager.

Both the UI and power consumption are two areas where significant progress is made in each device generation. It is thus likely that both will evolve in a way to enable devices that are light in weight, powerful, and yet have similarly low power consumption as today's devices. This will shift the dividing line and the type of use in the future that today still separate ultramobile devices such as smartphones and tablets on the one hand and nomadic devices such as netbooks on the other.

5.10   Feature Phone Operating Systems

The middleware between the mobile device hardware and applications is the operating system. It decides how applications can access device properties and network resources directly, how the UI looks like, and how the user can interact with the device. Operating systems thus have a significant impact on the usefulness, popularity, and market success of a mobile device. The following sections now describe the most popular mobile operating systems, APIs, and business models and discuss their potential evolution in the future.

For low-end mobile phones, also referred to as feature phones, most device manufacturers are using proprietary, or closed, operating systems today. Their main purpose is voice telephony and text messaging but some of these devices also offer basic Internet capabilities. The use of such phones in many parts of the world is on a steep decline. Nevertheless, it is likely that this device category will still be produced and used in countries with very low general income for quite some time to come, despite smartphones becoming cheaper with some basic smartphones having reached a price level of around €100 [66].

Feature phones typically have no or only limited multitasking support for user applications. Multitasking capabilities are usually restricted to running a single program in the background such as a music player application. Third-party developers have no access to the operating system and can only extend the functionality of such devices with Java programs that are executed in a Java Virtual Machine (JVM) of the Java Platform Micro Edition (ME) environment, originally developed by Sun Microsystems.

5.10.1 Java Platform Micro Edition

The advantage of the Java Platform Micro Edition, also known as Java 2 Micro Edition (J2ME), is that a program does not run on only one device or operating system but across a broad range of different devices and operating systems. Since JVM implementations and available Java packages differ slightly between devices, some adaptations are required for support of a broad range of devices. The downside of a JVM is that applications can only get a generalized access to the operating systems and have only limited capabilities to access data from other applications such as calendar and address book entries. Also, other applications cannot exchange data with Java applications, which makes it difficult, for example, to open a Web page from a link embedded in an application in a Java-based Web browser. This limitation, imposed by the Java sandbox concept, significantly reduces the usability and interaction between different applications on the device. From a security point of view, however, the sandbox ensures that the application cannot gain access to network functions such as sending an SMS or initiating a phone call without the consent of the user. This is of particular importance on mobile devices since, unlike on PCs, programs can accidentally or intentionally cause costs by accessing the network. The Java environment is open to developers and most developers choose to distribute their applications themselves, directly to the users. Users can then download the application file to the mobile phone via the cellular network, via Bluetooth from a PC or by transferring the application file to the device from a PC via a cable.

5.10.2 BREW

Another cross platform application runtime environment is Binary Runtime Environment for Wireless (BREW), developed by Qualcomm. It is mostly used in CDMA-based mobile devices in the USA and Japan. BREW developers have the choice between several programming languages, C, C++, and Java. Similar to the Java ME environment discussed above, BREW offers a cross-platform programming environment but without the restrictions imposed by the Java sandbox approach. To reduce potential security problems and to ensure the quality of applications, BREW applications have to pass rigorous tests in a certification laboratory before they can be distributed. This increases the time to market and reduces the number of developers since certification is not free. Furthermore, most CDMA network operators do not allow customers to install BREW applications themselves. Therefore, developers depend on network operators to distribute their applications. This makes developers dependent on network operators and requires negotiations with many network operators to reach a large customer base. Therefore, the BREW environment and ecosystem, which is controlled by the network operator, only attracts few developers compared with the Java ME environment, which is fully open.

5.11   Smartphone Operating Systems

At the publication of the first edition of this book in 2008, the most popular smartphone operating systems were Symbian and Windows Mobile and the use of features phones that were mainly used for voice telephony and SMS was prevalent. Accessing services and data on the Internet on mobile devices was only used by few people at the time. In only five years, this has changed significantly. Today, smartphones, tablet devices, and use of Internet-based services on those devices are well established in all parts of society and connected mobile devices are used by people of all ages. This has been enabled by more powerful hardware in mobile devices and new players entering the market with new operating systems, new ideas, new business models, and fewer limitations imposed by maintaining backwards compatibility. In only a few years, new players such as Apple and Google have completely reshaped the mobile operating system landscape. While the first edition of this book gave an in-depth description and reasoning of why these new operating systems might become popular in the future, this has become reality at the time of publication of the second edition. As a consequence, this section has been completely rewritten to describe the current state of the art and how this part of the industry might evolve in the future.

5.11.1   Apple iOS

Since the launch of the first Apple iPhone in 2007, this device and its successors have enjoyed tremendous success. The operating system, referred to as iOS, has also been developed by Apple and only runs on devices developed by the company. The limited number of device variants, the release of new phones only about once every 18 months, and full control over the development of the operating system have allowed the company to create a simple to use but powerful mobile operating system that works in the same way across different device generations.

Another important element of the iOS ecosystem is a centralized store to download third-party applications. Apple was not the first company to launch a centralized app store for users to download mobile applications to their device that are either available for free or paid for. Owing to their tremendous end user popularity and the resulting negotiating power with other players in the ecosystem, it was the first company that succeeded in doing so without network operator-imposed restrictions, local adaptations (except for languages), or revenue sharing with network operators. This was difficult and impossible for other companies to achieve, which were already established in the industry because of their dependency on network operator sales channels.

iOS is a closed and proprietary operating system and the source code is not publicly available. Apple also keeps tight control over third-party applications that can only be downloaded via the company's app store. To place an application in the app store, the developer has to adhere to Apple's developer guidelines and agree to a revenue-sharing model with Apple. This model applies for payments at the time of initial purchase via the app store and also later on for in-application purchases. The latter is a popular method for e-book and e-magazine publishers to monetize their content.

On the one hand, this strict control over applications allows the company to weed out malicious applications and thus to protect users by not approving apps in the first place or to remove them quickly from the store and on devices if deemed necessary later on. This tight control over the way applications have to behave and their central download place on the other hand make users completely dependent on the interests, moral viewpoints, and technical ideas of a single company. This has given rise to a number of solutions that offer ways to lift some of the restrictions imposed such as application downloads from outside the app store and the removal of "carrier locks," that is, the use of the device with only a SIM card of one network operator. Such solutions are often referred to as "jailbreak" toolkits.

In summary, because of its success in the marketplace, it seems unlikely that Apple will change its business model of tight control of every aspect of its iOS ecosystem in the next few years. With their focus on the ease of use rather than offering many options, it is likely that their platform will continue to be successful in the foreseeable future.

5.11.2 Google Android

A completely different business model for mobile devices to that of Apple is used by Google with their Android operating system that was launched with a first device in 2008. Except for a few proprietary hardware drivers, every aspect of the system is available as open source. Built on a mainstream Linux kernel that is also used for Linux-based desktop and server distributions such as Ubuntu, Red Hat, Debian, Suse, and so on, most of the additional software such as the mobile optimized UI and the Java-based Dalvik third-party application execution environment has been written by Google.

For third-party applications, Google maintains the "Google Play" app-store. Although applications that are deemed malicious can be removed, there are no restrictions on developers on what kind of applications can be put in the store. Also, there is no review process before applications can be made available. Developers are also free to publish their apps in other application stores or offer them for download on their own web pages. This makes the Google third-party ecosystem fully open in a similar way as programs for desktop PCs are available today. Access rights to different parts of Android such as the address book, calendar, network, GPS, and so on have to be requested by the application at installation time from the user. The user can then either grant all requested rights together or reject them in which case the application is not installed.

Like other Linux-based operating systems, Android applications are executed in user mode and thus do not have direct access to sensitive parts of the operating system to protect the system's integrity. An application, for example, has only access to a private directory and the external memory card, while all other parts of the file system are not accessible. While this is no obstacle and even desired for the majority of applications, there are applications that require more rights to directly access settings and protected parts of the system. An example of such an application is a network monitor program that can inspect and store data, which is exchanged with the network. Such rights are usually only granted to the administrator account of a system, which is referred to as "root" in the Linux world. Android per default does not have a mechanism for an application to request root privileges. There are, however, third-party applications that have found ways to enable the request of root privileges by applications. This is also referred to as "rooting" an Android device. Rooting a device does not give applications root rights by default. When an application requests root rights, there is still an interaction with the user required to allow or deny the request. It should also be mentioned at this point that there is a significant difference between Android device rooting and Apple jail breaking. While the former is a way to grant applications additional rights, the latter is intended to break out of Apple's walled garden business model. Details can be found in [67].

An important part of Google's business model is for Android to be closely linked with many of Google's online services such as the app store, the browser, search, maps, mail, talk (instant messaging), and so on. Although it is convenient, the downside is that users give very detailed personal and usage information to a single company. Owing to the openness of the system, however, many third-party alternatives to those programs are available that are not tied to the Google ecosystem.

Most parts of Android are available as open source and are governed by the Berkeley Software Distribution (BSD) and Apache software licenses (for details see Chapter 6), which allow any company to reuse the code for their purposes. This has given rise to modified versions ("mods") of Android for popular devices of third-party companies such as HTC and Samsung. An example is CyanogenMod [68] that is available for many different phone models. "Mods" have become popular among advanced users who would like to update to newer versions of Android more quickly than update cycles of the device manufacturers permit. Furthermore, such users often do not like the modifications made by the device manufacturers of the original Android UI and thus prefer "mods," which use the original Google UI. In addition, mods are usually already rooted, that is, applications can request more rights to access otherwise protected parts of the operating systems and storage space.

Another aspect of Android's license model is that any company can take the source code and produce an Android-based device outside the default Google development branch. This is referred to as "forking." There are a number of different Android forks such as Amazon's Kindle Fire, a tablet focusing on e-book reading [69].

As discussed earlier in this chapter, support of multitasking, that is, executing several applications simultaneously, is one of the big differences between mobile and desktop devices. Although both the iOS and Android kernels are capable of multitasking, iOS significantly restricts this capability for third-party applications. In effect, iOS only permits one third-party application to run at a time and the application is forced to exit when the user returns to the main screen. This is in contrast to Symbian, for example, that has had full multitasking capabilities for many years without compromising system stability or higher power consumption that are often quoted as being the reason for restricting multitasking on mobile devices. Android's multitasking capabilities are somewhere in between, with applications being permitted to run in the background if programed in a certain way. However, there are no guarantees that applications will continue to run in the background and it can be observed in practice that applications are often terminated after some time of inactivity in the background or when memory becomes scarce. This is quite different to Symbian and desktop operating systems, which can swap memory to the file system when more memory is required for active applications. As RAM in mobile devices continues to increase, this is likely to become less of a problem in the future and it is likely that Android's existing multitasking capabilities for third-party applications will continue to evolve over time to reach similar levels as Symbian and desktop based operating systems.

5.11.3 Android, Open Source, and its Positive Influence on Innovation

When Java was implemented on mobile devices many years ago, third-party companies for the first time had the possibility to write software that runs on many different mobile phones. Many other programming environments have followed over the years and the most popular ones are currently the native programming environment for the iPhone, Android, and Symbian. While those programming environments are very flexible, they nevertheless deliberately set one limitation: All third-party programs are shielded from the operating system. Applications can use the API offered to them but they are restricted from directly accessing any hardware or to interact directly with other parts of the operating system. This has partly changed, however, as the Android operating system is open source.

As Android does not only publish an API for applications running in the Dalvik virtual machine but the complete source code of the operating system, companies interested in offering functionality that requires interaction with the hardware or that directly extends the functionality of the operating system can do so relatively easily. This can be demonstrated with the following two examples:

- **GAN for Android**: Orange UK and T-Mobile US are selling Android-based phones with Kineto's GAN stack that tunnel GSM voice calls over Wi-Fi. Details can be found in [70, 71]. Developed many years ago, GAN depended on Nokia, Samsung, and others to integrate the GAN protocol stack into mobile devices as the programming environments of their phones did not allow third parties to dwell deep enough in their operating system.
- **NFC functionality**: NXP and G&D have worked on integrating NFC and SIM Secure Element functionality into the Android OS [72]. This is also clearly something that would have previously only been possible for device- or operating system manufactures. With Android being open source, this can now be done by third parties to enable their services.

Modifying or enhancing the operating system, however, has one disadvantage from a software distribution point: an application code running outside the environment provided by the API cannot be installed by users on non-rooted devices. In other words, such solutions have to be delivered as part of the Android firmware image of a particular manufacturer unless the code is included by Google in the default Android release cycle.

5.11.4 Other Smartphone Operating Systems

Apart from iOS and Android described above, there are a number of other mobile operating systems in use today. At the publication of the previous edition of this book, Symbian was used by a number of different companies for their smartphones. This has significantly changed with the arrival of iOS and Android for a number of reasons. Like Linux, the Symbian kernel and the various user interfaces such as S60, UIQ, and MOAP have a long history dating back to the first PSION PDAs. In later years, Symbian was developed in collaboration by established device manufacturers among which Nokia took the most active role. After the launch of iOS and Android however, Symbian's popularity saw a quick decline as the companies owning the OS had difficulties adapting the UI for touch input quickly enough. This is perhaps in part because of device manufacturers not being software driven and Internet based compared to Google and Apple, which resulted in slower innovation cycles. Also, their competition was free from legacy requirements and could thus develop ideas more quickly than the incumbent device manufacturers. In 2012, Symbian had evolved to a UI comparable to Android and iOS but by this time it had lost a lot of support in the community. A further contributing factor in this was Nokia's decision to hire a CEO who preferred Windows Phone as the new operating system for the company and who abolished the previous succession path from Symbian to Meego. Meego, a successor of Nokia's Memo and Intel's Moblin is based on a Linux kernel and got high praises by industry observers when launched on the N9 in a few countries. Despite its success, however, it was decided not to develop the system further until today. The future of Symbian and Meego is therefore uncertain and unless Nokia's course taken in 2011 is reversed, these operating systems are likely to disappear from the market.

Microsoft's Windows Mobile saw an equal drop in popularity at the end of the 2000s and Microsoft took the drastic step of designing a completely new mobile operating system, which became Windows Phone 7. With no backwards compatibility to Windows Mobile, Microsoft was free to innovate on various aspects of the operating system, especially on the UI, which is based on tiles unlike any other mobile operating system today. From a business model point of view, Microsoft has chosen a similar walled garden model as Apple with iOS, which means a closed-source operating system, an app store controlled by Microsoft without the possibility of loading third-party application from other sources. Also, multitasking capabilities are severely limited compared to Symbian and Android. Unlike Apple, however, Microsoft had initially chosen not to develop the hardware itself but to license the operating system to third-party companies. In 2012, companies such as HTC, Samsung, and LG produce Windows Phone 7 devices but only few models have appeared so far, perhaps because of the fact that Microsoft gives strict guidance on the hardware to be used for the devices and does not allow the manufacturer companies to modify the operating system. This approach has not led to an initial success and it remains to be seen how the OS will fare in the future. One possibility might be that the cooperation between Microsoft and Nokia may be intensified in the future to give Microsoft more control over the hardware itself. With Microsoft's decision to use a derivate of the Windows 8 desktop operating system for its new tablets, it is also not clear how the Windows Phone 7 operating system will evolve in the future as the evolution toward a combined smartphone and tablet operating system is thus in effect blocked. With smartphone hardware becoming more and more powerful and the desktop Windows operating system becoming more adapted to the tablet domain, there are speculations that at some point the desktop OS might also be used for smartphones.

5.11.5   *Fracturization*

Compared with the desktop computing world with its three main operating systems, Windows, Mac OS, and Linux, with relatively stable market shares, the mobile device landscape has been much more diverse and fluctuant over time. In addition to the operating systems and APIs discussed in this chapter, there are many other proprietary operating systems used in mid-tier and low-end mobile phones. This makes it difficult for developers to design mobile applications across a wide range of different devices. In recent years, two dominant platforms have emerged with iOS and Android, while other previously dominant platforms such as Symbian are being phased out.

This concentration on two or perhaps three major mobile operating systems in the future has made it easier for developers, and the situation is now similar to the PC world. iOS only runs on few device versions and thus is the easiest of all platforms to develop for from a software point of view. Android is used on very different devices in terms of screen resolution, processor, and graphics capabilities, and several versions of the operating system are in active use. Although most programs can be written in a generic way to accommodate for these differences, this environment is especially difficult for applications that are hardware dependent. 3D games are a typical example. Despite having a common API for 2D and 3D graphics programming as described above, the difference in processing speed of both the application processor and the graphics unit has to be taken into account.

On the positive side, diversity helps reduce the effect of malware. Although today the threat from malware such as viruses on mobile devices is still small, it is likely that this area will get more attention in the future as the number of users and devices increases. The more different operating systems and device combinations are on the market, the more difficult it is for a virus or other harmful program to propagate from one system to another.

It is interesting to note at this point that the market share of the major mobile operating systems is the reverse compared to their desktop counterparts. Microsoft Windows enjoys an overwhelming desktop market share and is followed at a distance by Apple. Linux only has a miniscule percentage of the overall desktop market share. In the mobile world, however, Linux-based Android is now the most popular mobile operating system. It is followed by Apple's iOS, which is in turn followed in the distance by Microsoft's Windows Phone operating system.

With services and applications now used on both mobile and desktop-like devices, it will be interesting to observe how this will influence market shares on both sides. Microsoft is hoping to replicate their success in the desktop market by developing versions of their desktop operating system for tablets and perhaps in the future for smartphone-like devices, thereby hoping to change the current market share situation in the mobile space to their advantage. Google tries to increase the take-up of web services that will run on any underlying operating system, thus securing their cloud-based business model. Apple's approach to control the ecosystem from end to end continues to attract many new users to their mobile devices and this popularity also had a positive effect on their desktop market share in the past few years. It is thus very likely that unlike in the PC world where Microsoft dominated the operating system market for many years, there will be at least two or three thriving ecosystems in the mobile space in the foreseeable future with the potential to change the balance of power in the desktop world as well.

5.12 Operating System Tasks

Today, operating systems for mobile devices have reached a level of functionality and complexity equal to operating systems for PCs and notebooks. With the introduction of Linux as an operating system for mobile devices, there is no longer even a difference from a practical point of view. The following section now takes a look at the basic building blocks and functionalities of a high-end mobile operating system such as Android, iOS, Symbian, and Windows Phone.

5.12.1 Multitasking

Operating systems of all mobile devices, from entry level to high-end devices, must be capable of multitasking as there are a number of tasks that need to be performed quasi simultaneously. The most important task of a connected mobile device, even while not communicating with the network, is to monitor periodic transmissions of broadcast information from the network. This is important to stay synchronized and to receive paging messages for incoming calls and SMS messages. In addition, the device also needs to react to user input and to execute the code for the required action, like, for example, updating the display as the user moves between applications or from one menu level to another. While, on simple devices, multitasking is limited and the execution of several user applications is not possible, smartphone operating systems such as Android and Symbian offer third-party application multitasking support, which includes the execution of several user applications in addition to all tasks required for staying connected with the network and dealing with all other external interfaces such as Wi-Fi, Bluetooth, USB, and so on.

While multitasking on the application layer is usually not time-critical, monitoring the network and making decisions about moving to another cell while in idle mode is a time-critical process and the processor has to be available at specific times to analyze incoming information. In smartphones, these tasks are usually managed by a dedicated processor. One processor thus runs a proprietary operating system and all software required to communicate with the cellular network. It is often referred to as the baseband processor, while the operating system visible to the user runs on one or more application processors.

Linux and other operating systems use pre-emptive multitasking. This means that a task cannot block other tasks from running as each is interrupted by the processor when its allocated time slice has been used up and the application has not yet returned control to the operating system.

5.12.2 Memory Management

Programs, also referred to as tasks, running in a multitasking environment do not only share the processor but also the available main memory (RAM), where programs and data are loaded from flash memory, sometimes also referred to as the flash disk, before they can be executed. Management of the main memory is therefore another important function of the operating system.

The operating system must ensure that programs can be executed no matter where they were loaded in memory. This is required since the order in which programs are started and stopped is not known to the operating system in advance. To make an unpredictable place in memory predictable for a program, virtual memory addresses are used in highend mobile operating systems in the same way that they are used in PC operating systems. When a program is prepared to run by the operating system for its timeslice, the microprocessor's memory management unit is instructed by the operating system how to map the virtual memory addresses known to the program to real memory addresses. While the program is executed the memory management unit of the processor transparently translates the virtual memory addresses used by the program into physical addresses for each command. This mapping has the additional benefit that a program cannot access the memory of another task since the memory management unit would never map a virtual memory address to a physical memory address belonging to another task. For additional security, the memory management unit can only be configured by the operating system, as code running outside the operating system's scheduler does not have the permission to access the unit.

As RAM is expensive and thus a scarce resource, most high-end operating systems have the ability to use a part of the flash disk as a swap space. If the operating system detects it is running out of memory, it starts removing parts of programs and data to the swap space, which cannot directly be accessed by the processor. If a program requires access to data that has previously been swapped out, the operating system interrupts the program and retrieves the data from the swap space. Afterwards, the program is allowed to continue. In practice this works quite well in a multitasking system because only a few applications are actively running, although many applications might be loaded into the main memory. Most applications are usually in a dormant state while the user does not interact with them. A practical example of this is a Web browser that the user starts once and then leaves running while using other applications like the calendar, the notes application or the photo application to take a picture. While the Web browser application is still in memory, it is not scheduled for execution by the operating system as it does not interact with the network or the user (assuming a static Web page without Java Script or flash content, both of which can run in the background). Thus, the application is completely dormant. If in such a situation the main memory is almost fully used, the operating system can start swapping out parts of the memory required by the Web browser to the flash disk and use it for another program. For this purpose the main memory is divided into pages. The operating system is aware when each page was last used and it can decide which pages to swap out to the flash disk once memory gets scarce. As memory is organized in pages, a program does not have to be fully dormant before the operating system can swap out some of its pages. Even if active in the background, there is usually always some program code or data that has not been used recently and can thus be swapped out in the hope that not being used recently also means that this part of memory will also not be used again soon.

It is interesting to note that while Symbian can make use of memory swapping to flash memory, this is not part of Android so far. Instead, it was decided to terminate background programs if memory was required.

5.12.3   File Systems and Storage

In the past few years the amount of internal storage space on mobile devices for applications and data such as pictures, videos, and music files has skyrocketed. Today, device internal memory for storage has reached 16–32 GB in high-end devices and this trend is likely to continue in the future. For a short time, hard disks were used in mobile devices to reach high capacities. With falling flash memory prices, the popularity of hard disks has decreased as flash memory is much smaller, requires less energy and is more robust against vibrations and shocks. Most devices also have an external memory slot for removable flash memory cards. As these cards can be used to exchange data with other devices including PCs and notebooks, the FAT (File Allocation Table) or FAT32 file systems are used on such cards, which were originally developed by Microsoft many years ago. While due to its age it is not the most sophisticated file system standard, it is supported by all major operating systems today including Microsoft Windows, Unix/Linux and Apple Mac OS. It is thus the best choice for use in mobile devices.

5.12.4   Input and Output

Like any other operating system, mobile operating systems are abstracting devices attached to the system for applications. The display is the first example that comes to mind as it is the main output device to interact with the user. The display is connected to the graphics device which in turn is connected to the processor via a bus system. The operating system then abstracts the function of the graphics card into an API that can be used by applications. These APIs offer a wide variety of functions to applications ranging from simple primitives of drawing lines and shapes, to printing text at a certain location on the screen to generating graphical menus and buttons. The keyboard is a typical input device, again abstracted for programs by the operating system. A keyboard driver receives keyboard input (a key was pressed) and the operating system is then responsible for passing that information to the currently active program.

Other external input and output devices in mobile devices are, for example, Wi-Fi and Bluetooth interfaces, GPS receivers, FM radios, cameras, TV video output, touch screens, and so on, as shown in Figure 5.4. To connect these external devices with the chipset, a number of different I/O bus systems are used.

Only a few years ago, a Universal Asynchronous Receiver/Transmitter (UART) interface was also used to connect mobile devices via a cable to the serial interface of a PC or notebook for exchanging address book entries and to establish data calls to the Internet, to another computer or to a FAX machine via the mobile phone. On the PC, the UART interface is limited to speeds of around 110 kbit/s. While sufficient for many applications, such a standard serial interface is no longer capable of transporting data exchanged via broadband wireless networks as data rates now exceed several megabits per second. In recent years, serial interfaces have been replaced by USB, which is also a serial bus system but capable of much higher speeds. Until recently, most mobile devices were equipped with a USB 1.1 interface with speeds of up to 11 Mbit/s. While almost increasing transfer rates by a factor of 10 and being sufficient for using the mobile device as a broadband wireless network interface for a PC or notebook, USB 1.1 quickly became a bottleneck for applications such as transferring music files, videos, pictures and maps between a mobile device and a PC. Consequently, high-end mobile devices now use USB 2.0 (also referred to as USB Hi-Speed) as an external interface. With data rates of up to 480 Mbit/s, the bottleneck has now moved from the transfer capabilities of the interface to how fast the processor and operating system can send and receive data and how fast that data can be written to the flash disk.

While USB 2.0 will be sufficient for connecting mobile devices to PCs for the next few years, USB 3.0 is now shipped as part of desktop and notebook computers with a peak data rate of 4 Gbit/s.

5.12.5 Network Support

As one of the main purposes of a mobile Internet device is to connect the user with other devices and people via a wireless network, support of different network types and interfaces and their abstraction for applications is another important task of mobile operating systems. Applications are usually not aware of the type of a network interface and instead request the creation of a Transmission Control Protocol (TCP) or User Datagram Protocol (UDP) IP connection to another device from the operating system. If no network connection is established at the time of the request, the operating system either decides on its own to connect to a network via one of the wireless interfaces or opens a dialog box to allow the user to select an appropriate network and configuration. Once a connection to a network is established, the operating system processes the application's TCP or UDP connection request and program execution continues. For many years now, the Internet community has been trying to migrate the current version of the IPv4 to IPv6 to counter the diminishing number of available IP addresses. This has proven to be a difficult process mainly on the network side and is due, in part, to a lack of IPv6-capable applications. Nevertheless, mobile operating systems such as Android and Symbian already support IPv6 and are shipped with IPv6 capable web browsers and other applications.

5.12.6 Security

In the days when connected mobile devices were only used for phone calls, there was little danger from external attackers gaining access to the mobile phone and the data inside. The reason for this was that the network itself isolated the devices from each other via a switching center and all commands exchanged were originated, terminated or filtered on the switching node. In addition, hardware and software of such devices were simple (cf. Figures 5.1 and 5.3) and thus offered few if any opportunities for external attacks. Today's sophisticated mobile devices are connected to the Internet, however, and are thus more and more exposed to the same kind of security threats as PCs and notebooks. It will therefore become increasingly important in the future for the operating system to defend the device against attacks or exploits. In practice there are many ways for malicious programs to gain access to a system:

- **Malicious programs** — a program should only be installed on a mobile device if its origin is known and trusted. It is therefore important that users realize that installing programs is a potential security risk and should only be done if the source is trusted.

Once a program is installed, operating systems like Linux protect the integrity of the system by executing the program in user mode, which prevents programs from making changes to the system configuration. A malicious program, however, still has access to the user's data and thus could potentially destroy or modify data without the consent of the user. Spyware, sometimes, also referred to as a Trojan horse, goes one step further and send private data it has found to a remote server on the Internet. The Symbian operating system uses a slightly different approach to application security. As described in Section 5.10.2, application developers have to get a certification for their program from an independent body before they can be distributed. Programs which do not need direct access to drivers and other lower layer components of the operating system can be distributed without being certified. The user is then informed during the installation process that the program has not been certified, which actions it wants to perform (e.g., access to the network, access to the file system, etc.) and that this presents a certain security risk. The user can then choose to abort the installation or to proceed. As most programs do not require access to lower-layer operating system services, this is the most common distribution method. Noncertified Symbian applications have similar capabilities as described for Linux applications in user mode and can thus also potentially corrupt or steal user data.

- **Peripheral software stack attacks**—another angle of attack is trying to break the software stack of network peripherals. Several well-known attacks on the Bluetooth protocol stack used malformed Bluetooth packets. In this way it was possible to access the calendar and address book on some devices without the consent or knowledge of the user. While such an attack is still possible today, fewer reports about successful attacks have been published recently and it appears that most mobile device manufacturers have done their homework. If a new vulnerability is found it is important that the manufacturer can and does react quickly and provides a patch, as is done in the PC world today.
- **Web browser attacks**—such attacks, which are quite common in the desktop PC world, try to exploit vulnerabilities of the browser software, for example, of the JavaScript implementation, to break out of the browser environment to execute system commands. Other attacks aim at potential vulnerabilities of plugins such as PDF or Flash, which can be tricked into executing infiltrated code or launch system commands with malformed documents or video files.
- **Attacks over the IP network**—in the PC world, attacks aimed at server programs waiting for incoming connections are also widespread. Like in the Web browser example above, such attacks exploit an operating system and processor weakness known as stack overflow, sometimes also referred to as buffer overflow. When one function in a program calls another, the return address is stored in a part of the memory referred to as the stack. Once the function has performed all its tasks the program returns to the previous function via the memory address stored on the stack. In addition to storing the return address, the stack is also used to store temporary data used by the called function. If the function does not ensure that the amount of memory is sufficient for incoming data, for example, from the network, then the return address and other variables can be overwritten by the incoming data. Under normal circumstances, this would result in a program fault as the program can no longer return to the previous function and the application would be terminated. This weakness, if not properly handled by the

program, can be used by malicious exploits to send a specific stream of data that overwrites the return address with a value that points to the data that was sent over the network. Instead of returning to the original function, control is given to the code which was contained in the data sent over the network. This code can then exploit further weaknesses in the operating system to gain higher operating system privileges to load further program code and to install itself in the system. While such exploits have mainly hit Microsoft's Windows operating system due to its widespread use, other operating systems such as Linux and Mac OS are by no means immune to this issue. One of the few operating systems immune to this kind of attack is Symbian as it uses descriptors that prevent buffer overflow attacks.

For the moment, there have only been a few reports about widespread or planned attacks on connected mobile devices. One reason for this is that their number compared with PCs and notebooks on the Internet is still small. Therefore, programs tailored to attack a specific type of mobile device would not find many targets yet. This also limits the spread of a virus from one mobile device to another as the virus would not work if it attacked a device such as a PC that runs a different operating system. As the number of mobile devices grows, however, there are no guarantees that mobile devices will not come under attack from viruses and other exploits in the future. Quick reaction from manufacturers to provide patches and systems automatically updating themselves is thus likely to become equally important on mobile devices as in the PC world today.

References

1. GSM Arena (2008) Panasonic GD 55 — Full Phone Specifications, http://www.gsmarena.com/panasonic _gd55-372.php (accessed 2012).
2. Asus eeePC Product Homepage, http://web.archive.org/web/20080622074016/http://usa.asus.com /products.aspx?l1=24&l2=164 (accessed 30 April 2012).
3. ARM (2008) http://www.arm.com/ (accessed 2012).
4. Krazit, T. (2007) ARM says it's Ready for the iPhone. Cnet Newsmaker, June 22, 2007, http://www.news .com/ ARM-says-its-ready-for-the-iPhone/2008-1006_3-6192601.html (accessed 2012).
5. HTC (2007) T-Mobile Unveils the T-Mobile G1–the First Phone Powered by Android, http://web .archive.org/web/20110712230204/http://www.htc.com/www/press.aspx?id=66338&lang=1033 (accessed 2012).
6. The Digital Electronics Blog (2008) Soft Macro Vs Hard macro? September 2008, http://digitalelectronics .blogspot.de/2008/01/soft-macro-vs-hard-macro.html (accessed 2012).
7. Smith, M. (2012) Orange Announces First Intel-Powered Android Phone for Europe, Codenamed Santa Clara, February 26, 2012, http://www.engadget.com/2012/02/26/orange-santa-clara-intel-medfield/ (accessed 2012).
8. Freescale Semiconductor (2006) i.2xx Platform Family Product Brief, document number I2XXPB, Revision 3, June 2006.
9. Sauter, M. (2006) *Communication Systems for the Mobile Information Society*, Chapter 1.7.3, John Wiley & Sons, Ltd, Chichester.
10. ARM (2008) ARM7TDMI, http://www.arm.com/products/CPUs/ARM7TDMI.html (accessed 2012).
11. ARM Google Android & Chrome OS Support, http://www.arm.com/community/software-enablement /google/index.php (accessed 30 April 2012).
12. Schimpi, A. (2012) Intel's Medfield & Atom Z2460 Arrive for Smartphones: It's Finally Here, January 10, 2012, http://www.anandtech.com/show/5365/intels-medfield-atom-z2460-arrive-for-smartphones (accessed 2012).
13. Qualcomm (2011) Snapdragon S4 Processors: System on Chip Solutions for a New Mobile Age, October 7, 2011, https://developer.qualcomm.com/download/qusnapdragons4whitepaperfnlrev6.pdf (accessed 2012).

14. ST Ericsson Novathor L9540, http://www.stericsson.com/products/L9540-novathor.jsp (accessed 30 April 2012).
15. ST Ericsson Novathor U8500, http://www.stericsson.com/products/u8500-novathor.jsp (accessed 30 April 2012).
16. Renesas Renesas Mobile Introduces First Integrated LTE Triple-Mode Platform Optimised for Full-featured, High Volume Smartphones, http://renesasmobile.com/news-events/news/news-20120215-LTE -Triple-Mode-Platform-Optimised-For-Full-Featured-High-Volume-Smartphones.html (accessed 30 April 2012).
17. Nvidia Tegra 2, http://www.nvidia.com/object/tegra-2.html (accessed 30 April 2012).
18. Mediatek MediaTek Launches MT6575 Android Platform, http://www.mtk.com.tw/en/News/news_content .php?sn=1052 (accessed 30 April 2012).
19. ARM (date unknown) Achieving Stronger SIM-lock and IMEI Implementations on Open Terminals Using ARM Trustzone Technology, http://infocenter.arm.com/help/topic/com.arm.doc.prd29-genc-009492c /PRD29-GENC-009492C_trustzone_security_whitepaper.pdf (accessed 2012).
20. Texas Instruments (2007) WiLink 6.0 Single-Chip WLAN, Bluetooth and FM Solutions, http://focus .ti.com/pdfs/wtbu/ti_wilink_6.pdf (accessed 2012).
21. Khronos Group OpenGL ES–The Standard for Embedded Accelerated 3D Graphics, http://www.khronos .org/opengles/ (accessed 30 April 2012).
22. Jacobs, D. (2010) iPhone Application Programming CS 193P Lecture 19 OpenGL ES, March 9, 2010, http://www.youtube.com/watch?v=_WcMe4Yj0NM (accessed 2012).
23. Becker, L. (2012) Marktforscher: Apple größter Smartphone-Hersteller 2011, February 15, 2012, http://www.heise.de/newsticker/meldung/Marktforscher-Apple-groesster-Smartphone-Hersteller-2011-1434905.html (accessed 2012).
24. Klug, B. and Shimpi, A. (2011) Qualcomm's New Snapdragon S4: MSM8960 & Krait Architecture Explored, July 10, 2011, http://www.anandtech.com/show/4940/qualcomm-new-snapdragon-s4-msm8960-krait-architecture/4 (accessed 2012).
25. Nvidia (2011) Variable SMP–A Multi-Core CPU Architecture for Low Power and High Per-formance, http://www.nvidia.com/content/PDF/tegra_white_papers/Variable-SMP-A-Multi-Core-CPU -Architecture-for-Low-Power-and-High-Performance-v1.1.pdf (accessed 2012).
26. Sauter, M. (2011) Video Stream Data Rates, December 1, 2011, http://mobilesociety.typepad.com/mobile _life/2011/12/video-stream-data-rates.html (accessed 2012).
27. Qualcomm Snapdragon Processors, http://www.qualcomm.eu/products/snapdragon (accessed 30 April 2012).
28. Globalfoundries 28 nm, http://globalfoundries.com/technology/28nm.aspx (accessed 30 April 2012).
29. Corning (2012) Corning Delivering Corning® Gorilla® Glass 2 for Next Generation Consumer Electronic Devices, February 27, 2012, http://www.corning.com/news_center/news_releases/2012/2012022701.aspx (accessed 2012).
30. Anscombe, N. (2000) Methanol Fuel Cells Seen as Mobile Phone Power Source. EETimes (Aug 2000), http://www.eetimes.com/story/OEG20000208S0036 (accessed 2012).
31. Litchfield, S. (2012) For Those Who Were Wondering about the Apple iPhone 4S's Camera vs the N8, October 5, 2012, http://www.allaboutsymbian.com/news/item/13334_For_those_who_were_wondering _a.php (accessed 2012).
32. Litchfield, S. (2010) Camera Nitty Gritty: Supplemental — EdoF, April 15, 2010, http://www.allabout symbian.com/features/item/Camera_Nitty_Gritty_Supplemental-EDoF.php (accessed 2012).
33. Sauter, M. (2012) Youtube Data Rates to Smartphones, February 17, 2012, http://mobilesociety.typepad .com/mobile_life/2012/02/youtube-data-rates-to-mobiles.html (accessed 2012).
34. Niccolai, J. (2007) Nokia Ships Its First GPS Phone, March 22, 2007, http://www.pcworld.com/article /130064/nokia_ships_its_first_gps_phone.html (accessed 2012).
35. Qualcomm (2011) GPS and GLONASS: "Dual-core" Location for Your Phone, December 14, 2011, http://www.youtube.com/watch?feature=player_embedded&v=gB9pXInwh3k (accessed 2012).
36. Layar Wikipedia. http://www.layar.com/layers/goweb3dwiki/ (accessed 30 April 2012).
37. Nokia Nokia N80 Support Pages, http://europe.nokia.com/find-products/devices/nokia-n80 (accessed 30 April 2012).
38. Nokia Nokia N95 Support Pages, http://europe.nokia.com/support/product-support/nokia-n95 (accessed 30 April 2012).

39. Wi-Fi Alliance Wi-Fi Direct, http://www.wi-fi.org/discover-and-learn/wi-fi-direct (accessed 30 April 2012).
40. Sauter, M. (2012) Droidsheep: Firesheep Moves to Android, January 4, 2012, http://mobilesociety.typepad .com/mobile_life/2012/01/firesheep-moves-to-android.html (accessed 2012).
41. Sports-Tracker Turn Your Mobile Into a Social Sports Computer, http://www.sports-tracker.com/ (accessed 30 April 2012).
42. FeliCA Networks KESAKA e-Key Service, http://www.felicanetworks.co.jp/en/cases/others.html (accessed 30 April 2012).
43. The NFC Forum FAQ, http://www.nfc-forum.org/resources/faqs (accessed 30 April 2012).
44. WG8 (2010) ISO/IEC 14443. *Proximity Cards (PICCs)*, October 25, 2010, http://wg8.de/sd1.html#14443 (accessed 2012).
45. Dawson, F. and Howes, T. (1998) vCard MIME Directory Profile, September 1998, http://tools.ietf.org/html /rfc2426 (accessed 2012).
46. NXP Semiconductors (2009) NFC Forum Type Tags, http://www.nfc-forum.org/resources/white_papers /NXP_BV_Type_Tags_White_Paper-Apr_09.pdf (accessed 2012).
47. Perez, S. (2012) Foursquare Adds NFC Support to its Android App, February 10, 2012, http://techcrunch.com/2012/02/10/foursquare-adds-nfc-support-to-its-android-app/ (accessed 2012).
48. Clark, S. (2012) EMVCo Picks GlobalPlatform Secure Element Compliance Programme, February 21, 2012, http://www.nfcworld.com/2012/02/21/313555/emvco-picks-globalplatform-secure-element -compliance-programme (accessed 2012).
49. Vétillard, E. (2012) NFC Payments 101, February 16, 2012, http://javacard.vetilles.com/2012/02/16/nfc -payments-101/ (accessed 2012).
50. Mobile Gazette (2008) LG HB620T with DVB-T, May 1, 2008, http://www.mobilegazette.com/lg-hb620t -08x05x01.htm (accessed 2012).
51. Blass, E. (2007) Hands-on with Texas Instruments' Cellphone Projector, Engadget, September 2007, http://www.engadget.com/2007/09/20/hands-on-with-texas-instruments-cellphone-projector/ (accessed 2012).
52. DLNA dlna Connect and Enjoy, http://www.dlna.org. (accessed 30 April 2012).
53. 3GPP (2012) User Equipment (UE) Radio Transmission and Reception (FDD). TS 25.101, Table 5.0, version 11.1.0.
54. 3GPP (2012) Evolved Universal Terrestrial Radio Access (E-UTRA); User Equipment (UE) Radio Transmission and Reception. TS 36.101.
55. Sauter, M. (2010) The Nokia N8–Pentaband UMTS, July 1, 2010, http://mobilesociety.typepad.com /mobile_life/2010/07/the-nokia-n8-pentaband-umts.html (accessed 2012).
56. Ziegler, C. (2006) T-Mobile Details 3G Plans, Engadget, October 2006, http://www.engadgetmobile.com /2006/10/06/t-mobile-details-3g-plans/ (accessed 2012).
57. Sorensen, C. (2008) Telus Considers Dumping its 'Betamax' of Wireless Networks, The Toronto Star, January 2008, http://www.thestar.com/Business/article/293353 (accessed 2012).
58. Varrall, G. (2007) RF Cost Economics for Handsets, RTT, May 2007, http://www.rttonline.com/Research /RF%20Cost%20economics-Handsets-white%20paper.pdf (accessed 2012).
59. Klug, B. (2011) HTC Thunderbolt Review: The First Verizon 4G LTE Smartphone, April 27, 2011, http://www.anandtech.com/show/4240/htc-thunderbolt-review-first-verizon-4g-lte-smartphone (accessed 2012).
60. Sauter, M. (2006) A Nokia N80 as WLAN Access Point–A Double Blow for Mobile Operators? August 2, 2006, http://mobilesociety.typepad.com/mobile_life/2006/08/a_nokia_n80_as_.html (accessed 2012).
61. Sauter, M. (2011) Android Wi-Fi Tethering–Great But Watch The Battery, April 5, 2011, http://mobile society.typepad.com/mobile_life/2011/04/android-wi-fi-tethering-great-but-watch-the-battery.html (accessed 2012).
62. Sauter, M. (2008) Wireless now Accounts for a Third of Austria's Broadband Connections, February 2008, http://mobilesociety.typepad.com/mobile_life/2008/02/wireless-now-ac.html (accessed 2012).
63. Bitkom (2012) Zeitenwende auf dem Handy-Markt, February 15, 2012, http://www.bitkom.org/de /presse/8477_71243.aspx (accessed 2012).
64. Bubley, D. (2011) Has Mobile Data Growth Flattened Off? Are Caps & Tiers Working Too Well? November 11, 2011, http://disruptivewireless.blogspot.fr/2011/11/had-mobile-data-growth-flattened-off.html (accessed 2012).

65. Google Introducing Chromebooks, http://www.google.com/intl/en/chrome/devices/ (accessed 30 April 2012).

66. Catanzariti, R. (2010) Huawei IDEOS (U8150) Android Smartphone, November 11, 2010, http://www.pcworld.idg.com.au/review/mobile_phones/huawei/ideos_u8150/363402 (accessed 2012).

67. Sauter, M. (2011) Rooting Does Not Equal Jailbreaking, June 1, 2011, http://mobilesociety.typepad.com/mobile_life/2011/06/rooting-does-not-equal-jailbreaking.html (accessed 2012).

68. CyanogenMod, http://www.cyanogenmod.com/ (accessed 30 April 2012).

69. Spurbeck, J. (2011) Three Android Forks that Exist Today, September 17, 2011, http://news.yahoo.com/three-android-forks-exist-today-135000744.html (accessed 2012).

70. Campbell, S. (2010) T-Mobile Offering WiFi Calling on Android Phones with UMA, October 7, 2010, http://fixed-mobile-convergence.tmcnet.com/topics/mobile-communications/articles/107057-t-mobile-offering-wifi-calling-android-phones-with.htm (accessed 2012).

71. Beren, D. (2010) Kineto Wireless Brings UMA to Android, Will T-Mobile Adopt? September 21, 2010, http://www.tmonews.com/2010/09/kineto-wireless-brings-uma-to-android-will-t-mobile-adopt/ (accessed 2012).

72. Clark, S. (2011) NXP and G&D Confirm Joint Android NFC Secure Element Interface, February 15, 2011, http://www.nfcworld.com/2011/02/15/36027/nxp-and-gd-confirm-joint-android-nfc-secure-element-interface/ (accessed 2012).

6

Mobile Web 2.0, Apps, and Owners

6.1 Overview

In addition to telephony services and mobile devices discussed in the previous chapters, Internet applications are another important driver for the evolution of wireless communication. After all, it is the use of applications and their demand for connectivity and bandwidth that drives network operators to roll out more capable fixed and wireless IP-based networks. This chapter looks at the application domain from a number of different angles.

In the first part of this chapter the evolution of the Web is discussed, to show the changes that the shift from "few-to-many communication" to "many-to-many" brought about for the user. This shift is often described as the transition from Web 1.0 to Web 2.0. However, as will be shown, Web 2.0 is much more than just many-to-many Web-based communication.

As this book is about wireless networks, this chapter then shows how the thoughts behind Web 2.0 apply to the mobile domain, that is, to mobile Web 2.0. Mobility and small-form factors can be as much an opportunity as a restriction. Therefore, the questions of how Web 2.0 has to be adapted for mobile devices and how Web 2.0 can benefit from mobility are addressed. During these considerations it is also important to keep an eye on how the constantly evolving Web 2.0 and mobile Web 2.0 impacts networks and mobile devices.

Native applications on mobile devices have also undergone a tremendous evolution in recent years and have become at least as important as web-based services for improving the way we communicate and gather information. Differences and commonalities to web-based applications are described and an introduction to programing in the Android environment is given.

In a world where users are no longer only consumers of information but also creators, privacy becomes a topic that requires special attention. It is important for users to realize what impact giving up private information has in the short and long term. Some Web 2.0 applications implicitly gather data about the actions of their users. How this can lead to privacy issues and how users can act to prevent this will also be discussed.

3G, 4G and Beyond–Bringing Networks, Devices and the Web Together, Second Edition. Martin Sauter.
© 2013 John Wiley & Sons, Ltd. Published 2013 by John Wiley & Sons, Ltd.

In practice, there are many different motivations for developing applications. Students, for example, create new applications because they have ideas they want to realize and can experiment without financial pressure or the need for a business model. Such an environment is quite different from the development environment in companies where deadlines, business models, and backwards compatibility rule during the development process. With this in mind this chapter will also discuss how the different environments shape the development of Web 2.0 vs the development of mobile Web 2.0 and examine the impact.

6.2 (Mobile) Web 1.0 — How Everything Started

For most users the Internet age started with two applications: e-mail and the Web. While the first form of e-mail dates back to the beginnings of the Internet in the 1960s and 1970s, the World Wide Web, or Web for short, is much younger. The first Web server and browser date back to the early 1990s. Becoming widespread in the research community by the mid 1990s, it took until the end of the decade before the Web became popular with the general public. Popularity increased once computers became powerful enough and affordable for the mass market. Content proliferated and became more relevant to everyday life, as shops started to offer their products online, banks opened their virtual portals on the Web, companies started to inform people about their products and news started being distributed on the Web much faster than via newspapers and magazines. Furthermore, the availability of affordable broadband Internet connections via DSL (Digital Subscriber Line) and TV cable since the early 2000s helped to accelerate the trend. While the Web was initially intended for sharing information between researchers, it got a different spin once it left the university campus. For the general public, the Internet was at first a top-down information distribution system. Most people connected to the Internet purely used the Web to obtain information. Some people also refer to this as the "read-only" Web, as users only consumed information and provided little or no content for others. Thus, from a distribution point of view, the Internet was very similar to the "offline" world where media companies broadcast their information to a large consumer audience via newspapers, magazines, television, movies, and so on. Non-media companies also started to use the Web to either advertise their services or sell them online. Amazon is a good example of a company that quickly started using the Web not only to broadcast information but also as a sales platform. However, what Amazon, and other online stores, had in common with media companies was that they were the suppliers of information or goods and the user was merely the consumer. Note that this has now changed, to some degree, as will be discussed in the next section.

In the mobile world, the Web had a much more difficult start. First attempts by mobile phone manufacturers to mobilize the Web were a big disappointment. In the fixed line world the Internet had an incubation time of at least a decade to grow, to be refined and fostered by researchers and students at universities before being used by the public, who already had sufficiently capable notebooks, PCs, and a reasonably priced connection to the Internet. In the mobile world, things were distinctly different when the first Web browsers appeared on mobile phones around 2001:

- Mobile Internet access was targeted at the general public instead of first attracting researchers and students to develop, use, and refine the services.

- Unlike at universities, where the Web was free for users, companies wanted to charge for the mobile service from day 1.
- It was believed that the Web could be extended into the mobile domain solely by adapting successful services to the limitations of mobile devices, rather than looking at the benefits of mobility. That is like taking a radio play, assembling the actors and their microphones in front of a camera, and broadcasting them reading the radio play on TV [1].
- Little, if any, appealing content for the target audience was available in an adapted version for mobile phones.
- Mobile access to the Internet was very expensive so only a few were willing to use it.
- Circuit-switched bearers were used at the beginning, which were slow and not suitable for packet-switched traffic.
- The mobile phone hardware was not yet powerful enough for credible mobile Web browsers. Display sizes were small, screen resolutions not suited for graphics, there was no color, not enough processing power and not enough memory for rendering pages.
- The use of a dedicated protocol stack (the Wireless Application Protocol, (WAP)) instead of HyperText Markup Language (HTML) required special tools for Web page creation and at the same time limited the possibilities to design mobile and user-friendly Web pages.

Any of the points mentioned above could have been enough to stop the mobile Web in its tracks. Consequently, there was a lot to overcome before the Internet on mobile devices started to gain the interest of a wider audience. This coincided with the emergence of the Web 2.0 and its evolution into the mobile domain, as described in the next section.

6.3   Web 2.0 — Empowering the User

While the Web 1.0 was basically a read-only Web, with content being pushed to consumers, advances in technology, thinking, and market readjustment (with the bursting of the dot com bubble at the beginning of the century) have returned the Internet to its original idea: exchange of information between people. The ideas that have brought about this seismic shift from a read-only Web to a read/write Web are often combined into the term Web 2.0. Web 2.0, however, is not a technology that can be accurately defined; it is a collection of different ideas. With these ideas also being applicable to the experience of the Web, and the Internet in general, on mobile devices, it makes sense to first discuss Web 2.0 before looking at its implications for mobile devices and networks. The following sections look at Web 2.0 from a number of different angles: from the user's point of view, from a principal point of view and from a technical point of view.

6.4   Web 2.0 from the User's Point of View

For the user, the Web today offers many possibilities for creating as well as consuming information, be it text-based or in the form of pictures, videos, audio files, and so on. The following section describes some of the applications that have been brought about by Web 2.0 for this purpose.

6.4.1 Blogs

A key phenomenon that has risen with Web 2.0 is blogging. A Blog is a private Web page with the following properties:

- Dynamic information — Blogs are not used for displaying static information but are continuously updated by their owners with new information in the form of articles, also referred to as Blog entries. Thus, many people compare Blogs with online diaries. In practice, however, most Blogs are not personal diaries accessible to the public, but platforms on which people share their knowledge or passions with other people. Companies have also discovered Blogs as a means of telling their story to a wider audience in a semi-personal fashion. Blogs can also be valuable additions to books, giving the author the possibility to interact with his readers, go into details of specific topics, and to share his thoughts. Figure 6.1, for example, shows the Blog that complements this book.
- Ease of use — Blogs are created, maintained, and updated via a Web-based interface. No Web programing skills are required. Thus, Blogs can be created and used by everyone, not only technically skilled people.
- Blogs order content in a chronological fashion with the latest information usually presented at the top of the main page.
- Readers of a Blog can leave comments, which encourages discussion and interaction.
- Other people can subscribe to an automated news feed of a Blog. This way they can easily find the Blog again (bookmarking functionality) and be automatically informed when the author of the Blog publishes a new entry. This is referred to as aggregation and is discussed in more detail in Section 6.6.3.
- A Blog is often the central element for the online activities of a user. It may be used to link to other online activities, for example, links to accounts at picture sharing sites, the user's pages in social networks and so on. Readers of the Blog can thus easily discover additional information from or about the owner of the Blog.

6.4.2 Media Sharing

Blogs can also be used to share nontextual content such as pictures and videos. In many cases, however, it is preferable to share such content via dedicated sharing sites such as Flickr [2] for pictures, YouTube [3] for videos, and so on. This has the advantage that users looking for a video or picture about a specific subject can go to such a sharing site and obtain a relevant list of videos that other people have made available to share. In private Blogs, links can then be used to point Blog readers to the content. It is also possible to embed pictures and videos from sharing platforms directly in Blog entries. Thus, no redirection is required for Blog readers, while people who are unaware of the Blog can still find the content.

6.4.3 Podcasting

Podcasting is another important form of media sharing. The word itself is a combination of the words iPod and broadcasting. Podcasting combines audio recording and making

Figure 6.1 The author's blog.

the recording available on Blogs, Web pages, and via automated feeds. Automated feeds allow interested users to be informed about a new podcast in a feed and connected MP3 players can automatically download new podcasts from feeds selected by the user. Thus, distributing audio content is no longer an exclusive domain of radio stations. Radio stations, however, have also discovered the value of podcasting and today many stations offer their content as podcasts after the initial traditional broadcast. The advantage for listeners is that radio shows can now be downloaded and consumed at any time and any place.

While Web sites exist that offer podcast directories and podcast archives, many podcasters host the audio files themselves and only use podcast directories to make others aware of their podcasts.

6.4.4 Advanced Search

Being a publisher of information is only useful when a potential audience can find the content (Blog entries, pictures, videos, etc.). This is made possible by advanced search engines such as Google, Bing, Yahoo, Technorati, and others, who are constantly updating their databases. The ranking of the search results is based on a combination of different parameters such as the number of other sites linking to a page and their own popularity, when the page was last updated and algorithms which are the well guarded secrets of search companies. While search engines can analyze text-based information, automated analysis of images and videos is still difficult. To help search engines find such nontextual information, users often add text-based tags to their multimedia content. Tags are also useful to group pieces of information together. It is thus possible to quickly find additional information on a specific topic on the same Blog or sharing site.

6.4.5   User Recommendation

In addition to ensuring a certain quality in reporting news, traditional media, such as newspapers and magazines, select the content they want to publish. Their selection is based on their understanding of user preferences and their own views. Consequently, a few people select the content that is then distributed to a large audience. Furthermore, mass media tailors content only for a mass audience and are thus not able to service niche markets. The Web 2.0 has opened the door for democratizing the selection process. User recommendation sites, such as Digg [4], let users recommend electronic articles. If enough people recommend an article it is automatically shown on the front page of Digg or in a section dedicated to a specific subject. This way the selection is not based on the preferences of a few but based on the recommendation of many.

6.4.6   Wikis — Collective Writing

Wikis are the opposite of Blogs. While a Blog is a Web site where a single user can publish their information and express their views, Wikis let many users contribute toward a common goal by making it easy to work on the same content in a Web-based environment. The most popular Wiki is undoubtedly the Wikipedia project. Within a short time the amount of articles and popularity has far surpassed other online and offline lexica of traditional media companies. Today, Wikipedia has hundreds of thousands of users helping to write and maintain the online encyclopedia. Participating is simple, since no account is needed to change or extend existing articles. The quality of individual articles is usually very good since people interested in a certain topic often ensure that the related articles on Wikipedia are accurate. As anyone can change any article on Wikipedia, entries on controversial topics sometimes go from one extreme to the other. In such cases, articles can be put under change control or set to immutable by users with administrator privileges. This shows that, in general, the intelligence and knowledge of the crowd is superior to the intelligence and knowledge of the few, but that the concept has its limits as well.

It is also possible to subscribe to Wiki Web pages in a similar fashion as subscribing to Blogs and podcasts. Thus, changes are immediately reported to interested people.

Apart from Wikipedia, a wide range of other Wikis exist on the Web today that are dedicated to specific topics. Starting a Wiki is just as easy as starting a Blog, since there are many Wiki hosting services on the net where new Wikis can be created by anyone with a few minutes to spare. Figure 6.2 shows a Wiki dedicated to the topic of how to access the Internet with prepaid Subscriber Identity Module (SIM) cards of 2G and 3G network operators. Started by the author of this book, many people have since contributed and added information about prepaid SIM cards and Internet access in their countries. As is the nature of a Wiki, articles are frequently updated when people notice that network operators have changed their offers.

Wikis are also finding their way into the corporate world, where they are used for collaboration, sharing of information or to help project teams to work together on a set of documents.

Figure 6.2 A small Wiki running on a server of a Wiki hosting service. (Reproduced by permission of Wetpaint.com, Inc., 307 third Avenue S., Suite 300, Seattle, WA 98104, USA. Photograph reproduced from Martin Sauter.)

6.4.7 Social Networking Sites

While Blogs, Wikis, and sharing sites make it easy to publish, share, and discover any type of content, social networking sites are dedicated to connecting people and making it easy to find other people with similar interests. Famous social network sites are Facebook [5] in the private domain and LinkedIn [6] for business contacts. It is interesting to note that the popularity of early social networking sites such as Myspace [7] has as quickly diminished as it rose initially. This shows how dynamic this area of technology continues to be. Being a member of a social networking site means sharing and exchanging private information with others. Many different types of social networks exist. Some focus on fostering professional contacts and offer few additional functionalities, while others focus on direct communication between people, for example, by offering blogging functionality and automatically distributing new entries to all people who the user has declared as friends on the site. The Blogging behavior on social networking sites is usually different to dedicated Blogs, since entries are shorter, usually more personal and dedicated to the people in the friends list rather than a wider audience. Many social networking platforms also allow users to create personalized Web pages on which they present themselves to others.

6.4.8   Web Applications

In the days of Web 1.0 most programs had to be installed on a device and the Web was mostly used to retrieve information. Advanced browser capabilities, however, have brought about a wide range of Web applications which do not have to be installed locally. Instead, Web applications are loaded from a Web server as part of a Web page. They are then either exclusively executed locally or are split into a client and server part, with the server part running on a server in the Internet. Google has many Web applications, a very popular one being Google maps. While the maps application itself is executed in the browser, as a JavaScript application, the "maps and search" databases are in the network. When users search for a specific location, or for hotels, restaurants, and so on at a location, the application connects to Google's search database, retrieves answers, and displays the results on a map that is also loaded from the network server. The user can then perform various actions on the map, like zooming and scrolling. These actions are performed locally in the browser until further mapping data is required. At this point the map's application running on the Web browser asynchronously requests the required data. During all these steps the initial Web page on which the map's application is executed is never left. The application processes all input information itself, updates the Web page and communicates with the backend server.

Today, even sophisticated programs such as spreadsheets and word processors are available as Web applications. Documents are usually not stored locally, but on a server in the network. This has the advantage that several people at different locations can work on a document simultaneously. Also, a user can work on documents via any device connected to the Web, without taking the document with him. Another benefit of Web applications is that they do not have to be deployed and installed on a device. This makes deployment very simple and changes to the software can be done seamlessly, when the application is sent to the Web browser as part of the Web page. The downside of Web applications and Web storage is that the user becomes dependent on a functioning network connection and relies on the service provider to keep their documents safe and private.

6.4.9   Mashups

Mashups are a special form of Web application. Instead of a single entity providing both the application and the database, mashups retrieve data from several databases in the network via an open Application programing Interface (API) and combine the sources in a new way. An example of a mashup is a Web application that uses cartographic data from the Google maps database to display the locations of the members of a user's social network, where data about the members is retrieved from the social networking site of the user. This is something neither Google maps nor the social networking site can do on their own. The crucial point for mashups is that other Web services allow their data to be used without their own Web front end. This is the case for many Web services today, with Google maps just being a prominent example. Also important for mashups is that the interface provided by a Web application does not change, otherwise the mashup stops working. Mashups also depend on the availability of their data sources. As soon as one of the data sources is not available the mashup stops working as well.

6.4.10   Virtual Worlds

Another way how the Internet connects people is the concept of virtual worlds. The most prominent virtual world is Second Life by Linden Labs [8] even though their popularity has diminished somewhat in recent years. Virtual worlds create a world in which real people are represented by their avatars. Avatars can look like the real person owning them or, more commonly, how that person would like to look. Avatars can then walk through the virtual world, meet other avatars and communicate with them. Avatars can also own land, buy objects, and create new objects themselves. While virtual worlds might have initially been conceived as pure games, companies have experimented with the concept and have opened virtual store fronts. Avatars of employees work as shop assistants and interact with customers. Also, some universities have experimented with virtual worlds for online learning by holding classes in the virtual world, which are attended by real-life students, who visit the classroom with their avatars. Communication is possible via instant messaging but also via an audio channel. It should be noted at this point that most virtual worlds require a client application on the user's device. Therefore, they are not strictly a part of the Web 2.0, as they are not running in a Web browser. Nevertheless, in everyday life most people count virtual worlds as part of the Web 2.0.

6.4.11   Long-Tail Economics

Web 2.0 services enable users to move on from purely being consumers to also become creators of content, which in turn considerably increases the variety of information, viewpoints, and goods available via the Web. By using search engines or services such as eBay, Amazon, iTunes, and so on, this information, or these goods, can also be found and consumed by others without having to be promoted by media companies and advertisements. The ability to find things "off the beaten path" also facilitates the production and sale of goods for which, traditionally, there has been no market, because people were not aware of them. Chris Anderson has described this phenomenon as long-tail economics in [9]. The term long-tail is explained in Figure 6.3. The vertical axis represents the number

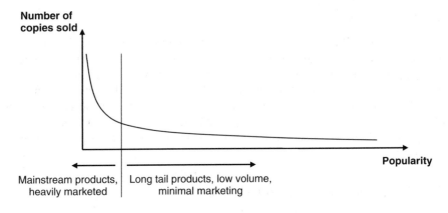

Figure 6.3   The long tail.

of copies sold of a product, for example, a book, and the horizontal axis shows its popularity. Very popular items start out on the left of the graph, with the long tail beginning when it is economically no longer feasible to keep the items in stock, that is, when only limited space and local customers are available.

While still making a fair percentage of their revenue with mainstream products, companies like Amazon today are successful because they can offer goods which only sell in quantities too small to be profitable when they have to be physically distributed and stored in many places. This in turn again increases the popularity of the site since goods are available which cannot be bought at a local store where floor space is limited and interest in stocking products which sell in small quantities is not high. As there are many more products sold in small quantities compared with the few products sold in very large quantities, a substantial amount of revenue can be generated for the company running the portal. eBay is another good example of long-tail economics. While not stocking any goods itself, eBay generates its revenue from auctions of goods from the long tail and not from those sold at every street corner. Whether it is possible to be profitable by producing goods or content on the long tail, however, is another matter [10]. For many, however, generating revenue is not the goal of providing content on the long tail, as their main driver is to express their views and give something back for the information, produced by others, that they have consumed for free.

6.5   The Ideas behind Web 2.0

Most of the Web 2.0 applications discussed in the previous section have a number of basic ideas behind them. Tim O'Reilly, who originally coined the term Web 2.0, has written an extensive essay [11] about the ideas behind Web 2.0. Basically, he sees seven principles that make up Web 2.0 and points out that, for applications to be classified as belonging to Web 2.0, they should fulfill as many of the criteria as possible. This section gives a brief overview of these principles as they form the basis for the subsequent analysis, that is how these principles are enhanced or limited by mobile Internet access and if the mobile Web 2.0 is just an extension of Web 2.0 or requires its own definition.

6.5.1   The Web as a Platform

A central element of Web 2.0 is the fact that applications are no longer installed locally but downloaded as part of a Web page before being executed locally. Also, the data used by these applications is no longer present on the local device but is stored on a server in the network. Thus, both the application and the data are in the network. This means that software and data can change and evolve independently and the classic software release cycle which consists of regularly upgrading locally installed software is no longer necessary. As software and data change, Web 2.0 applications are not packaged software but rather a service.

6.5.2   Harnessing Collective Intelligence

User participation on Web 2.0 services is the next important element. Services that only exist because of user participation are, for example, Wikipedia, Flickr, and Facebook. While the organizations behind these services work on the software itself, the data (Wiki entries, pictures, personal entries, etc.) is entirely supplied by the users. Users submit their information for free, working toward the greater goal of creating a database that everybody benefits from. While in the traditional top-down knowledge distribution model, the classification of information (taxonomy) was done by a few experts, having a countless number of people working on a common database and classifying information is often referred to as folksonomy. Classifying information is often done by tagging, that is by adding text-based information (catch words) to anything from articles to pictures and videos. This way it is possible to find nontextual information about a certain topic and quickly correlate information from different sources.

Collective intelligence also means that software should be published as open source and distributed freely so everybody can build on the work of others. This idea is similar to contributing information to a database (e.g., Wikipedia) that can then be used by others.

Blogs are also a central element of the Web 2.0 idea, as they allow everyone with a computer connected to the Internet to easily share their views in articles, also referred to as Blog entries. Blog entries are usually sorted by date so visitors to a Blog will always see the latest entries first. In contrast to the above services, however, Blogs are not collecting information from several users but are a platform for individuals to express themselves. Therefore, powerful search algorithms are required to open up this "wisdom," created by the crowds, to a larger audience as users first of all need to discover a Blog before they can benefit from the information. Some Blogs have become very successful because users have found the information so interesting that they have linked to the source from their own Blogs. When this is repeated by others a snowball effect occurs. As one input parameter for modern search algorithms is the relevance of a page based on the number of links pointing to it, this snowball effect gives such Web pages a high rating with search engines and thus moves them higher in the search result lists. This in turn again increases their popularity and creates more incoming links.

Less frequented Blogs, however, are just as important to Web 2.0 as the few famous ones. Many topics, such as mobile network technology, for example, are only of interest to a few people. Before Blogs became popular, little to no information could be found about these topics on the Web, since large media companies focus on content that is of interest to large audiences and not niche ones. With the rise of Web 2.0, however, it has become much simpler to find people discussing such topics on the net. Blogrolls, which are placed on Blogs and contain links to Blogs discussing similar topics, help newcomers to quickly find other resources.

As many Blogs are updated infrequently and thus interesting information is spread over many different sources, a method is required to automatically notify users when a Blog is updated. This is necessary since it is not practical to visit all previously found interesting Blogs every day to see if they have been updated. Automatic notification is done with feeds, to which a user can subscribe to with a feed reader. A feed reader combines all feeds and shows the user which Blogs have updated information. The Blog entries are then either read directly in the feed reader or the feed reader offers a link to the Blog.

6.5.3 Data is the next Intel Inside

While users buy standalone applications like, for example, word processors because of their functionality, Web 2.0 services are above all successful because of their database in the background. If services offer both information and the possibility for users to enhance the database or be the actual creator of most of the information the service is likely to become even more popular, due to the rising amount of useful information that even the most powerful company could not put together. An example of this is a database of restaurants, hotels, theaters, and so on. Directories assembled by companies will never be as complete or accurate as directories maintained by the users themselves. Control over such user-maintained databases is an important criterion for them to become successful, as the more information is in the database the harder it gets for similar services to compete. To stimulate users to add content it is also important to make the database accessible beyond the actual service, via an open interface. This allows mashups to combine the information of different databases and offer new services based on the result. This can in turn help to promote the original service. An example is Google maps. It allows other applications to request maps via an open interface. When mashups use maps for displaying location information (e.g., about houses for sale, hotels, etc.), the design of the map and the copyright notice always point back to Google.

6.5.4   End of the Software Release Cycle

As software is no longer locally installed, there are no longer different versions of the software that have to be maintained so users no longer need to upgrade applications. Errors can thus be corrected very quickly and it enables services to evolve gradually instead of in distinctive steps over a longer period of time. This concept is also known as an application being in perpetual beta state. This term, however, is a bit misleading as beta often suggests that an application is not yet ready for general use.

Running applications in a Web browser and having the database and possibly some processing logic in the network also allows the provider of the service to monitor which features are used and which are not. New features can thus be tested to see if they are acceptable or useful to a wider audience. If not, they can be removed again quickly, which prevents rising entropy that makes the program difficult to use over time.

Web 2.0 services often regard their users as co-developers, as their opinions of what works and what does not can quickly be put into the software. Also, new ideas coming from users of a service can be implemented quickly if there is demand and deployed much faster than in a traditional development model, in which software has to be distributed and local installations have to be upgraded. This shortens the software development cycle and helps services to evolve more quickly.

6.5.5   Lightweight Programing Models

Some Web 2.0 services retrieve information from several databases in the network and thus combine the information of several information silos. Information is usually accessed either via Real Simple Syndication (RSS) feeds or a simple interface based on Hypertext Transport Protocol (HTTP) and Extensible Markup Language (XML). Both methods allow loose coupling between the service and the database in the network. Loose coupling means that the interface has no complex protocol stack for information exchange, no service description and no security requirements to protect the exchange of data. This enables developers to quickly realize ideas, but of course also limits what kind of data can be exchanged over such a connection.

6.5.6   Software above the Level of a Single Device

While in a traditional model, software is deployed, installed, and executed on a single device, Web 2.0 applications and services are typically distributed. Software is downloaded from the network each time the user visits the service's Web page. Some services make extensive use of software in the backend and only have the presentation layer implemented in the software downloaded to the Web browser. Other Web 2.0 software runs mostly in the browser on the local machine and only queries a database in the network.

Some services are especially useful because they are device-independent and can be used everywhere with any device that can run a Web browser. Web-based bookmark services, for example, allow users to get to their bookmarks from any computer, as both the service and the bookmarks are Web-based.

Yet another angle to look at software above the level of a single device is that some services become especially useful because they can be used from different kinds of devices and not only computers. Instant messaging and social networks, for example, can be enhanced when the user does not only have access to the service and data when at home or at the office, but also when he roams outside and only has a small mobile device with him. As both the service and the data reside in the network and are used with a browser, no software needs to be installed and use of the service on both stationary and mobile devices is easy. This topic will be elaborated in more detail in the next section on mobile Web 2.0.

Some companies have also combined Web 2.0 services, traditional installable software and mobile devices to offer a compelling overall service to users. Apple, for example, offers iTunes, which is a traditional program that has to be installed. The media database it uses, however, is not only created by Apple and media companies but also includes a podcast catalog entirely managed by users. To make the service useful, a mobile device is sold as part of the package, to which content can be downloaded directly over the network or via the software installed on the computer.

6.5.7   Rich User Experience

Web 2.0 services usually offer a simple but rich user experience. This requires methods beyond static Web pages and links. Modern browsers support JavaScript to create interactive Web pages in addition to XHTML and CSS for describing Web page content. The XML is used to encapsulate information for the transfer between the service and the database in the network. This way, standard XML libraries can be used to encapsulate and retrieve information from a data stream without the programers having to reinvent data encapsulation formats for every new service. All methods together are sometimes referred to as Asynchronous JavaScript and XML (AJAX). Asynchronous in this context means that the JavaScript code embedded in a page can retrieve information from a database in the network and show the result on the Web page, without requiring a full page reload. This way it is possible for services running in a browser to behave in a similar way to locally installed applications and not like a Web page in the traditional sense.

6.6   Discovering the Fabrics of Web 2.0

The previous sections have taken a look at Web 2.0 from the user's perspective and which basic ideas are shared by Web 2.0 services. This section now introduces the technical concepts of the most important Web 2.0 methods and processes.

6.6.1   HTML

A central idea of the World Wide Web is that web pages are not sent from a Web server to a Web browser as they are presented on the screen. Instead, a description of a web page is downloaded in a language referred to as HyperText Markup Language and then rendered into the web page presented to the user.

Markups are instructions in the HTML document that tell the Web browser how to present (render) different parts of a web page. A typical example are markup instructions that tell the Web browser that a certain part of the text is to be shown in bold, while another part is to be shown to the user in a certain color or different size from the rest of the text. Other elements of the page such as text input boxes or buttons are also described as a markup and not downloaded as an image from the Web server. One of the most important markup instructions is doubtlessly the "referrer" with which links to other web pages can be included in a web page.

Figure 6.4 shows how the HTML markups and text look like for a very basic web page. In practice, most web pages are obviously far more complex and contain many more markup commands. To visualize the HTML code of a page, Web browsers usually have an option to show the "source code" of the page. Firefox, for example, shows the HTML code when the mouse pointer is placed over a web page after pressing the right mouse button and selecting "View Page Source" in the context menu.

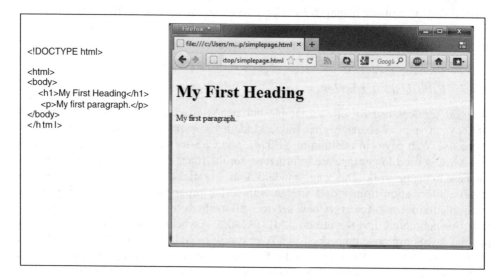

Figure 6.4   HTML code and the corresponding Web page in a Web browser.

The address of a web page that usually leads to a HTML document usually only contains the skeleton of a web page and just references many other objects to be included on the page such as pictures and embedded objects such as videos. These have then to be requested separately from the same or a different Web server and are included in the overall web page.

A web page (the document) usually consists of many parts (objects) that are either part of the HTML description or loaded separately such as text, pictures, input boxes, buttons, and so on, which are all described in HTML. The Web browser parses the HTML description of the page and assembles what is referred to as the document object model (DOM) tree to render the page object by object. The DOM tree is then also used to enable the Javascript code embedded in the web page to modify objects on the page. This can be used, for example, to validate input text from the user and return feedback whether the input is correct or not before the information is returned to the web server for processing. Further details on Javascript are discussed in the following.

Over the years, HTML was significantly extended and the latest version, HTML5, is an industry-wide harmonization approach to ensure interoperability between different browsers. In addition to describing how web pages look like, HTML5 is now also capable to describe how to include the following functionalities in a web page:

- Video and audio content which has previously been done by including the proprietary Adobe Flash player in a web page.
- A local access API for Javascript. This way, scripts that are part of a web page can access local resources such as storage, location information, a built-in web cam, and so on.
- Drag and drop support.
- 2D and 3D graphics.

Many functionalities of HTML5 have been designed specifically with Web browsers on mobile devices in mind. Details on the different HTML5 elements and support in mobile Web browsers can be found in [12].

6.6.2 AJAX

In Web 2.0 the Web browser is the user interface for services. The more capabilities Web browsers have, the better the user experience. In Web 1.0 most Web pages were static. Whenever the user, for example, put text into an input field or set a radio button and pressed the "ok" or "continue" button, the information was sent to a Web server for processing and a new Web page with the result was returned. The user experience of such an approach is relatively poor compared with local applications where the reaction to user input is displayed on the same screen without the typical reload effect of one Web page being replaced by another.

The solution to this problem comes in three parts. The first part is the support of JavaScript code on Web pages by the Web browser. The JavaScript code can interact with the user via the Web page by reading user input such as text input or when the user clicks on buttons on the Web page and so on. Unlike in the previous approach, where such actions resulted in immediate communication with the Web server in the network and the transmission of a new Web page as a result of the action, the JavaScript code can process the input locally and change the appearance of the Web page without the page reload effect.

The second part is allowing a JavaScript application on a Web page to send data to a Web server and receive a response without impacting what is shown on the Web page. The JavaScript can thus take the user input and send it to the Web server in the background. The Web server then sends a response and the JavaScript application embedded in the Web page will alter the appearance of the page without the need for loading a new Web page. Since this exchange of data is done in the background, it is also referred to as being asynchronous, since the exchange does not prevent the user interacting with the Web page (scrolling, pressing a button, etc.) while the JavaScript application is waiting for a response. The JavaScript application embedded in a Web page can modify the page in a similar way as a program running locally is able to modify the content of its window. Thus, for the user the behavior is similar to that of a local program.

The third part is a standardized way of exchanging information between the JavaScript application running in the Web browser and the program running on a Web server on the web. A format often used for this exchange is XML, which is also used as a descriptive language for Atom and RSS feeds, as shown in Figure 6.5. XML is a tag-based language that encapsulates information between tags in a structured way. The class used in JavaScript to exchange information with the backend includes sophisticated functions for extracting information from an XML formatted stream. This makes manipulation of the received

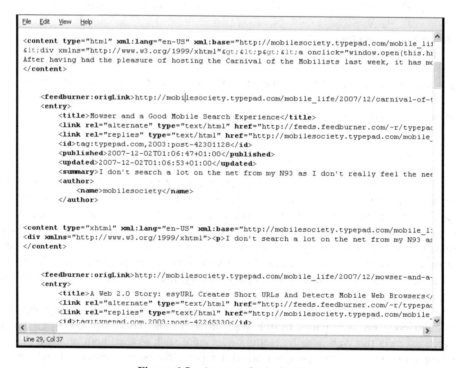

Figure 6.5   An atom feed of a blog.

```
00  <script language="JavaScript"
01     type="text/javascript">
02  // <![CDATA[
03  var XMLHTTP = null;
04
05  XMLHTTP = new XMLHttpRequest();
06
07  function Output Content(){
08
09     var xml = XMLHTTP.responseXML;
10
11     if(XMLHTTP.readyState == 4){
12        var d = document.getElementById("data");
13        d.innerHTML += xml.documentElement.getAttribute("title");
14     }
15  }
16
17  window.onload = function(){
18     XMLHTTP.open("GET",
"getfeed?feed=mobilesociety.typepad.com/feed");
19     XMLHTTP.onreadystatechange = OutputContent;
20     XMLHTTP.send(null);
21  }
22  // ]] >
23  </script>
24
25  <body>
26     < p id="data">Data received from server: </p>
27  </body>
```

Figure 6.6 A Web page with a simplified embedded JavaScript Application.

data very simple for a JavaScript application embedded in a Web page. All three parts taken together are commonly referred to as Asynchronous JavaScript and XML for short.

An example of a very simple JavaScript application embedded in a Web page communicating asynchronously with an application hosted on the Web server is shown in Figure 6.6. The actual content of the Web page is very small and is contained in lines 25–27 between the <body> tags. The JavaScript code itself is embedded in the Web page before the visible content from line 0 to 23. On line 5 the JavaScript code instantiates an object from class XMLHttpRequest. This class has all the required functions to send data back to the Web server from which the Web page was loaded via the HTTP, asynchronously receive an answer and extract information from an XML formatted data stream. The XMLHTTP object is first used in line 18 where it is given the URL to be sent to the Web server. In this example, the JavaScript application sends the URL of a Blog feed. The application on the Web server then interprets the information and returns a result, for example, it retrieves the Blog's feed and returns what it has received back to the JavaScript application running in the Web browser. This is done asynchronously as the send function on line 20 does not block until it receives an answer. Instead a pointer to a function is given to the XMLHTTP object, which is called when the Web server returns the requested information. In the example, this is done in line 19 and the function which is called when the Web server returns data is defined starting from line 7.

The JavaScript application therefore does not block and is able to react to other user input while waiting for the server response. Functions handling user input, however, are not part of the example in order to keep it short.

When the Web server returns the requested information, in the example the XML-encoded feed of a Blog, the "OutputContent" function in line 7 is called. In line 12 the text from line 26 of the Web page is imported into a variable of the JavaScript application. In line 13, the "getAttribute" function of the XMLHTTP object is used to retrieve the text between the first <title> tags of the feed. This text is then appended to the text already present in line 26 and put on the Web page without requiring a reload.

While the JavaScript application shown in Figure 6.6 is not really useful, due to its limited functionality, it nevertheless shows how AJAX can be used in practice. More sophisticated JavaScript applications can make use of the asynchronous communication to download much more useful information and draw graphics and other style elements on the Web page based on the data received.

6.6.3 Aggregation

The glue that holds Web 2.0 together is aggregation, or the ability to automatically retrieve information from many sources for presentation in a common place or for further processing. Blog or feed-reading programs, for example, are based on aggregation. The idea of Blog or feed readers is to be a central place from which a user can check if new articles have been published on Blogs or Web pages supporting aggregation. For the user, using a feed reader saves time that has otherwise to be spent on visiting each Blog in a Web browser to check for news. Figure 6.7 shows Mozilla Thunderbird, an e-mail and feed reader program. On the left side, the program shows all subscribed feeds and marks those in bold which have new articles. On the upper right, the latest feed entries of the selected Blog are shown. New entries are marked in bold so they can be found easily. On the lower right, the selected Blog entry is then shown. The link to the article on the Blog is also shown, as it is sometimes preferable to read the article on the Blog itself rather than in the feed reader, as sometimes no pictures or only scaled down versions are embedded in the feed.

From a technical point of view, feed readers make use of Blog feeds, which contain the articles of the Blog in a standardized and machine readable form. When a user publishes a new article on a Blog, the feed is automatically updated as well. Each time a user starts a feed reader, the feeds of all sources, the user is interested in are automatically retrieved with a HTTP request, just like a normal Web page, and analyzed for new content, which is then presented in the feed reader. In practice, there are two different feed formats, and feed readers usually support both:

- RSS, specified in [13], and
- ATOM syndication format, specified in [14].

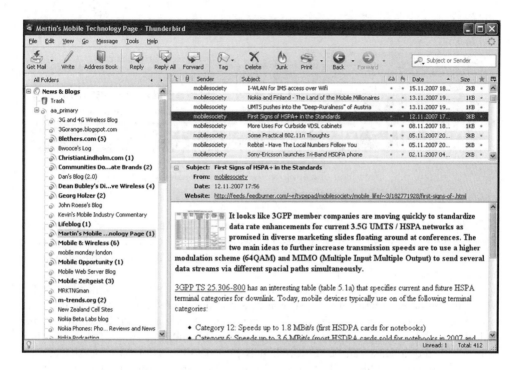

Figure 6.7 Mozilla Thunderbird used as a feed reader.

Both feed formats are based on XML, which is a descriptive language and a generalization of the HTML, used for describing Web pages.

Figure 6.5 shows an extract of an Atom Blog feed. Information is put between standardized tags (e.g., <title> and </title>); so feed readers or other programs can search XML feeds for specific information. Besides the text of Blog entries, a lot of additional information is contained in feeds, such as the date an entry was created, information about the Blog itself, name of the author, and so on. The text of the Blog entries can be formatted as HTML text and can thus also contain references to pictures embedded in the article or links to external pages. The feed reader can then request the pictures from the Blog for presentation in the Blog entry and open a Web browser if the user wants to follow a link in the article to another Web page.

In practice, users do not have to deal with the XML description delivered by an XML feed directly. The usual method to import a feed is by clicking on the feed icon that is shown next to the URL of the Web page, as shown in Figure 6.8. The Web browser then shows the URL of the feed which the user can then copy and paste into the feed reader.

Feeds are not only used for aggregating Blog feeds in a Blog reader. Today, other types of Web pages also offer RSS or Atom feeds; so content from those pages can also be viewed in a feed reader program. Picture-sharing sites such as Flickr, for example, offer feeds for individual users or tags. Each time the feed reader requests updated information from a Flickr feed, Flickr includes the latest pictures of a user in the feed or the pictures for the specified tags.

Figure 6.8   Feed icon of a Blog on the right of the URL of the Web page.

Feeds are also used by applications to automatically aggregate a user's information from different places. An example is social networking sites. Pictures from picture sharing sites or new Blog entries are thus automatically imported into the user's page on a social networking platform.

Feeds are also used in combination with podcasts. Apple's iTunes is a good example, which among other functionalities also works as a podcast directory. A podcast directory is in essence a list of podcast feeds. The feeds themselves contain a description of the podcasts available from a source, information about the audio file (e.g., size) and a link from which the podcast can be retrieved. It is also possible to use a podcast feed in a feed reader program, which will then present the textual information for the podcast and present a link from which the audio file can be retrieved. Most people, however, prefer programs like iTunes for podcast feeds.

6.6.4   Tagging and Folksonomy

While analyzing textual information on Blog entries and Web pages is a relatively easy task for search engines, classifying other available media such as pictures, videos, and audio files (e.g., podcasts) is still not possible without additional information supplied by the person making the content available. A lot of research is ongoing to automatically analyze the content of nontextual sources on the Web and a result of this is, for example, the face recognition feature on pictures uploaded to Facebook [15]. However, for the time being, search engines and other mechanisms linking content still rely on additional textual information. The most common way of adding additional information is by adding tags, that is, search words. As this form of classification is done by the users and not by a central instance, it is sometimes also referred to as folksonomy, that is, taxonomy of the masses.

Flickr, an image hosting and sharing Web service, is a good example of a service that uses tagging and folksonomy. Tags can be added to pictures by the creator, describing the content and location as shown in Figure 6.9. Tags can also contain other information like, for example, emotion, event information, and so on. Tags can also contain geographical location tags (latitude and longitude), which were generated automatically by the mobile device with which the picture was taken because it was able to retrieve the GPS position from a GPS device (internal or external) at the time the picture was taken. The tags are then used by the image-sharing service and other services for various purposes. The

Figure 6.9 Tags alongside a picture on Flickr, an image-sharing service.

picture sharing service itself converts the tags into user clickable links. When the user clicks on a tag the service searches for other pictures with the same tag and presents the search result to the user. Thus, it is easy to find pictures taken by other users at the same location or about the same topic.

The picture sharing service treats the geographic location tags in a special way. Instead of showing the GPS coordinates, which would not be very informative for the user, it creates a special "map" link. When the user clicks on the "map" link a window opens up in which a map of the location is shown. The user can then zoom in and out and move the map in any direction to find out more about the location where the picture was taken. The picture sharing site also inserts the location of other pictures the user has taken in the area which is currently shown and on request presents pictures other users have taken in this area, which are also stored together with geographical location tags. This functionality is a typical combination of the use of tags to find and correlate information, of AJAX for creating an interactive and user-friendly Web page and of open interfaces which allow information stored in different databases to be combined (pictures and text in the image database and the maps in a map database on the network).

The tags and geographical location information alongside images are also used by other services. Search engines such as Yahoo or Google periodically scan Web pages created by Flickr from its image/tag/user database. It is then possible to find pictures not only directly in Flickr but also via a standard Internet search. This is important since Flickr is not the only picture-sharing service on the net and searching for pictures with a general Web search service results in a wider choice, as the search includes the pictures of many sharing sites. It is important to note at this point that without tags the value of putting a picture online for sharing with others is very limited, since it cannot be found and correlated with other pictures.

6.6.5   Open Application Programing Interfaces

Many services are popular today because they offer an open API, which allows third party applications to access the functionality of the service and the database behind it. Atom and RSS feeds are one form of open API to retrieve information from Blogs or Web pages. Requesting the feed is simple, as it only requires knowledge of the URL (Universal Resource Locator, for example, http://mobilesociety.typepad.com/feed). Analyzing a feed is also possible since the feed is returned as an Atom or RSS formatted XML stream. How the XML file can be analyzed is part of the open RSS and Atom specifications. In Figure 6.7, Thunderbird, a locally installed feed reader was shown. There are also Web 2.0 feed readers which run as JavaScript applications in Web pages and which get feed updates and store information in a database in the network (e.g., which feeds the user has subscribed to, which Blog entries have already been read, etc.).

While feeds only deliver information and leave the processing to the Web service running on a user's computer, remote services can also share a library of functions with a JavaScript application running in the local Web browser. Examples of this approach are the APIs of Yahoo [16], Google maps [17] and OpenStreetMap [18]. These APIs allow other Web 2.0 services to show location data on a map generated by these services. A practical example is a Web statistics service that logs the IP addresses from which a Web site was visited. When the owner later on calls the statistics Web page, the service in the background queries an Internet database for the part of the world in which the IP addresses are registered. This information is then combined with that of the mapping service and a map with markers at the locations where the IP addresses are registered is shown on the Web page. As the map is loaded directly from the server of the mapping service it is interactive and the user can zoom and scroll in the same way as if he had visited the map service directly via the mapping portal.

Figure 6.10 shows how this is done in principle. The statistics service comprises both a server component and a front-end component, that is a program or script running on the Web server and a JavaScript application executed in a Web page. The backend component on the Web server is called when people visit a Web site which contains an image that has to be loaded from the statistics server. Requesting the image then invokes a counting procedure. It is also possible to trigger an HTTP request to the statistics server for counting purposes with a tiny JavaScript application that is embedded in the Web page. The counting service on the statistics Web server processes the incoming request to

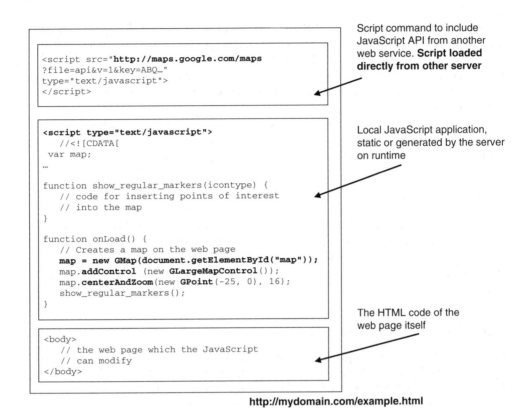

```
<script src="http://maps.google.com/maps
?file=api&v=1&key=ABQ..."
type="text/javascript">
</script>
```
Script command to include JavaScript API from another web service. **Script loaded directly from other server**

```
<script type="text/javascript">
  //<![CDATA[
 var map;
 ...

function show_regular_markers(icontype) {
  // code for inserting points of interest
  // into the map
}

function onLoad() {
  // Creates a map on the web page
  map = new GMap(document.getElementById("map"));
  map.addControl (new GLargeMapControl());
  map.centerAndZoom(new GPoint(-25, 0), 16);
  show_regular_markers();
}
```
Local JavaScript application, static or generated by the server on runtime

```
<body>
  // the web page which the JavaScript
  // can modify
</body>
```
The HTML code of the web page itself

http://mydomain.com/example.html

Figure 6.10   Remote JavaScript code embedded in a Web page.

retrieve the origin of the request and stores it in its database. When the owner of the Web site later on visits the statistics service Web page, the following actions are performed:

- After the user has identified himself to the service running on the Web server, the IP addresses from which the user's Web site was visited in the past are retrieved from the statistics database. The service running on the Web server then queries an external database to get the locations at which those IP addresses are registered.
- Once the locations are known the statistics service generates a Web page. At the beginning of the Web page, a reference to Goggle's mapping API is included. It is important to note that this is just a reference to where the Web browser can retrieve the API, that is, the Web browser loads the API directly from Google's server and not from the Web statistics server.
- Next, the JavaScript code of the statistics service is put into the Web page by the server application. As the source code is assembled at run time, it can contain the information about where to put the location markers on the map either in variables or as parameters of function calls. In the "onLoad" function shown in Figure 6.10, the JavaScript application embedded in the page then calls the JavaScript API functions of the mapping service that have been loaded by the script command above.

- As the API functions were loaded from the mapping service Web server, they have permission to establish a network connection back to the map server. They can thus retrieve all information required for the map.
- The map API functions also have permissions to access the local Web page. Thus, they can then draw the map at the desired place and react to input from the user to zoom and move the map.
- In the example above the local "show_regular_markers" function is called afterwards to draw the markers on to the map with further calls to API functions. Note that the implementation of the function is not shown to keep the example short.

6.6.6   Open Source

In his Web 2.0 essay [11], Tim O'Reily also mentions that a good Web 2.0 practice is to make software available as open source. This way the Web community has access to the source code and is allowed to use it free of charge for their own projects. There are many popular open source license schemes and this section takes a closer look at three of the most important ones.

6.6.6.1   GNU Public License (GPL)

Software distributed under the GNU Public License (GPL) [19], originally conceived by Richard Stallman, must be distributed together with the source code. The company distributing the software can do this for free or request a fee for the distribution. The GPL allows anyone to use the source code free of charge. The condition imposed by the GPL is that in case the resulting software is redistributed this also has to be done under the GPL license. This ensures that software based on freely received open source software must also remain open source.

The GPL open source principle — to make the source code of derivate work available — only applies when the derivate work is also distributed. If open source software is used as the basis for a service offered to others, the GPL does not require the derivate source code to be distributed. The following example puts this into perspective: a company uses open source database software licensed under the GPL (e.g., a database system) and modifies and integrates it into a new Web-based e-mail service to store e-mails of users. The Web-based e-mail service is then made available to the general public via the company's Web server. Users are charged a monthly fee for access to the system. As it is the service and not the software that is made available to users (the software remains solely on the company's server), the modified code does not have to be published. If, however, the company sells or gives away the software for free to other companies, so they can set up their own Web-based e-mail systems, the distributed software falls under the terms of the GPL. This means that the source code has to be open and given away free of charge. Other companies are free to change the software and to sell or distribute it for free again. The idea of the GPL is that freely available source code makes it easy for anybody to build upon existing software of others, thus accelerating innovation and new developments. The most successful project under the GPL license is the Linux operating system. The business model of companies using GPL software to develop and distribute their own software is not usually based on the sale of the software itself. This is why all Linux distributions are free. Instead, such companies are typically selling support services around the product such as technical support or maintenance.

Many electronic devices such as set-top boxes, Wi-Fi access points and printers with built-in embedded computers are based on the Linux operating system. Thus, the software of such systems is governed by the GPL and the source code has to be made available to the public. This has inspired projects such as OpenWrt [20], which is an alternative operating system for Wi-Fi routers based on a certain chipset. The alternative operating system, developed by the Web community, has more features than the original software and can be extended by anyone.

6.6.6.2   The BSD and Apache License Agreements

Software distributed under the Berkeley Software Distribution (BSD) license agreement [21] is also provided as source code and the license gives permission to modify and extend the source code for derivate work. The big difference to source code distributed under the GPL license is that the derivate work does not have to be redistributed under the same licensing conditions. This means that a company is free to use the software developed by a third party under the BSD license within its own software and is allowed to sell the software and keep the copyright, that is to restrict others from redistributing the software. Also, it is not required to release the source code.

A license agreement similar to BSD is the Apache license [22], which got its name from the very popular Apache Web server — the first product to be released under this license. In addition to the BSD license, the Apache license requires software developers to include a notice when distributing the product that the product includes Apache licensed code.

Google's Android operating system, discussed in the previous chapter, makes use of the Apache license for applications created in the user space and the GPL license for the Linux kernel [23]. This means that companies adopting the Android OS for their own developments do not need to publish the code for the software running on the application layer of Android if they do not wish to do so. It is likely that this decision was made in order to attract more terminal manufacturers to Android than would be the case if the whole system was put under the GPL, which would force companies to release their source code.

6.7   Mobile Web 2.0 — Evolution and Revolution of Web 2.0

The previous sections have focused on the evolution of the Web as it happens today on PCs and notebooks. With the rising capabilities of mobile devices, as discussed in Chapter 5, the Web also extends more and more into the mobile world. The following sections now discuss how Web 2.0 services can find their way to mobile devices and also how mobility and other properties of mobile devices can revolutionize the community-based services aspects of Web 2.0 and the possibilities for self expression.

As the extension of Web 2.0 into the mobile domain is both an evolution and revolution, many people use the terms Mobile Web 2.0 or Mobile 2.0 when discussing topics around the Internet and Web-based services on mobile devices.

6.7.1   The Seven Principles of Web 2.0 in the Mobile World

In Section 6.5 the seven principles of Web 2.0 as seen by Tim O'Reily [11] were discussed. Most of these principles also apply for Web 2.0 on mobile devices:

6.7.1.1   The Web as a Platform

As on PCs and notebooks, services or applications can be used on mobile devices either via the built-in (mobile) Web browser or via local applications. Local applications can run entirely locally and store their data on the device. In this case they are "non-connected" applications and do not come into the Web 2.0 category. If local applications communicate with services or databases in the network and in addition incorporate several of the other principles, they can be counted as Web 2.0 applications.

A few years ago, Java and a standardized mobile device API was the most popular way to program local applications as programs could be run on different mobile operating systems. The disadvantage was, however, that those programs could not use device-specific features or adapt to the "look and feel" of the user interface of a particular device or operating system. With the rise of iOS and Android, most applications today are thus developed in the programing language and the API of a particular operating system.

As in the Web 2.0 world, many mobile Web 2.0 services use the Web browser as their execution and user interaction environment. While in the past, there were a wide range of different browsers in the mobile space with significantly different capabilities and feature implementations, this has improved over time. Today, the Google and Apple mobile Web browsers are used in the majority of mobile devices and both strive for interoperable support of HTML 5 and JavaScript. Their capabilities have significantly advanced compared to browsers available when the first edition of this book was published in 2008 and it can be observed that web pages are now designed with large buttons and Java Script code for use on the desktop, on tablet devices, and also on smartphones.

When considering the Web as a platform for a service it has to be kept in mind that mobile devices are not always connected to the network when the user wants to use a service. When possible and desirable from a user's point of view, a service should have an online component but also be usable when no network is available. A distributed calendar application is a good example of a service that requires an online and an offline component. It is desirable to integrate a calendar application with a central database on a Web server so people can share a common calendar—even for a single user a distributed calendar with a central database in the network is interesting as many users today use several devices—but the calendar must also be usable on a device even when no network is available. In the future, there will certainly be fewer places where no network is available and therefore, an offline component will become dispensable for some applications, while for others, such as calendars, it will remain an important aspect due to the required instant availability of the information, at any time and in any place. A number of different approaches are currently under development to make Web applications available in offline mode. This topic is discussed further in Section 6.7.3.

Another scenario which has to be kept in mind when developing Web-based mobile services is that a network might be available but cannot be used for a certain service due to the limited bandwidth (e.g., General Packet Radio Service (GPRS) only) or high costs for data transfers. While checking the weather forecast is likely to cause only minimal cost no matter what kind of connection is used, streaming a video from YouTube should be avoided without a flat rate cellular data subscription if outside the coverage area of a home or office Wi-Fi network.

Services could in theory be aware of the connections they can use and which they cannot in terms of available bandwidth and cost and adapt their behavior accordingly. This is referred to as "bearer awareness". This term is somewhat inaccurate however, as it is not only the bearer technology (UMTS, LTE, etc.) that sets the limits but rather the cost for the use of the bearer set by the network operator. Therefore, the neutral term "connection" is used in this chapter instead of "bearer."

It should be noted at this point however, that today, most mobile applications are unaware of the type of network connectivity available and its cost. At the time of publication, there are no large-scale efforts underway to focus on this; so it is likely that bearer awareness will remain an academic concept for the foreseeable future. It is thus left to the user to control the data usage of applications himself, for example, by only using a video streaming application such as a mobile Youtube viewer at home over Wi-Fi if his cellular data subscription only includes a small monthly data volume.

6.7.1.2   Harnessing Collective Intelligence

Many Web 2.0 applications and services enable users to share information with each other and break up the traditional model of top-down content and information distribution. This applies for mobile devices as well and is moreover significantly enhanced since mobile devices offer access to information in far more situations than desktop computers or notebooks, which rely on Wi-Fi networks and sufficient physical space around the user. Furthermore, users carry their mobile devices with them almost everywhere and access to information is therefore not limited to times when a notebook is available. Thus, it is possible to use the Internet in a context-sensitive way, for example, to search for an address or to get background information about a topic in almost any situation.

Mobility also simplifies the sharing of content, as mobile devices are used to capture images, videos, and other multimedia content. Downloading content from a mobile device to the desktop computer or notebook before publishing is complicated and content is not shared at the time of inspiration. Connected mobile devices simplify this process as no intermediate step via a computer is necessary. Furthermore, users can share their content and thoughts at the point of inspiration, that is, right when the picture or the video was made or when a thought occurred. This will be discussed in more detail in the following sections.

On the software side, harnessing collective intelligence describes using open source software and making new developments available as open source again for others to base their own ideas on. While at the publication of the first edition of this book, only few manufacturers were using open source software, this has changed significantly in the meantime because of Google's Android operating system. The operating system kernel is based on Linux and the GPL open source license and the application environment is distributed under the Apache open source license. In the same way as on the desktop, developers can now modify all layers of the software stack of mobile devices and are no longer confined to the API of an operating system.

6.7.1.3 Data is the next Intel Inside

Another attribute of Web 2.0 is a network-based database. The database becomes more valuable as more people use it and contribute information. In the mobile world, network databases are even more important since local storage capacity is limited. Furthermore, databases supplying location-dependent information (e.g., restaurant information or local events) and up-to-date information from other people are very valuable in the mobile space as mobile search is often related to the user's location.

6.7.1.4 End of the Software Release Cycle

The idea behind the end of the software release cycle is that applications are executed in the Web browser and have a Web-server-based backend and database. This way, software modifications can be made very quickly and new versions of an application are automatically distributed to a device when it loads the Web page of the service. This applies to mobile devices as well but, as discussed above for "the Web as a platform" principle, many mobile applications have to include local extensions as network access might not always be available.

6.7.1.5 Lightweight Programing Models

Easy to use APIs are very important to foster quick development of applications using the services (APIs) of other Web-based services. As in the desktop world, Javascript applications running in mobile Web browsers and XML-based communication with network databases and services provide a standardized way across the many different devices of different manufactures to create new services. Whenever confidential user data is transmitted over the network, a secure connection (e.g., secure HTTP, HTTPS (Hypertext Transport Protocol Secure)) should be used to protect the transmission. This is especially important when Wi-Fi hotspots are used, as data is transferred unencrypted over the air which makes it easy for attackers to intercept the communication of other hotspot users. While security awareness is a growing trend, there were still instances of popular services in 2012 that did not use adequate security measures to protect the exchange of data. Details can be found in [24].

6.7.1.6 Software above the Level of a Single Device

Most services are not exclusively used on a mobile device but always involve other devices as well. A good example is a Web-based Blog reader application such as Google Reader [25], which can be used both on the desktop and also via a Web browser of a mobile device. Such Web 2.0-based applications have a huge advantage over software that is installed on a device and keep information in a local database, as the same data is automatically synchronized over all devices. An example, while at home a user might read their Blog feeds with a Web-based feed reader and mark Blog entries as being read or mark them for later on. When out of the home, the user can use a version of the application adapted to mobile devices and continue from the point where he stopped reading on the desktop. Articles already read on the desktop will also appear as read on the mobile device since both Web-based front ends query the same database on the Web server. All actions performed on the mobile device are also stored in the network so the process also works vice versa.

Another example of software (and data) above the level of a single device is a music library and applications which enable the use of the music library via the network from many devices. If the music library is stored on a mobile device which allows other devices to access the library, the music files can be streamed to other devices over the network. This could be done over Bluetooth, for example, and a mobile device could output the music stream via a Bluetooth connection to a Bluetooth enabled hi-fi sound system. Music can also be streamed over the local Wi-Fi network to a network enabled hi-fi sound system which is either Universal Plug and Play (UPnP) capable or can access the music library of a mobile device via a network share.

Yet another example is streaming audio and video media files via the network to a mobile device. The Slingbox [26] is such a device and adapts TV channels and recorded video files to the display resolution of mobile devices and sends the media stream over the network to the player software on a mobile device. Such services also have to take the underlying network into account and have to adapt the stream to the available network speed.

6.7.1.7   Rich User Experience

Early Web-capable mobile phones suffered from relatively low processing power and screen resolutions which made it difficult to develop an appealing user front end. Since then, however, display sizes and screen resolutions have significantly improved and enough processing power is available to run sophisticated operating systems and appealing graphical user interfaces. The user interface of Apple's iPhone is a good example of a mobile device with a rich user experience and fast reaction to user input. At the same time, the device is small enough to be carried around almost anywhere and battery capacity is sufficient for at least a full day of use.

6.7.2   Advantages of Connected Mobile Devices

The Internet cannot and should not be replicated piece by piece from the desktop onto mobile devices. This is partly because of the limitations of small devices, such as the need to scroll to see more than a few lines of text, a small keypad, or virtual keypad, which makes it difficult to input text, no mouse for easy navigation, and a smaller screen size then on the desktop. However, it is also because mobile devices are game changing, that is they are much more then just the "small" Internet. Tomi Ahonen sees the Internet on mobile devices as the seventh mass media and explains in an essay that instead of looking at the disadvantages one should rather explore the unique elements of connected mobile devices and how they can be used to create new kinds of applications [1].

The current mass media channels are:

- print media (e.g.. books, newspapers, magazines);
- various forms of discs or tapes (recording and music industry);
- movies and documentaries for entertainment;
- radio broadcasting;
- television broadcasting; and
- personal computers and the Internet.

Each channel has unique elements, with the Internet being a bit of an exception since it universally embraces all other mass media types and adds interactivity and search.

New forms of mass media have been able to establish themselves alongside already existing media because they offered something the previous channels did not have. One of the advantages of radio broadcasting over print media is, for example, that news can be spread much faster than would ever be possible with newspapers.

The emergence of a new mass media usually does not lead to a complete demise of previous types of mass media, as some of their properties are not shared by the new media. Instead, it can be observed that media types usually adapt to the arrival of new media. In the case of the print media, newspapers adapted to the fact that they were no longer the source of breaking news once radio and television broadcasting became popular. They have still retained a roll in the media landscape, however, due to their ability to cover news in much more depth and because they are much more suitable for delivering background information. Even with the emergence of the Internet, the print media is still alive and well, as in some circumstances it is still more convenient to read an article in the newspaper than on a computer screen.

Ahonen describes the following advantages of connected mobile devices over previous mass media channels.

6.7.2.1   Mobile is Personal

Previous mass media channels were not personal. A single copy of a newspaper, a book, a movie, a CD, or a television set is potentially used by more than one person. Therefore it is difficult to establish a direct relationship with a customer through a single copy or a single device. Even connected desktop PCs or notebooks at home are often not personal, as the device is usually used by several family members. The connected mobile device on the other hand is highly personal, as it is not shared with friends or even family members. For content creators one device therefore equals one user, which is ideal for assembling statistics about the use of a service, for marketing purposes, and also as a sales channel. As the number of personal devices per user increases, the downside is that the marketer can no longer assume that one device represents one person.

6.7.2.2   Always On

Connected mobile devices are the first type of mass media that is always online. Thus, users can be informed or alerted about events even more quickly than via radio or television — which users do not watch or listen to all of the time. Mobile devices are rarely switched off. A service that has capitalized on this is mobile e-mail, for example, which can be delivered instantly to mobile devices. RIM was probably the first company to fulfill this need with its Blackberry mobile e-mail devices.

6.7.2.3   Always Carried

Always on is so valuable because users tend to carry their mobile devices with them most of the time. Users can be informed instantly about breaking news and they can get access to information at any time. Studies show that many people even keep their phones switched on at night and have them on their bedside table, mostly to serve as alarm clocks. Equally, users have the ability to get or search for information at any time and at any place. As a consequence, search engines are now taking into account that searching for information on a topic while on the move with a mobile device is usually different than searching for information via a desktop or notebook.

The idle screen or the screen saver of connected mobile devices is also an ideal place to display content. Opera's mobile widgets were one of the first programs to explore this area by displaying content retrieved from the Web, sending updates to users in real time from their favorite news feeds [27]. Others have followed in the meantime and popular information to be displayed on the idle screen ranges from weather forecasts to Facebook status updates.

6.7.2.4   Built-In Payment Channel

Most of the traditional mass media channels use advertisements as a source of revenue. Companies advertise via mass media channels in the hope that people like a product and will buy it later on. The issue for advertisers is that there is a gap between users reacting to the advertisement and the opportunity to actually buy the product. This gap has been shortened considerably by advertising on the Internet but users usually still have to type in credit card details before the product is sold. As connected mobile devices are personal and as connecting to a cellular network requires identification and authentication, even this final step of typing in credit card details can be removed. The process from advertising something and giving the user the possibility to buy the product thus becomes a single click, or "single click to buy." This approach is used on Web portals and online stores of network operators and third-party companies such as those of Google, Apple, Microsoft, Nokia, and others to sell games, music, ebooks, magazines, and so on. The user can browse a catalog and when deciding on a game, for example, can pay instantly by clicking on a purchase button. No identification is necessary as the mobile is personal and the transaction is performed in the background and either deducted immediately from the user's prepaid account, via his telephone bill, via credit card information stored in the shop's database, and so on.

6.7.2.5   At the Point of the Creative Impulse

As connected mobile devices are personal and continuously carried, they are a unique tool to capture thoughts and impressions at the point of inspiration. An advantage of this is the capture of pictures and videos that would otherwise never have been recorded, as other nonconnected and single-purpose devices such as digital photo cameras are not always at hand. In addition, by being connected, mobile devices enable users to instantly send that picture, video, or thought to a picture or video-sharing platform, to a social network, to an instant messaging service, to a microblogging service like Twitter, or to a personal Blog. This helps to inform others of big and small events as they break. Thoughts are also quickly recorded for personal use as, unlike PCs, mobile devices do not need a minute to boot before an application is available for note taking. Also, connected mobile devices simplify the sharing process as it is no longer necessary to first transfer pictures and videos from a digital camera to a PC and then upload the files to Web-based services. By simplifying this process it is much more likely that people will be willing to share content, as it can be made available with much less effort.

6.7.2.6 Summary

When looking at the unique properties of connected mobile devices, it becomes clear that their importance as a new mass media channel will continue to grow and that other channels will have to adapt, as print media had to, following the emergence of radio broadcasting at the beginning of the last century.

6.7.3 Access to Local Resources for Web Apps

On the desktop, many Web-browser-based JavaScript applications exist today with well-designed user interfaces and connectivity to services and databases on the Web. Web-based applications such as Gmail (e-mail reader), Google Reader (Blog reader), Google Docs (text processor, spreadsheet), and so on are now even taking over some of the functionality of locally installed programs. On mobile devices, this trend is growing but not as widespread yet mainly because of two reasons:

- Applications need to be available even without network coverage. This is crucial for applications such as calendars, address books, and so on, which have to be accessible at any time and any place.
- Many applications require access to the file system, the camera, and other services of the device. This is important for applications that upload information, such as pictures, to the Web or for applications that use the device's built-in GPS unit, in order to include information relevant to the user's current location with uploaded content.

To address these shortcomings, a number of different initiatives have developed solutions for these shortcomings. By 2012, many results had been incorporated in the HTML5 standard and mobile browser support was rising. A good overview is given in [12] and the following paragraphs describe some of the key HTML 5 functionalities now available in mobile browsers.

To address the requirement that a web application must remain usable even when a device is currently not connected to a network, HTML5 describes mechanisms to store web pages and data on the device. This is done by declaring in the main Web page of the

Web app which resources are to be stored in a locally. Later on if the user invokes the Web app and no network connectivity is available, the browser recognizes that the Web app and its data are available locally and loads it from local storage.

To store data locally a JavaScript application get access to local Web storage [28] and a local SQL database [29] unless the user takes steps to prevent this for privacy reasons. A calendar application, for example, could store and modify a copy of the user's calendar entries that are usually held in a web-based database locally and synchronize the modified data with the web-based database once connectivity is restored.

As synchronizing the local and the remote data depository after going online could take considerable time, it is important that such activities do not block the execution of the JavaScript application on the Web page. Thus, HTML5 has defined the "web workers" API that can be used by a JavaScript application embedded in a web page to spawn background threads to perform actions that are not directly related to interacting with the user [30].

HTML5 also enables access to local resources and information such as:

- GPS and network location information,
- motion sensors such as accelerometers, gyroscopes, and magnetometers built into every smartphone today,
- network connection type info (Wi-Fi, cellular), and
- access to the audio, image, and video capture capabilities of the mobile device.

If and how access to these resources are granted depends on the implementation of the Web browser. Security- and privacy-aware browsers therefore actively request the user's consent before access to such resources is granted to Web apps.

It should be noted at this point that not all functionalities listed in this section are necessarily supported in all mobile browsers. Consequently, developers are either restricted to a subset to be cross-platform compatible or have to focus on specific browsers or operating systems. Owing to the broad acceptance of HTML5 as a common standard, however, most functionalities described above are already implemented by mainstream mobile Web browsers and future enhancements are likely to further improve cross-platform compatibility.

6.7.4   2D Barcodes and Near Field Communication (NFC)

In the desktop world, URLs of new Web pages found in print magazines, books, and so on can usually be quickly entered into the Web browser via the keyboard. In the mobile world, however, this is much more difficult as there is usually just a numeric keypad available. The use of T9 (text on nine keys), a predictive text input technology to speed up the writing process of text on numeric keypads, is difficult to use with URLs as names of Web sites are often composed of words that do not exist in the local language. Even if short Web site names are chosen, entering the Web site name into a mobile Web browser is still time-consuming. This tends to keep users from viewing new Web pages on their mobile device unless there is a very strong motivation to do so.

Most mobile phones, however, are equipped with a camera which can be used to simplify the process. In the future, mobile phones may become powerful enough to include text recognition algorithms which are able to extract URLs from a picture the user has

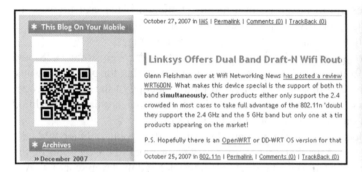

Figure 6.11   A two-dimensional barcode.

taken from an advertisement or a Web site URL in a print magazine. The extracted URL can then be forwarded automatically to the mobile Web browser and the page opened without any further interaction with the user.

A current alternative is two-dimensional (2D) barcodes. 2D barcodes are similar to standard one-dimensional Universal Product Code (UPC) barcodes which have been in use for a long time on everyday products. These are used in combination with cash registers in supermarkets, for example, to speed up the checkout process. Two-dimensional barcodes have an advantage over one-dimensional UPC barcodes in that much more information can be encoded in them. Figure 6.11 shows a 2D barcode that can typically be found today on advertisements, in magazines, and shop windows. Instead of typing a URL into a mobile web browser, barcode reader apps can now be downloaded to smartphones and used to scan a 2D barcode and to automatically decode the content as shown on the right in Figure 6.11. After decoding, the barcode application forwards the URL to the mobile browser, where it is opened automatically.

Another emerging technology to bridge the gap between an advertisement in the physical world and an online Web presence is Near Field Communications (NFCs) as described in the previous chapter. If posters and advertisements contain an Radio Frequency ID (RFID) tag in addition to a 2D barcode and the mobile device is NFC capable, a URL or other information can be obtained by simply holding a device close to the poster or advertisement. This is simpler and faster than starting a 2D bar code scanning application as the process of detecting the RFID tag is fully automatic. This is because in the default configuration an NFC-capable device is scanning for RFID tags continuously while the screen is not locked.

Another advantage of 2D barcodes in magazines, on posters, on advertisements, and so on is the strong message to the user that the Web site behind the barcode can be viewed with a mobile device and that no PC or notebook is required.

6.7.5 Web Page Adaptation for Mobile Devices

Today, the majority of Web pages are still designed for desktop-based screen resolutions and often it is also assumed that the user is connected via a broadband connection, that is, that pictures and other content can be downloaded almost instantly. This is an issue for mobile Web browsing since the screen is usually much smaller. Advanced Web browsers on mobile devices are "full screen" Web browsers as they can render the page exactly as intended for the desktop and only show part of the Web page on the screen. The part that is shown on the screen is either a particular area of the Web page or the text from a particular area, reformatted to fit into the mobile device's display.

While this works quite well, there are a number of disadvantages to the approach depending on the situation. The first disadvantage is the processing power required to render a standard Web page. Long pages with a lot of content take some time to be rendered and require the user to wait for some time. Fortunately, because of increasing processing power available on mobile devices, this disadvantage is about to disappear more and more. A second disadvantage is the time required to download all parts of a page if only a slow connection to the network is available.

In practice, a number of different solutions exist to compensate for these shortcomings, which are discussed in the sections as follows.

6.7.5.1 Mobile-Friendly Pages

Some Web sites and Blogs are available in a full desktop browser format and also with a mobile layout. The mobile friendly layout usually does not use JavaScript and Flash, the page is formatted differently and advertisements are either missing or inserted in a different way, that is, not on a side bar since the screen of a mobile device is too small for Web pages to have sidebars. Mobile-friendly Web pages can be requested from the Web server in a number of different ways. One approach is to have different URLs for the desktop Web site and the corresponding mobile Web site. How this is done is not standardized. Google, for example, offers a mobile-friendly search engine page via http://google.com/m. Others put the indication that a Web site contains mobile content at the beginning of the URL such as http://mobile.domain.org or http://m.domain.org.

Since these methods are not standardized, the mobile industry created a new top-level domain, ".mobi." URLs like http://martin.mobi indicate that the Web pages on this Web site are specifically tailored for mobile devices. The creation of the mobi domain was quite controversial at the time as the top-level domain now indicates the type of device for which it is intended. No other top-level domain has done this before and many had been against such a solution as they were in favor of the "one Web" (for all devices) approach to prevent the World Wide Web from splitting into a desktop and a mobile sphere. One of the advocates of the "one Web" movement was Sir Tim Berners Lee, inventor of the world wide Web. In the article "New Top Level Domains.mobi and .xxx Considered Harmful" [31] he gives some details of why he fears that a mobile-specific top-level domain might have a detrimental effect on the Web. In the end he did not prevail and the .mobi domain was created.

6.7.5.2 Web Servers with Content Adaptation

An approach to web page adaptation that has become quite popular in recent years is that Web sites themselves analyze the browser identification string in an HTTP request message and deliver a standard Web page for desktop browsers and an adapted version for mobile devices. From a theoretical point of view, this approach is superior as:

- The Web is not fractured and the "one Web" paradigm is fulfilled.
- No external content adaptation services are needed and pages get rendered according to the capabilities of the requesting device.

Unfortunately the disadvantage of this approach in practice is that users have no way of determining from the URL that the content will be delivered in a mobile-friendly format and thus might not consider visiting the Web site from their mobile phone if they are not informed by other means like, for example, a "mobile-friendly" logo on the Web site itself.

In addition, this approach does not take the type of network connection and the associated costs into account. A mobile Web browser might be capable of rendering sophisticated Web pages and thus the Web server can send the standard desktop page, but the user might prefer a scaled down version due to the high transmission costs.

6.7.5.3 Web Browser Content Adaptation Proxies

Yet another way to adapt Web pages for mobile devices is to combine a light-weight mobile Web browser with a content adaptation proxy in the network. Opera very successfully uses this strategy with its Opera Mini web browser, a Java, and native

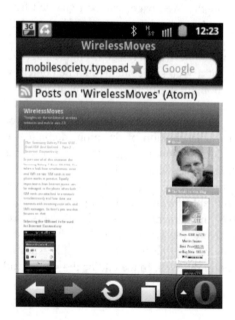

Figure 6.12 Opera Mini with network-based proxy and light-weight browser. (The screen-shot of OperaMini and the Opera logo is reproduced with kind permission from: Opera Software ASA, Norway.)

application on iOS, Android, and Symbian available for most mobile phones and other mobile devices [32]. The Opera Mini Web browser on the mobile phone only communicates with the Web proxy, which requests the page on the user's behalf, modifies the page, and leaves it to Opera Mini to display it. Since the proxy and the Web browser work together, the resulting adaptation is usually excellent. This way it is also possible for the browser to show an overview of the full page without having to load the full content of the page. Figure 6.12 shows how this content adaptation approach looks in practice on an entry-level Android device.

6.7.5.4 Web-Based Transparent Proxies

Many network operators today use transparent proxies to reduce the amount of data transferred for a Web page and decrease the page loading times. This is done, for example, by compressing images and removing unnecessary HTML code from Web pages. As the proxy is transparent, the Web browser and Web server do not need to be modified. While the approach has some advantages, the downside is that the quality of the images embedded in a Web page is significantly reduced. As current 3G networks and future B3G networks are fast enough to serve full Web pages, most operators allow the user to deactivate this kind of content adaptation. Deactivation is done either via a Web page, or via proprietary software which was shipped together with the wireless network card or by sending a modified HTTP header [33].

6.7.5.5 Conclusion

In practice, a number of factors determine how the user can or prefers to surf the Web with their mobile device:

- the capabilities of the device itself (processing power, screen resolution, etc.),
- the mobile Web browser delivered with the device,
- the speed of the network connection, and
- the cost of a particular network connection.

In theory, the "one Web" approach is a good idea, but it does not work well in some usage scenarios since not all Web sites supply their Web pages in formats adapted to specific types of devices and wireless networks. Thus, users that only have access to slow networks, for example, during their commute, in remote areas, or in developing markets have to be aware of which sites support optimization techniques and which do not and consequently less ideal approaches than "one Web" are sometimes still preferable in practice. Fortunately, devices have become more powerful in recent years, costs for using cellular networks have decreased significantly, and network coverage of high-speed wireless broadband networks has increased; so content adaptation is required in fewer and fewer usage scenarios. These trends are likely to continue further. In developed markets, server side content adaptation and wireless broadband network coverage will at some point be widespread enough; so users will no longer require Web browser content adaptation proxies or other means of adapting to slow networks and low-end mobile devices with limited browsing capabilities.

6.8   (Mobile) Web 2.0 and Privacy and Security Considerations

The inherent goal of many (mobile) Web 2.0 social networking applications is to let users share private information with the rest of the world or with a group of other people. Privacy on this level can be controlled by the user, that is, by consciously deciding what kind of information they want to share with others. When making a decision of what to share and what not to share, one should not only think about the immediate implications of sharing private thoughts, ideas, and so on, but also about what happens to this information over time. Everything that is shared publicly will remain on the Internet for an indefinite time and can be found via search engines by anyone and not only by those for whom the information might have initially been intended. Everything shared with a group of users is also shared with the company running the social networking site and so this information can be used for various things such as directed advertising or later on sold to other companies. Where the boundary lies between what to share and what to keep private is an individual decision and can be controlled at the time the information is shared.

There are, however, many mechanisms at work behind the scenes which track user behavior and collect sensitive private information without the direct consent of users at the time this information is gathered and stored. Thus, users cannot make individual and conscious decisions about such processes. Instead, users are required to know about these mechanisms and to actively take countermeasures in case they do not agree to this kind of private data collection and analysis. The following examples show some of the Web techniques used by companies to collect information about user behavior and what users can do protect their privacy if they not agree with these methods. Unfortunately, these countermeasures require a fair amount of technical knowledge which the majority of Internet users do not have.

6.8.1   On-Page Cookies

The oldest form of tracking users is a mechanism known as "HTTP cookies." When requesting a Web page, a Web server can, in addition to the requested Web page, return a cookie (a text string) to the Web browser. The Web browser stores the cookie in an internal database and returns it in future requests to the same Web server. This is one of the most popular mechanisms used today by Web sites to track a user through their Web pages. It is a useful tool during an order process, for example, which requires the user to go through several pages. Without cookies or similar technologies, the Web server could not correlate the information the user has supplied in the different pages. Cookies are also useful for online services such as Web-based e-mail, blogging tools, or social networking sites. Here, a Web-based environment is used which is spread over many different Web pages and it would not be practical to require the user to identify himself on each new page.

Cookies are also used by many companies to correlate user actions during site visits in order to specifically alter the content of a page based on previous behavior. Amazon is a good example. When browsing through the Amazon store the system tracks which items have been looked at and presents previous items as part of the new page. The system also tracks which articles the user has viewed during a session. Based on this knowledge, other users will later on get suggestions like "other users who have viewed this item have also viewed the following item ...." There are a number of privacy concerns with such

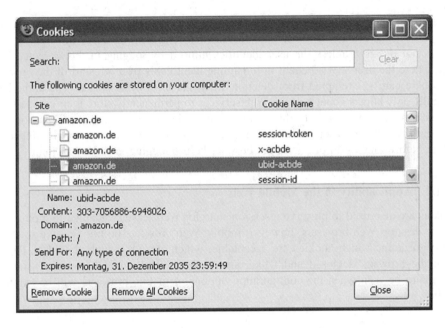

Figure 6.13   Cookie information stored in the browser.

behavior. First, the shop owner stores and uses information not only about which items a user has bought in the past, but also which items she has only looked at. Over time the shop owner can thus collect a huge amount of information about customers which can then be used to deduce preferences and to place specifically targeted ads on a Web page. Also, if this private information is sold to a third company or stolen, private data is revealed without the knowledge or consent of the user.

Second, cookies are stored in the Web browser's database for an indefinite amount of time by default and are stored even when the browser is closed. This is shown in Figure 6.13 for a cookie that has set its expiry time to the year 2035. By default, cookies are still present in a Web browser's database and are resent to the Web server when the user revisits the Web site at some later date. The advantage for the user is that he does not need to log in again as they are immediately recognized. This is useful in some instances, for example, for frequently used services such as Web-based e-mail or blogging, as the user does not have to identify himself again. A potential downside side, however, is that user behavior can be tracked without a gap. Again, Amazon is a good example of a company that tracks user behavior across sessions. As soon as a user revisits the Amazon store, the user's browsing record is retrieved from a database and information and ads are presented based on the items the user has bought and viewed during their previous visits. The shop or service software logic in the background then also tracks the user's current visit and stores their browsing record for future analysis.

6.8.1.1   How can Cookie Use be Controlled?

By default, Web browsers today accept cookies from all Web sites and keep them as long as requested by the Web servers. It is possible, however, to manually ban cookies from certain Web sites and to configure the browser to delete all cookies at the end of the session from the remaining sites. This ensures that long-term behavior tracking by Web sites is not possible unless the user identifies himself by logging in. For selected Web sites, where the user sees a benefit in remaining identifiable over a restart of the browser, the cookie can be exempted from deletion when the browser is closed. The author, for example, only allows about a dozen Web services to permanently store their cookie while the rest get deleted. To ensure that shopping Web sites do not use the information they have gathered about a user while browsing the shop, the user should manually delete the cookie before logging in or restart the browser before logging in to buy an item.

6.8.1.2   Applicability to the Mobile Web

Cookies are also used to preserve a session state for Web services adapted to mobile use. Unlike desktop Web browsers, however, mobile Web browsers do not yet give users the ability to define which cookies are acceptable, which are not, which should be deleted when the browser is closed and which are allowed to be stored between sessions. It remains to be seen when the configuration options of mobile browsers will be extended to allow this.

6.8.2   Inter-Site Cookies

Cookies can also be used to track a user over several Web sites. All that is required for this is that Web sites include a picture or other element from the same third-party Web server. When this picture or page element is retrieved from the third-party server the cookie previously stored when visiting another site will be sent alongside the request. This can be used for marketing purposes, for example, to track how the user navigates through the Web. A precondition is, of course, that the Web sites the user visits use the same external marketing partner. As there are only a few large online marketing companies, inter-site user tracking and use of the gathered data for targeted marketing and other purposes is easily achieved.

Inter-site cookies can be controlled in the same way as described above for single-site cookies that are only used on one Web site, since from the Web browser point of view there is no difference. Some browsers also offer an option to disable third-party cookies. In addition, Web browser plug-ins such as Adblock Plus reduce the amount of advertising shown on Web pages loaded from third-party servers and their corresponding cookies.

In general, third-party cookies are also applicable for mobile Web browsing where privacy issues today are greater than for standard Web browsing, as most mobile Web browsers do not allow the user to view and manage cookie information properly.

6.8.3   Flash Shared Objects

Interactive Web page content and animations based on Adobe's Flash environment [34] are found on many Web pages for various purposes today. These range from social Web site flash plug-ins to advertisement banners. The Flash environment allows applications to store information on the local hard drive. Information is not removed when a Web page is left or when the browser session ends. In effect, Flash animations/applications have a cookie-like mechanism which cannot be controlled via the browser. The same advantages and disadvantages discussed for HTTP cookies thus also apply for Web page Flash applications.

6.8.3.1 How to Control Shared Objects

Depending on the kind of Flash player installation, the environment can be configured to deny applications the right to store local information. Another possibility to prevent Flash applications from tracking user behavior is to use freely available Web browser plug-ins such as Adblock [35]. With such plug-ins the user can control which Flash applications and other page elements are loaded and executed.

6.8.3.2 Applicability to Mobile Web Surfing

While Apple has banned the use of Flash on web pages on its smartphones and tablets, it is supported on many other mobile operating systems and mobile web browsers, most notably Android. Leakage of private information via Flash local storage is thus also a concern on mobile devices.

6.8.4 Session Tracking

One of the big benefits of Blogs is that readers can discuss articles with the author and other readers by leaving a comment on the Web page. In the past, some companies have extended this principle to allow background discussions on any Web page. This worked by installing a browser plug-in so that for every Web page the user visits, an indication is provided of whether other people have left a comment for this page. While this functionality is quite interesting, the issue with this approach is that the plug-in queries a database in the background for every Web page the user visits, to find out if other people have left a "virtual" comment. In effect the database in the network can thus record every step made by all users who have installed the plug-in. From a privacy point of view this is problematic as it potentially allows an external service to track a user's every move on the Web.

6.8.4.1 Countermeasures

The only effective countermeasure is not installing Web browser plug-ins which contain such a functionality, even if it is only a part of their overall functionality. Also users should be careful when installing plug-ins from an unknown author as in recent history, such functionality was attempted to be included in browser plug-ins such as, for example, ShowIP [36]. This was quickly discovered by users however and pointed out on the download page of the plugin-in on the Mozilla web site.

6.8.4.2 Applicability to Mobile Web Surfing

Most mobile Web browsers do not yet support plug-ins and thus the threat is small to mobile Web surfing today.

6.8.5   HTML5 Security and Privacy Considerations

As discussed in Section 6.7, HTML5 allows a Web application to store data locally or to even access local information beyond the browser cache. This has a number of security and privacy implications, which should be considered from a development and from a user point of view. The main issue with having a local application and data cache in the Web browser is that JavaScript applications could use the local cache for user tracking. Advertisement services could use the local interface to write a JavaScript program that is included in all Web pages of their advertising partners, which records the pages from which it has been invoked in the local cache. Each time the same JavaScript application is executed from a different page, it reports the last page back to the advertisement server, which can then generate a user profile. As most users are unlikely to agree with such methods, the user should be informed that an application wants to store information locally and what kind of privacy issues this could bring with it. The user should then have the choice to allow or deny the use of local storage. If allowed, then this should be either temporarily or, where the JavaScript application is trusted, more permanently.

Access to local device information outside the browser cache is even more sensitive, as malicious JavaScript applications could read the user's private data and send it to a server in the network. Therefore, the user has to be informed if a JavaScript application tries to access local data and give the user the opportunity to allow or deny the request. The user then has to decide whether he trusts the application and thus allows access to local resources. Again, the user should have the choice to allow access temporarily or permanently as trusted applications should be able to access local data without further queries to the user once consent has been given. Otherwise, usability of the application would suffer.

At the time of writing, many PC and mobile Web browsers do not yet offer fine grained control of web storage settings. One exception is Firefox which explicitly asks the user for permission if a Web app wants to store data. Also, users can later on verify which Web apps use local storage and can revoke storage rights for individual web apps as shown in Figure 6.14.

6.8.5.1   Countermeasures

Depending on the Web browser, local storage can be disabled globally or by rejecting individual requests.

6.8.5.2   Applicability to Mobile Web Surfing

As this functionality was specifically designed for mobile devices, the feature is available in most mobile browsers today. Most of them, however, do not offer fine grained control beyond the possibility to occasionally delete local storage data of all web apps.

6.8.6   Private Information and Personal Data in the Cloud

The purpose of many Web services is to offer social services to connect people with each other. This usually requires that personal information is stored on a server in the

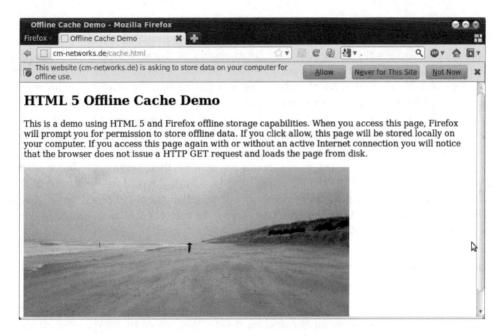

Figure 6.14   Control of local storage for Web apps in Firefox.

network ("the cloud"), which is then combined with data of other users. The risk of this is that once personal data leaves the personal devices of a user, control is lost over the data. Privacy declarations of cloud-based services describe for what purposes the user's data can be used by the service and many cloud-based service providers such as Google and Facebook grant themselves far reaching rights concerning monetization. Furthermore, cloud-based services are usually required by law of their country of origin to provide detailed information about usage of the service and data stored about a user to law enforcement agencies ranging from the police to homeland security. This extends far beyond personal information and also includes, for example, any kind of files a user has uploaded to a cloud-based services such as, for example, a remote backup server. If data is stored unencrypted or in an encrypted format with the encryption key also being stored by the service provider puts users at risk that their data might fall into the hands of hackers, which can then be used for all kinds of purposes. Examples are stolen credit card records and social security numbers that can be used for performing illegal actions such as unauthorized monetary transactions and identity theft for other purposes. For companies, storing confidential information on external servers opens the door for industrial espionage unless the company takes care that all data is encrypted before it is sent to the remote storage server.

6.8.6.1   Countermeasures

Protection of data sent to social network services such as Facebook is not possible as the purpose of such services is to share information. Users should be aware that they are not actually the customer of social network services as it is usually not them who pay for the service. In reality, companies paying for the presentation of advertisement are the customers of social networking services. Users should be aware that they loose control over what their data will be used for by the service today and in the future and should thus be cautious on what kind of information is released into the public domain. Also, users should be aware that many social services collect information about the location via their IP address, Wi-FI SSID (Service Set ID), GPS, and other means. This has to be kept in mind when interacting with such services and only data should be shared that the user would be willing to share with the public.

For cloud-based data storage, users should ensure that their service of choice offers true pre-Internet encryption (PIE) [37], that is, the data is encrypted on the device of the user with their personal encryption key that is not stored on the cloud-based server. Even if the data stored on the cloud server is later stolen, the attacker will have gained nothing as the key is not part of the compromised data.

6.8.6.2 Applicability to Mobile Web Surfing

The privacy considerations just described for cloud-based social networks and cloud-based storage are also applicable for services used on mobile devices as both types of services are now typically available on mobile devices as well. In many cases, services are offering a Web site for small screens and desktop-based browsers.

6.9 Mobile Apps

The advantages of connected mobile devices compared to desktop-based devices such as always on, always carried, a built-in payment channel, and so on, as described in Section 6.7.2, are not only realized with services running in a Web browser but also with native applications. While lacking some Web 2.0 characteristics such as the end of the release cycle and not running in the Web browser, they share many others such as harnessing the collective intelligence, data as the next Intel inside, and the rich user experience paradigm as described in Section 6.7.1. Furthermore, many apps in use today extensively use databases in the network ("the cloud") and are updated and enhanced on a regular basis with semi- or fully automatic procedures.

Some Web 2.0 services such as Facebook, for example, have expanded into the mobile domain and are now widely used on mobile devices. Like on the PC, Facebook is available as a Web-browser-based application. As mobile Web browsers still have some limitations compared to their desktop counterparts and over native applications in terms of speed and access to local resources, Facebook decided to also implement native applications on several mobile platforms. These behave in a similar way as the Web application but in addition they have access to local information such as the address book and location information via GPS, network cell-id's, and by identifying nearby Wi-Fi networks. It will be interesting to see if Facebook will over time make use of increasing browser capabilities and HTML5 extensions for local access or if the current trend of providing local application that is superior to their Web app counterparts continues.

The following sections now take a look at a number of aspects of local applications beginning from describing the different local ecosystem approaches, security and privacy implications, an introduction to programing mobile applications for the Android operating system, and how applications should behave to optimize power consumption and interaction with the network.

6.9.1 App Stores and Ecosystem Approaches

Third-party applications on mobile devices have been available for many years and many mobile operating systems offered the possibility to install programs downloaded directly from the Internet or via a PC. The lack of Internet access on mobile devices, small screens, and limited user interfaces, however, kept the popularity and use of such applications rather low. This changed with the emergence of the iPhone and Android-based devices that brought the following changes to the mobile ecosystem:

- Large displays,
- An easy to use touch-based graphical user interface,
- A central market place from which mobile applications could be downloaded,
- Easy access for developers to these markets, also referred to as app stores,
- Many free of charge apps either written by enthusiasts or monetized through advertisement banners.

While there is a central app store for each mobile operating system, their philosophies and terms of use for application developers and users are very different. On the one hand, there are app stores that enforce a walled garden ecosystem. The Apple iPhone environment is a typical example of this category as users can only install third-party applications via Apple's app store. Application developers can thus only distribute their applications under Apple's control and by adhering to Apple's developer guidelines. These guidelines significantly restrict the developer's freedom and monetization capabilities outside Apple's control. Also, developers do not have access to the source code of the operating system and no API exists to modify the system's behavior. As a consequence, developers cannot innovate outside the narrow confines of the API and users are limited to applications that comply with Apple's moral and monetary view.

On the other hand, there are mobile application stores and environments that are completely opposite from the walled garden approach described above. Google's Android is an example of such an open garden approach in which the user is free to download applications not only from the central application store but from any source they like. Also, the operating system is open, the source code is available, and many API functions are available to change the behavior of the operating system. This is as far reaching as allowing developers to replace the complete user interface look and feel [38].

Applications wishing to access local and private information such as calendar and address book entries, location information, network access, and so on have to request proper permissions at installation time. This allows the users to detect when applications would like to access information they have no need for and helps to stop potential privacy breaches and loss of personal information. An alarm clock application, for example, does not require access to location information and the network. If still requested, it is likely that the application is either not well programed or wants to access this information to send it to the network, for example, for selecting advertisements relevant to the user's location as a form of monetization. The user can then decide if he is willing to give up his privacy in exchange for a free application or not. It should be noted that many users may not review the permissions requested by an application during the installation process, which is somewhat a security issue in itself. However, this approach has worked for many decades in the personal computer world, where users are free to install any kind of software from any source desired. From the author's point of view, having a choice and the freedom to make an informed decision is preferable to a walled garden environment.

6.10 Android App Programing Introduction

After a general introduction to mobile applications and central market places in the previous section, the following section gives the reader a high-level introduction on how software is written for mobile devices in practice and how the API provided by a mobile platform is used. Google's Android is taken as an example because all tools are freely available and excellent documentation is available online. Also, no application certification or complex mobile device setup is required to test the program during the development process.

6.10.1 The Eclipse Programing Environment

Most Android apps run in a virtual machine referred to as "Dalvik" and are programed in Java. The main benefit of using a virtual machine is that it makes applications independent from the CPU architecture. To interact with the user, the device hardware, and the network, the Android API is used. The API and the virtual machine have been developed by Google; so Dalvik programs will work on devices of all manufacturers using Android as the operating system for their devices. This is a significant difference to the previous Java approach that became very fragmented over time and developers often having to write different versions of their application for phones with different Java API implementations.

Applications for mobile devices are usually programed on a desktop or notebook computer. There are different development environments that can be used. Google recommends the use of the open-source Eclipse development environment together with the Android Software Development Kit (SDK) and describes how it can be installed in [39]. Eclipse can either be installed directly in Windows, Mac OS, or a Linux distribution. While a high screen resolution and a fast processor increase the efficiency during the development and debugging process, it is also possible to use Eclipse on netbooks with a small screen and low processing power as shown in Figure 6.15. The major drawbacks are that less information and fewer windows of the development environment can be seen on the screen and that it is better to use a real mobile device for testing and debugging the application as the Android device simulator requires significant resources to run sufficiently fast.

6.10.2 Android and Object Oriented Programing

In essence, an Android application is based on two major concepts. The understanding of these is the basis for the subsequent discussion of a simple application. The first concept

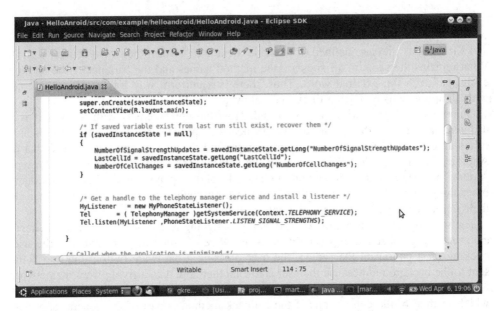

Figure 6.15   The Eclipse programing environment running on a netbook.

of every Android App is that it is programed in Java, which is an object-oriented language. The core of object orientation can be described in two parts:

First, object orientation means that every line of code of an app is part of a class. On the highest level of abstraction, every app has at least one class and one method in the class that contains the code. This method is called when the app is started by the user. That method then creates what the user sees, that is, it arranges text, buttons, and other things on the screen. As an app usually performs more than just a startup procedure, additional methods are defined which are then called when necessary.

The question that arises at this point is how Android knows which method to call when the App is launched. This can be answered by looking at the second idea of object orientation. Classes are a kind of blueprint and to make them come alive, they have to be instantiated. This can be thought of in terms of real objects. The abstract term "car" is a good example. The "car" class contains the basic description of what all cars have in common such as a steering wheel, doors, and so on. A particular car is then instantiation of the abstract term car. There can be several instances of a car, like, for example, John's car and Debby's car. In object orientation, an instance is called an object. In other words, an object is an instance of a class. What Android does with object orientation is describing the basic principles of how an app is supposed to work and defines the names and parameters of the methods that are called by the OS when specific events occur.

When the app is started, a specific method of an object is called. To make this approach useful, the app only uses the method definition, that is, how the method looks to the outside but defines what is inside the method itself. This is called "overriding." So if there are two apps running, they both have the same method that is called by the OS when the app is launched. The content of that method, however, is defined by the programer of each app.

The method that is called when the app is started is only one of many that are already contained in the basic app class and the application programer is free to either use methods and variables as they are defined or override them and perform their own actions with it. What remains the same in all apps are the method names and the parameters the method exchanges with the external word. Thus, there is a uniform way for the OS to interact with an app as will be further discussed below.

Object orientation is a means to an end. How the app interacts with the world is a different matter. For that purpose, the OS needs to provide an API. Apple's iOS also offers an object-oriented interface to programing but the API is quite different to that of Android. To display an "OK" button on the screen, for example, an object of the type "button" is instantiated from the button class. The text displayed on the button and many other things are then defined by the app by calling methods of that object that then do their job inside the class to bring the button to live.

The second major concept of Android app programing is call back methods. Whenever an event occurs that the app should become aware of, such as, for example, the user clicking on a button on the screen, the user pressing a hardware button, the app being sent to the background because the user starts another app, data arriving from the network for the app, and so on, a callback method is called by the OS. But how can the OS know which method of the app to call? If the callback method is part of the general app class described above, then there is nothing special to do because the callback methods are predefined and the programer can simply override the existing methods for the different events. Most elements a user interacts with such as buttons, for example, are not part of the general Android app class. As described above, the user instantiates a button object from the button class and then the app object contains a button object. To be informed about the user clicking on the button, the programer needs to define a method in the app class to receive the event. A reference to that method is then given to the button object. When the user then clicks on the button, the button object calls the given method.

Most apps that interact with the user are fully callback controlled. When the app starts, everything is put on the screen and callback methods are put in place. Then, while the user is not interacting with the app, no further processing power is needed for the app and no code is executed. Only once the user interacts with the app by, for example, clicking on a button, a callback method is called. The callback method can then perform an action when the button is clicked, for example, a calculation is made and an output is made to the screen, and then go back to sleep until a new event arrives.

And finally, in addition to callback methods, an app can also spawn background threads for actions that are continuously performed without user interaction or for actions that take a long time to complete. A separate execution thread has to be used for this; so the main thread with the callback methods can be called by the OS at any time the user interacts with the app.

6.10.3 A Basic Android Program

After the general introduction to the Eclipse programing environment and the basic concepts of object orientation and how it is used, this section now takes a look at how Android programs are implemented in practice. For this purpose, some parts of an application are shown that retrieves information about the network information such as cell-ids, signal strength, and other data that Android provides to apps via its API. The full source code and the packaged app can be found in the Android store or at the web site to this book [40].

As described in the previous post, Android apps use an event-driven execution approach. When the app is started by the user, Android loads the app into memory and then jumps to the "onCreate" method. The following code excerpt shows the beginning of the onCreate method of the app:

```
/* Called when the activity is first created. */
@Override
public void onCreate(Bundle savedInstanceState) {
    super.onCreate(savedInstanceState);
    setContentView(R.layout.main);

    /* If saved variable exist from last run still exist, recover them */
    if (savedInstanceState != null) {
    [...]
    }
```

As the app has a number of specific actions to perform, the onCreate method is "overridden." This means that while the original input and output parameters of the method are used in the app to allow Android to call the method in a defined way, the code in the method itself is unique for this application. In some cases, it makes sense to not only have individual functionality in overridden methods but to also execute the code of the original method. This is done in the first line of the method by calling "super.onCreate."

The onCreate method receives a parameter that references data that the app has saved while it was last executed. This way, the app can populate all necessary variables at startup; so it can continue from its previous state. This is quite helpful as Android closes an app automatically for various reasons such as, for example, when the system runs out of memory. An app is also closed and then automatically reopened when the user rotates the device and the screen orientation changes, something that happens quite frequently. As a consequence, the app uses this reference to populate a number of variables, such as the number of cell and location area changes that have occurred before (not shown in the code excerpt above).

Once the variables are initialized, the onCreate method performs a number of preparations to get access to the network information. The app is mainly interested in the cell-id, the location area code, the signal strength, and when one of these parameters change. On a machine that only runs a single program, a loop could be used to periodically check if these parameters have changed and then display them on the screen. In practice, however, this would not work as Android is a multitasking operating system; so a single program must not lock-up the device by running in a loop. Therefore, apps are event driven, that is, a handler method is called when something happens. The app then performs an action such as updating the display and then goes back to sleep until another event happens and the handler is called again. So the way to get to the network status information and to be informed about changes is to create a listener class with listener methods that Android calls when certain cellular network events such as signal strength changes occur. This is done with the following code:

```
/* Get a handle to the telephony manager service and install a
   listener */
MyListener = new MyPhoneStateListener();
Tel        = (TelephonyManager)
              getSystemService(Context.TELEPHONY SERVICE);

Tel.listen(MyListener, PhoneStateListener.LISTEN SIGNAL STRENGTHS);
```

In the first line, an object is instantiated from a class that that has been defined further below in the source code, the "MyPhoneStateListener" class. In the second line, a reference is retrieved to the service object that manages the cellular network connectivity. In Android, this object is called the "Telephony_Service," which is a bit misleading. In the third line of the code, the listener object is then installed as a listener of the telephony system service. This way, whenever the signal strength or other properties of the cellular network connectivity changes, the app is automatically notified and can then perform appropriate actions such as displaying the updated signal strength, cell-id, and other information on the screen.

The code for the listener object is part of the app and is a private class that is only visible to the app but not to the outside world. The following code excerpt shows how it looks like:

```
private class MyPhoneStateListener extends PhoneStateListener {
  /* Get the Signal strength from the provider,
     each time there is an update */
  @Override
  public void onSignalStrengthsChanged(SignalStrength signalStrength) {
  [...]

  try {
      /* output signal strength value directly on canvas of
         the main activity */
      outputText = "v7, number of updates: " +
      String.valueOf(NumberOfSignalStrengthUpdates);
      NumberOfSignalStrengthUpdates += 1;

      outputText += "\r\n\r\nNetwork Operator: " +
                    Tel.getNetworkOperator() +
                    " "+ Tel.getNetworkOperatorName() + "\r\n";
      outputText += "Network Type: " +
                    String.valueOf(Tel.getNetworkType()) +
                    "\r\n\r\n";

  [...]
```

To keep things simple, the app only uses one listener method in the listener object, "onSignalStrengthsChanged." As the name implies, this method is called whenever the signal strength of the current cell changes. It is "overridden" as the Android default actions should not be executed. Instead the method should only perform the actions in this instance of the class. When this method is called, it also retrieves other network status information and compares it to previously received values. If, for example, the cell-id has changed, it increases a corresponding counter variable accordingly.

Reading network status information only works while the mobile device receives a network. If network coverage is lost and the method is invoked, status query methods return an error, known as an "exception" in Java. If the exception is not handled, the application will be terminated with an error message presented to the user. To prevent this from happening, all network status queries are performed in a "try-catch" construct. If an exception occurs in the "try" part of the code, execution continues in the "catch" part where the exception can be handled. In the case of this app, the code for the "catch" part is rather short as it does not do anything with the exception and the method just exits gracefully without the app terminated by the operating system.

The following methods provided by the Android API are used by the demo app:

```
TelephonyManager.getNetworkOperator()
TelephonyManager.getNetworkOperatorName()
TelephonyManager.getNetworkOperatorType()
SignalStrength.getGsmSignalStrength() --> works for UMTS as well...
GsmCellLocation.getCid()
GsmLellLocation.getLac()
```

There are also objects and methods in the API to get information about neighboring cells. Unfortunately, this functionality is not implemented consistently across different devices and network technologies. Only one out of several device models tested would return correct neighboring cell information and then only for GSM but not for UMTS. Here is the code for it:

```
/* Neighbor Cell Stuff */
List<NeighboringCellInfo> nbcell = Tel.getNeighboringCellInfo ();
outputText += "Number of Neighbors: " +
                String.valueOf(nbcell.size()) + "\r\n";
Iterator<NeighboringCellInfo> it = nbcell.iterator();
while (it.hasNext()) {
  outputText += String.valueOf((it.next()getCid())) + "\r\n";
}
```

And finally, the output text that has been generated needs to be presented on the screen. There are several ways to do this and the app uses the most straightforward approach:

```
/* And finally, output the generated string with all the
   info retrieved */
TextView tv = new TextView(getApplicationContext());
tv.setText(outputText);
setContentView(tv);
```

Seeing the current location area code and cell-id on the screen is a first step, but having a record of cell and signal strength changes would be even better for further analysis later on. In other words, the data needs to be written to a file that can then be retrieved and further processed on the PC. The best place to store the file is on the flash disk, which can then be either accessed via a USB cable or by removing it and connecting it via an adapter to a PC. Here is to code for this operation:

```
try {
    File root = Environment.getExternalStorageDirectory();
    if (root.canWrite()){
        File logfile = new File(root, filename);
        FileWriter logwriter = new
            FileWriter(logfile, true); /* true=append */
        BufferedWriter out = new BufferedWriter(logwriter);

        /* now save the data buffer into the file */
        out.write(LocalFileWriteBufferStr);
        out.close();
        }
    }
catch (IOException e) {
    /* don't do anything for the moment */
    }
```

Writing a file could fail, in which case Java would throw an exception and terminate the program if unhandled. Hence the need for the "try" construct that has already been introduced. In the first line of the code excerpt given above, the directory name of the flash disk is retrieved with the "getExternalStorageDirectory()" method, which is part of the Android API. If the disk is present and writable, a file is opened in append mode, that is, if it already exists, all new data is appended. Then, the standard Java "out.write()" and "out.close()" methods are used to write the data to the file and then to close it again.

To ensure that no data is lost when the flash card is suddenly removed while the file is open, data is written and then the file is closed again immediately. This has been done as it is entirely possible that the flash disk is removed while the app is running, either physically or when a USB cable is connected to the PC. To reduce the wear on the flash memory because of frequent file operations, the app uses a cache string in memory and only writes to the file after several kilobytes of data have accumulated. The downside of this is that the app has to be aware when it is closed and restarted, for example, when the user turns the device and the screen is switched from portrait to landscape mode or when the app is sent to the background.

And finally, the app includes a menu; so it can show an "About" text to reset counters and toggle the debug mode. This is done in Android as follows:

```
@Override
public boolean onCreateOptionsMenu(Menu menu) {
    menu.add(0, RESET COUNTER, 0, "Rest Counters");
    menu.add(0, ABOUT, 0, "About");
    menu.add(0, TOGGLE DEBUG, 0, "Toggle Debug Mode");
```

```
      return true;
  }

  @Override
  public boolean onOptionsItemSelected (MenuItem item) {

      switch (item.getItemId()) {
        case RESET COUNTER:

            NumberOfCellChanges = 0;
            [...]

            return true;

        case ABOUT:
            AlertDialog.Builder builder = new
                  AlertDialog.Builder(this);
            [...]

            AlertDialog alert = builder.create();
            alert.show();

            return true;

        case TOGGLE DEBUG:
            /* Toggle the debug behavior of the program
              when the user selects this menu item */
            if (outputDebugInfo == false) {
               outputDebugInfo = true;
            }
            else {
               outputDebugInfo = false;
            }

        default:
            return super.onOptionsItemSelected(item);

      }
  }
```

As mentioned before, the code excerpts shown in this section are not the full program and the reader is invited to go the web site of this book at www.wirelessmoves.com to download the full source code and the executable program that can be installed and run directly from the Web page on Android-based devices.

6.11 Impact of Mobile Apps on Networks and Power Consumption

When GSM networks were first launched in the early 1990s, voice and Short Message Service (SMS) messaging were the main services. The network and mobile devices were designed to be as power and resource efficient as possible. Standby time of devices reached several weeks and talk (usage) times of 5 h and more became possible. This has changed significantly with smartphones, and more importantly with the arrival of Internet connectivity on these devices. It is not now uncommon for smartphone users to recharge their device once a day when used for more than just voice calls and text messaging. A number of factors contribute to this trend and are described in the following paragraphs.

A major contributing factor are the larger, brighter, and higher resolution displays that draw significantly more power than the small screens with low resolutions used in mobile devices only a few years earlier. In combination with the power required for maintaining an active communication link to the network during activities that exchange data over the network such as Web surfing, device batteries tend to only have enough capacity to support such activities for 2 or 3 h at most. Advances in battery technology are slow and overall capacity has only slowly risen over time, while power consumption in mobile devices because of the ever increasing number of usage scenarios has increased significantly.

One countermeasure taken by mobile devices and networks in recent years is to optimize the network state switching during an active communication session to:

- reduce power consumption,
- minimize the number of simultaneous users per cell in a fully active state, and
- reduce the amount of signaling traffic required to maintain connections.

As described in Chapter 2, the radio link between a subscriber and the base station can be in a number of different states. In UMTS, for example, data can be sent and received instantaneously and at a very high speed in the fully active state, referred to as Cell-DCH (Cell Dedicated Channel), at the expense of high power consumption. When no data has been transferred for some time, usually between 15 and 20 s, the connection to a device is usually put to the Idle state. Very little power is required in this state as the mobile device only occasionally monitors the network; so it can change between cells when the user is moving and receive incoming communication requests on the paging channel. The disadvantage of the idle state is that when renewed communication takes place, for example, when the user has clicked on a link on a Web page, it takes between 2.5 and 3 s before the link is reestablished. This is clearly noticeable to the user. This is why many network operators do not send the mobile device into the Idle state anymore but a state referred to as Cell-PCH or URA-PCH (UTRAN Registration Area — Physical Channel). These states are similar to the Idle state but the mobile keeps its radio identity and ciphering parameters. Reestablishing the communication link is thus accomplished in around 700 ms, which significantly improves the response time to user input. In most networks, mobile devices are not transferred from the fully active state to the almost fully dormant state immediately. Instead, a state in the middle referred to as Cell-FACH (Forward Access Channel) offers a compromise between power consumption, response time, and data transfer speeds.

Programers wishing to optimize the power consumption of their mobile applications should be aware of these state transitions and use network connectivity accordingly if possible. Many applications, for example, are advertisement banner supported and reload banners on a frequent basis, such as, for example, once or twice a minute. This does not only incur a significant amount of signaling between the mobile device and the network to transition between the different status of the radio link but also increases power consumption as the radio link is almost never in a dormant state. Power and network-optimized advertisement supported applications; therefore load several advertisements at once and cache them locally instead of loading each advertisement banner individually at the time it is needed. A more detailed discussion of this and how this can be implemented on different mobile operating systems can be found in [41].

Another contributing factor to higher power consumption are data transfers taking place in the background even when the user is currently not interacting with the device. Most email applications, for example, are continuously keeping a connection to the server in the network; so incoming messages can be pushed to the device immediately. This requires keep-alive messages being sent to the server in the network, usually in the order between 15 and 20 min, that is, three to four times an hour. Also, instant messaging clients rely on a continuous connection to the server in the network and thus have a similar behavior in addition to the messages that are received in the background. This background behavior has a significant impact to the mobile device's standby time as the device establishes a connection to the network several times an hour. In essence, there are two measures to decrease power consumption. How often a keep alive message is required to keep the connection to the server in the network in place depends on the Transmission Control Protocol (TCP) connection timeout at the gateway between the mobile network and the Internet and the server in the network. A power-efficient connection implementation should adapt the keep-alive interval to the maximum time experienced in a particular network. A second method to significantly reduce power consumption for sporadic background traffic is for the device's radio protocol stack to implement a feature referred to as "fast dormancy." Instead of waiting for the network to change the radio connection through the different states as described above, the device can signal to the network that no further data is expected to be sent and that the connection can therefore be set to an energy-efficient state right way. How the device determines that no further data is to be sent is implementation specific. Often, devices invoke the fast dormancy mechanism when the display is not active and no data has been received for a few seconds. Further details of this mechanism from a network point of view are described in Chapter 2. From the mobile device's point of view, the mechanisms and how they can be used in a mobile application are also described in [41].

Another shortcoming from an efficiency point of view often observed in mobile applications is that content is only adapted for the mobile device after it has been received. In some cases as described in [42], it has been observed that applications download images in their full resolution, which requires downloading several megabytes of data only to scale them to a fraction of their size to display them. Such behavior not only increases the users waiting time until content is presented but also leads to a significant amount of unnecessary data traffic that the user has to pay for. It is much better, therefore, to download content that is already adapted for mobile viewing on smaller screens.

6.12 Mobile Apps Security and Privacy Considerations

Today, mobile devices have reached a level of complexity, flexibility, and connectivity that equals that of desktop devices. As in the PC world, only a few operating systems make up the majority of connected devices, which is likely to lead over time to similar malware and virus issues as can be found on desktop systems today.

Outside attacks on mobile devices are especially worrying because of the increasing amount of private data stored on mobile devices such as contacts, email addresses, calendar entries, private images, credit card information, and so on that could be stolen. If attackers manage to make hidden phone calls or send SMS messages to premium numbers, they can incur significant financial loss that can often only be detected when the next phone bill is delivered to the customer.

The following sections give examples of how Internet connectivity and access to personal data and payment information on mobile devices have already been exploited and how users can protect themselves from similar attacks in the future.

6.12.1 Wi-Fi Eavesdropping

One of the most straightforward security and privacy attacks that can be easily implemented even by people with little technical knowledge is Wi-Fi eavesdropping in public Wi-Fi hotspots. Data packets in Wi-Fi hotspots are usually not encrypted to enable users to easily join the network. Proper authentication and encryption is thus left to individual applications. Even in 2012, however, many popular applications on notebooks and mobile devices were still sending their data without any encryption. Facebook and Whatsapp [43] were two examples of apps that used nonsecure communication, and a number of programs had appeared on the market to eavesdrop on the communication of other users in Wi-Fi networks. Two examples are Firesheep, a plugin for the Firefox desktop Web browser [44] and Droidsheep, a standalone application for Android devices [24]. These programs set the network card into a special mode that forward all packets it receives to the IP protocol stack of the device independent on whether those packets are destined for the device or not. Firesheep and Droidsheep then read these data packets and in case Facebook content is found, they display the content to the user in a way that a single click suffices to take over the connection and impersonate another user that is in the same Wi-Fi hotspot. Such eavesdropping could easily be prevented if the service used secure http (https) instead of http for communication and it is likely that over time, services will adopt this measure as their use on mobile devices becomes even more popular. As it cannot be seen from the outside whether an application transmits data in encrypted form, users should be very careful when using public Wi-Fi hotspots. Web apps and local apps should only be used if it has been ascertained that proper encryption is used.

6.12.2 Access to Private Data by Apps

While on notebooks and other desktop-based devices, private data such as contacts, calendar information, email, and so on is administered by different programs and a variety of programs exist for each purpose. On mobile devices, however, this kind of data is usually administered by applications that are delivered as part of the operating system and offer a standardized API for other applications to access the information. On the one hand, this offers the tremendous advantage that third-party programs such as, for example, an alternative mobile email program such as K9 mail [45] can access contact information from the device's phonebook application to retrieve email addresses. On the other hand, such standardized APIs also open the door for third-party applications to access the local database and export private information over the network to their remote central database. While popular social networking apps offer the possibility to deactivate this feature, this is not straightforward to perform for the average user. Malicious programs that perform such actions without asking for the user's permission before retrieving and uploading private information have also been observed in practice.

On the Android mobile platform, applications have to ask for the user's consent before they can use functions to access private data or resources ranging from contact information, getting access to the Internet to the retrieval of location information. This is done once at installation time and the user can either grant all permissions requested or reject the installation of the app entirely. On other mobile platforms, no such measures are used and the user has to trust the app store screening policy to check applications for malicious or privacy breaching actions before listing them. Central app stores on the other hand make it easier to counter malware threats as once detected, malicious apps can be removed quickly. Already installed apps can even be deleted from mobile devices by the app store owner and this has been done in the past [46]. While there is certainly a good side to this, it should also be noted that an external party has significant control over mobile devices, which is perhaps not in everyone's interest.

6.12.3 User Tracking by Apps and the Operating System

Many people are not aware that once they have entered their a username and password into an mobile device to benefit from synchronization and other network-based services offered by the device or operating system manufacturer, private information is collected in the background across different programs and combined centrally in a single account. The following list gives some examples:

- Web searches in the Web browser are recorded in the network,
- A background task records and forwards GPS location information,
- Wi-Fi and cell-ids are observed and combined with GPS location information while the user is moving,
- The use of online mapping and navigation applications leaves traces of locations the user was searching for or has visited,
- The app market stores information about which apps have been installed,
- Calendar and address book information is uploaded to centralized servers, and
- Instant messaging clients send information to centralized servers on when the user has locked and unlocked his phone to show other contacts that the user might not be reachable.

Further details can be found in [47]. All of these separate pieces of information are combined and stored under a single account, often in addition to the traces left by online activities of the user on his desktop devices. This gives the company collecting the information incredibly detailed information about their users, their preferences, whereabouts, and activities. This information is not even anonymized as many users also give their full name, address, phone number, and credit card information for online purchases to at least one of the services all administered under the same user account.

Mitigating this loss of privacy is difficult as per default this information is sent once the username and password are known to the device unless the user explicitly disables this functionality in each application separately. Also, loss of privacy can be reduced using alternative programs such as a different Web browser or an offline navigation application from an independent company.

6.12.4   Third-Party Information Leakage

While users still have some means to control information leakage on their own devices, no control can be gained of one self's private information released by other people. As many apps on mobile devices scan and export the data contained in the contact list to the centralized server, social networking services usually also have information on members not giving personal details to the service and also nonmembers who have never interacted with the service such as name, address, phone number, and their relationship with other people. In addition, features such as automatic face recognition allow social network services to identify persons once one has been identified by someone else on a picture. Some social networks offer the possibility for a user to restrict such information to be given out to others but it is likely that the internal database is still aware of which pictures a person is on and may make use of it for various purposes.

6.13   Summary

Despite the security and privacy considerations mentioned above, sophisticated mobile devices have been adopted by the masses and have doubtlessly enriched the way we interact and communicate with each other. For privacy and security conscious users, remedies and alternatives are available and this is likely to be the case in the future as well as long as open-source-based mobile devices exist on the market that can be adapted by the community.

Over the years, there have been many debates about the advantages and disadvantages of Web-based vs. native applications and which approach will eventually dominate the mobile space. Until now, native apps are used for more purposes than Web apps, but both types of apps have found their place in the current mobile ecosystem, while each side is trying to increase its usability to extend into the domain of the other and into uncharted territory altogether.

On the Web app side, the most advanced approach is Google's Chromebook [48], which aims at giving users a fully Web-based environment on a desktop device. All applications are downloaded from a Web server and are executed in the Web browser. Also, all data is stored in the network instead of being saved locally, which makes the device only useful while network coverage is available. This is a major weakness of the approach, which will however decrease over time because of coverage extensions of mobile broadband networks and software enhancements that enable local caching of Web apps and data while no network is available.

On the native mobile application side, a trend is visible toward the use of tablets, which have taken over tasks for which many people would have previously used a desktop PC or notebook. Using a tablet with dedicated apps for Youtube video streaming, Web browsing, writing emails, and viewing documents is often more convenient on a tablet than on desktop-based devices and it is going to be interesting to observe how the gap between desktop-based devices and tablets will shrink over time as tablets overcome their shortcomings such as limited multitasking support, no physical keyboard and operating systems being not as flexible, and versatile as those on desktop devices today.

References

1. Ahonen, T. (2007) Mobile: The 7th Mass Media is to Internet Like TV is to Radio, Communities, Dominate Brands Blog, February 2007, http://communities-dominate.blogs.com/brands/2007/02/mobile_the_7th_.html (accessed 2012).
2. Flickr (2008) www.flickr.com (accessed 2012).
3. YouTube (2008) www.youtube.com (accessed 2012).
4. Digg (2008) Digg/all News, Videos & Images, http://www.digg.com (accessed 2012).
5. Facebook (2008) http://www.facebook.com/ (accessed 2012).
6. LinkedIn (2008) Relationships Matter, http://www.linkedin.com/ (accessed 2012).
7. Myspace (2008) http://www.myspace.com/ (accessed 2012).
8. Linden Lab (2008) Makers of Second Life & Virtual World Platform Second Life Grid, http://lindenlab.com/ (accessed 2012).
9. Anderson, C. (2004) The Long Tail. Wired Magazine (Oct. 2004), http://www.wired.com/wired/archive/12.10/tail.html (accessed 2012).
10. Iskold, A. (2007) There's No Money in the Long Tail of the Blogosphere, ReadWriteWeb, http://www.readwriteweb.com/archives/blogosphere_long_tail.php (accessed 2012).
11. O'Reily, T. (2005) What is Web 2.0–Design Patterns and Business Models for the Next Generation of Software, September 2005, http://www.oreillynet.com/pub/a/oreilly/tim/news/2005/09/30/what-is-web-20.html?page=1 (accessed 2012).
12. Firtman, M. (2012) Mobile HTML5 Compatability Overview, http://mobilehtml5.org/.
13. The RSS Advisory Board (2007) RSS 2.0 Specification, October 2007, http://www.rssboard.org/rss-specification (accessed 2012).
14. Nottingham, M. and Sayre, R. (2005) RFC 4287. *The Atom Syndication Format*, The Internet Society, http://tools.ietf.org/html/rfc4287 (accessed 2012).
15. Mitchel, J. (2011) Making Photo Tagging Easier, June 30, 2011, http://blog.facebook.com/blog.php?post=467145887130 (accessed 2012).
16. Yahoo (2008) Yahoo! Maps Web Service–AJAX API Getting Started Guide, http://developer.yahoo.com/maps/ajax/index.html (accessed 2012).
17. Google (2012) Google Maps API Reference, https://developers.google.com/maps/documentation/javascript/ (accessed 2012).
18. OpenStreetMap (2012) http://www.openstreetmap.org/ (accessed 2012).
19. The Free Software Foundation (2007) GNU General Public License, Version 3, http://www.gnu.org/licenses/gpl-3.0.html (accessed 2012).
20. OpenWrt (2012) http://www.openwrt.org/ (accessed 2012).
21. The Open Source Initiative (OSI) (2006) The BSD License, October 31, 2006, http://www.opensource.org/licenses/bsd-license.php (accessed 2012).
22. The Apache Software Foundation (2004) Apache License, Version 2, January 2004, http://www.apache.org/licenses/LICENSE-2.0.txt (accessed 2012).
23. Google (2012) Licenses, http://source.android.com/source/licenses.html (accessed 2012).
24. Sauter, M. (2012) Droidsheep : Firesheep Moves to Android, January 2012, http://mobilesociety.typepad.com/mobile_life/2012/01/firesheep-moves-to-android.html (accessed 2012).
25. Google (2012) Google Reader, http://reader.google.com (accessed 2012).
26. Slingbox (2012) http://www.slingmedia.com/ (accessed 2012).

27. Belic Dusan (2007) Opera Widgets on New 3G Phones from KDDI, October 2007, http://www.intomobile
    .com/2007/10/25/opera-widgets-on-new-3g-phones-from-kddi/ (accessed 2012).
28. W3C (2012) Web Storage, April 2012, http://dev.w3.org/html5/webstorage/ (accessed 2012).
29. W3C (2010) Web SQL Database, November 2010, http://www.w3.org/TR/webdatabase/ (accessed 2012).
30. W3C (2012) Web Workers, April 2012, http://dev.w3.org/html5/workers/ (accessed 2012).
31. Sir Berners-Lee, T. (April 2004) New Top Level Domains .mobi and .xxx Considered Harmful.
    http://www.w3.org/DesignIssues/TLD (accessed 2012).
32. Opera (2008) Opera Mini–Free Mobile Web Browser for Your Phone, http://www.operamini.com/
    (accessed 2012).
33. Sauter, M. (2007) Deactivating the Vodafone Websession Compression Proxy, July 2007, http://mobile
    society.typepad.com/mobile_life/2007/07/deactivating-th.html (accessed 2012).
34. Adobe Flash Player (2008) http://www.adobe.com/products/flashplayer (accessed 2012).
35. Adblock   (2008)   https://addons.mozilla.org/en-US/firefox/addon/adblock-plus/?src=search#id=1865
    (accessed 2012).
36. Sauter, M (2012) Privacy Breached and Adware With an Update, May 2012, http://mobilesociety.typepad
    .com/mobile_life/2012/05/privacy-breached-and-adware-with-an-update.html (accessed 2012).
37. Gibson, S. and Laporte, L. (2011) Security Now! #307, June 2011, http://www.grc.com/sn/sn-307.txt
    (accessed 2012).
38. Bogawat, A. (2011) Customize Every Aspect of Your Android Experience, April 2011, http://android
    .appstorm.net/roundups/customize-every-aspect-of-your-android-experience/ (accessed 2012).
39. Google Android Developer Guide, http://developer.android.com/guide/developing/index.html (accessed 30
    April 2012).
40. Sauter, M. (2011) Cell-Logger Source, http://www.wirelessmoves.com (accessed 30 April 2012).
41. The GSM Association (2012) Smarter Apps for Smarter Phones, April 2012, http://www.gsma.com
    /technicalprojects/smarter-apps-for-smarter-phones/ (accessed 2012).
42. Hunt, T. (2011) Secret iOS Business; What you Don't Know about Your Apps, October 2011, http://www
    .troyhunt.com/2011/10/secret-ios-business-what-you-dont-know.html (accessed 2012).
43. Tyson, M. (2012) WhatsApp Chat Spying? There's an App for that! May 2012, http://hexus.net/mobile
    /news/general/38813-whatsapp-chat-spying-theres-app-that/ (accessed 2012).
44. Sauter, M. (2010) Firesheep and Hotspot Hacking, November 2010, http://mobilesociety.typepad.com
    /mobile_life/2010/11/firesheep-and-hotspot-hacking.html (accessed 2012).
45. The K9 Project k9mail for Android, http://code.google.com/p/k9mail/ (accessed 28 May 2012).
46. Wallen, J. (2011) Google Deletes Malware-Infected Android Apps from users' Phones, March
    2011, http://www.techrepublic.com/blog/smartphones/google-deletes-malware-infected-android-apps-from
    -users-phones/2375.
47. Sauter, M. (2011) Android Calling Home–Part 2, May 2011, http://mobilesociety.typepad.com/mobile
    _life/2011/05/android-calling-home-part-2.html.
48. Google Introducing Chromebooks, http://www.google.com/intl/en/chrome/devices/ (accessed May 2012).

7

Conclusion

This book has looked at mobile networks, mobile devices, and the mobile Web 2.0 from a number of different perspectives and this chapter now gives a summary and outlook of the areas discussed.

The mobile Web 2.0 both extends and improves Web 2.0 applications from the desktop world. Like Web 2.0, mobile Web 2.0 is also about user participation and new services and information is no longer only distributed top-down but is created by the users themselves. This is possible because of a decentralized approach, because of the openness of services to be included in other new services, freely available information in network databases, open interfaces, and open source software. Furthermore, applications and services are continuously evolving and the user is part of the development process. Changes to services and products can be made quickly and global reach ensures there are enough users contributing and sharing information to create a critical mass for the service. The main driver for users to share information on Blogs, Wikis, picture-sharing sites, social network platforms, and so on, is the wish to communicate the need for self-expression and in some cases the desire to return something to the community for using other services and information for free. Most Web services generate revenue by including advertisements on Web pages or do not require revenue at all because they are driven by enthusiasts. New services are often introduced to the Web community without a firm business plan in mind at first. Most properties described above apply to both Web 2.0 and the mobile Web 2.0. In addition, mobile devices are transforming Web 2.0 due to their unique properties such as always on, always carried, being available at the point of inspiration, and so on.

In recent years, both native applications and Web-browser-based services have become popular on mobile devices. On the one hand, apps can be easily downloaded from centralized app stores and both open and walled garden approaches exist in practice each with their particular advantages and disadvantages. When given permission, apps can access private and sensitive information on mobile devices. This has important privacy and security implications but it is up to the users of how much data they want to share with services, other people, and so on. This is often challenging in practice as options to restrict access or use of private data are often not presented to the user in an easily accessible fashion.

3G, 4G and Beyond–Bringing Networks, Devices and the Web Together, Second Edition. Martin Sauter.
© 2013 John Wiley & Sons, Ltd. Published 2013 by John Wiley & Sons, Ltd.

An interesting aspect to note about Web-browser-based applications is that the environment in which they can operate is continuously evolving because of increasing HTML5 support in mobile Web browsers such as offline storage and better support to access local resources such as location information. Apart from making such advanced functionalities available in mobile Web browsers, managing the privacy- and security-related questions arising from the use of such new functionalities will be one of the challenges in the years ahead.

Services deployed by traditional mobile network operators, on the other hand, have little in common with mobile Web 2.0 services described in this chapter. A main requirement for operator deployed services is to generate revenue from day one, which requires a business model and a significant customer base from the start. This is difficult to achieve in practice and there are only a few services that can develop successfully under such a business model. Services that might prosper under these conditions are voice-based services of the IP Multimedia Subsystem as described in Chapter 4. Creating successful voice-centric applications is a difficult task, especially in the wireless domain, since the network layer has a profound impact on the quality of service (QoS) of this type of application. An important aspect in this regard is that telephony is a convenience service, that is, it must be very easy to use, work under all circumstances, and be instantly available at any time. Unlike Internet companies, wireless network operators have a lot of experience running and optimizing their networks, in particular for circuit-switched voice services. Control of the network layer and QoS mechanisms will help to achieve the ambitious goal of moving circuit-switched voice to the packet-switched domain with a similar quality in terms of voice quality, call drop rates, and sound quality. This might, however, also raise questions concerning network neutrality because of the QoS preference for network operator voice packets over other data including packets containing voice data of Internet-based services. Today, migrating cellular voice service to packet-switched next generation wireless networks is still in its infancy, and IMS services are still not very well tested or widely deployed. It is therefore likely that some sort of intermediate technology such as Unlicensed Mobile Access, VoLGA, or CS-Fallback, all introduced in Chapter 4, will have an interesting role to play in fixed wireless convergence before IMS-based voice services may one day take over as a long-term solution.

In the future it is most likely that the fixed and mobile operators who will be the most successful will be those that are able to combine fixed-line and wireless access networks in an intelligent way to deliver a smooth end to end user experience. Furthermore, successful operators will take advantage of the fact that, while they can offer some services themselves, there will be countless others which are developed and run by Internet companies and private individuals. In the end, both types of applications will generate revenue for them if they can find a balance between:

- being a network provider for the user and thus the enabler of popular Internet-based applications and applications in the long tail,
- providing operator-centric services such as voice-based applications themselves,
- getting the right combination of cellular coverage complemented with very high-speed Wi-Fi "bubbles" at homes and offices to offload traffic from the cellular network layer.

On the cellular network side, a number of significant advances in throughput per user and overall network capacity have been made in recent years and this trend is likely to continue in the future. Networks have or are in the process to evolve to HSPA+, LTE, and LTE-Advanced, and network capacity has increased and will continue to increase using more spectrum, densification of the macrolayer, smaller cells in dedicated areas, devices with advanced receivers, interference cancellation mechanisms, and multiantenna technologies.

While still a novelty when the first edition of this book was published, a Wi-Fi interface has become a standard feature of any mobile device today. Switching between cellular and Wi-Fi is now performed automatically by devices once configured by the user. This way, the cellular network is only used when Wi-Fi is not available, for example, when the user is not at home or in the office, which helps to reduce the load in cellular networks where transferring data is much more expensive than over Wi-Fi for network operators and users. Switching to Wi-Fi is often encouraged by tiered subscription models that include a limited amount of data volume per month after which the maximum throughput is throttled to a low value. This model is likely to be used in the future as well and will be potentially refined such as, for example, by enabling users to buy an additional data volume during the month if they have consumed more than their included monthly data allowance.

A topic often discussed is the seamless handover of data connections between cellular networks and Wi-Fi. Despite many approaches having been presented over time, few have been implemented in practice and none has so far gained wide spread acceptance. One of the main reasons for this is that cellular networks and private Wi-Fi hotspots at home, in the office, and in public places do not share a common core network infrastructure. Therefore, IP connections are broken when switching from one network technology to the other as a new IP address is assigned each time. For most applications, this is not a problem. For some exceptions such as ongoing Voice over IP calls, client-based solutions such as delaying the switch over from one technology to the other will likely be the only mitigation used in the foreseeable future.

Perhaps an even bigger evolution has been seen in mobile devices in recent years. High-end devices are now equipped with multicore embedded processors clocked at speeds of 1.5 GHz and beyond and touch screens with display resolutions at which individual pixels are no longer visible to the user are likely to be incorporated also in mid-tiered devices in the foreseeable future. The capabilities of dedicated graphics processing units are advancing impressively as well, mostly for the benefit of 3D games. Recently, Near Field Communication (NFC) support has improved in mobile devices and even entry-level smartphones now have an NFC chip included. Like desktop PCs, mobile devices allow running many applications simultaneously. Applications for email, instant messaging, and so on, run in the background and frequently communicate with servers in the network. This means that smartphones and other mobile devices have become truly always-connected devices, that is, they are assigned an IP address continuously and exchange information with servers in the network several times an hour in the background.

The downside of these trends is that power consumption of mobile devices keeps increasing while battery capacities are only slowly increasing. For many power users, it has become difficult in practice to use a device without having to recharge it at some time during the day. Charging a connected device once a day, for example, overnight has become the norm for many rather than an exception.

Areas in which little progress has been seen in recent years are other hardware aspects such as device size, usability of hardware because of the lack of hardware buttons, and less choice for users when it comes to full keyboards and other hardware features that manufacturers have experimented with in the past. It will be interesting to observe if the monotonous landscape of bar-shaped mobile devices with few buttons will persist or if manufacturers will produce a larger variety of hardware variants again once the smartphone market hits a saturation point, and advances in software will no longer be enough to increase the market share.

Increasing processing power and graphics capabilities might, however, give rise to another kind of hardware evolution, the potential dual use of smartphones in combination with additional external hardware. While an operating system such as Android is used on the smartphone itself, other more desktop-oriented operating systems could run in parallel when the device is used in combination with additional components such as larger displays and input devices such as a hardware keyboard and mouse. Such combinations could potentially be an alternative to standalone tablets or netbooks used today.

While developments on the network infrastructure side have been incremental, the development on the mobile operating systems has been of a very disruptive nature in recent years. While in 2008, Symbian and Windows Mobile were two of the dominant smartphone operating systems, both have quickly vanished in recent years because of the strong competition from newly emerging operating systems such as Apple's iOS and Google's Android. Unlike former operating systems that were developed by telecommunication companies and often tailored to the wishes of network operators, the new operating systems were developed by software companies from the IT domain with little input from traditional telecommunication companies. Owing to their ease of use and Apple's popularity, significant pressure was created on network operators to sell devices not designed to their specifications. While Apple has doubtlessly initially opened the floodgate, Android has taken this momentum forward. It is interesting to note that this evolution is now also having a profound effect on desktop operating systems as their user interfaces are also evolving to be usable not only on traditional desktops and notebooks but also on devices such as tablets and potentially also on even smaller devices in the future. Some companies are trying to create user interfaces usable on devices of different categories but it remains questionable if a "one type fits all" user interface that spans all device categories from smartphones to desktops will be a success in practice.

As this book has shown, there is a very dynamic relationship between networks, mobile devices, and the Web 2.0. With new developments announced almost daily, readers of this book are invited to keep up to date by subscribing to articles published by the author on this book's Blog at http://www.wirelessmoves.com.

Index

16QAM, 30, 50
2D barcode, 329
2D graphics, 253
3D graphics, 254
64QAM, 34, 50

Absolute Radio Frequency Channel
    Number, 229
Accelerometers, 265
Aggregation, 314
AJAX, 311
All over IP, 6
Analog networks, 1
Android, 283
Android programming, 342
Antennas, 107
Apache license, 321
APN, 179
Application Programing Interface,
    318
ARFCN, 229
ARM architecture, 243
Asynchronous Transfer Mode, 11
ATM, 140
Average Revenue per User, 99

Backhaul, 138
Battery capacity, 108
Beamforming, 102
Bearer establishment, 13
Bearer Independent Core Network, 43,
    151

BICN, 152
Bluetooth, 267
BREW, 281
BSD license, 321
Busy hour, 117

Call Data Record, 16
Carrier aggregation, 68
CDMA, 19
CDR, 16
Cell capacity, 100
Cell Update, 31
Cell-DCH, 23, 26, 41
Cell-FACH, 22, 26, 40
Cell-PCH, 23, 26, 349
Channel bundling, 81
Charging, 96
Chipset, 257
Circuit switched interworking, 192
Circuit switching, 13, 150
CMOS process, 259
Compass, 263
Continuous Packet Connectivity, 37
Cookies, 334
Coordinated Multi-Point Operation, 70
CPC, 37
CS Fallback, 232
CSMT flag, 234

DCH, 24
Development Cycle, 308

3G, 4G and Beyond–Bringing Networks, Devices and the Web Together, Second Edition. Martin Sauter.
© 2013 John Wiley & Sons, Ltd. Published 2013 by John Wiley & Sons, Ltd.

Differential Binary Phase Shift Keying, 78
Digital Signal Processor, 247
Direct Link Protocol, 84
Distributed Coordination Function, 83
DLNA, 272
Doppler Effect, 53
DSLAM, 142
DTM, 233
DTMF, 195
Dual Tone Multiple Frequency, 195
Dual Transfer Mode, 233
Dual-carrier, 35

E-1, 11
E-DCH, 31
E-paper, 260
EAP-AKA, 204
EAP-SIM, 204
Eclipse, 342
EDCA, 89
Emergency Calls, 170, 216
eNodeB, 45
Equipment refresh, 126
Evolved Packet System, 9

Femtocells, 223
FFT, 50
File system, 290
Flash objects, 336
Folksonomy, 316
Frame, 54
Frame aggregation, 81
Frequency bands, 105
Frustum, 255

G.711, 190
Gateway GPRS Support Node, 18
Generic Access Network, 228
GGSN, 18, 154, 174
Global Positioning System, 263
GNU public license, 320
GPL, 320
GPS, 263
GPU, 249
Graphics acceleration, 253

Graphics Processing Unit, 249
GSM Switch-Off, 125
GTP, 18
Guard time, 81

Handover, 25
HARQ, 30
HDMI, 273
Header compression, 216
Hetnet, 73
HLR, 19
Home Location Register, 19
Home Subscriber Server, 47, 177
HSDPA, 28
HSPA, 28
HSS, 47, 177
HSUPA, 28, 31
HTML, 310
HTML-5, 322
Hypertext Markup Language, 310

I-CSCF, 177
I-WLAN, 204
IARI, 213
ICSI, 213
IFFT, 50
IM, 214
Image sharing, 214
IMS, 169, 173
IMS API, 211
IMS Application Reference ID, 213
IMS Centralized Services, 202
IMS Communication Service ID, 213
IMS multi-party calls, 181
IMS security, 184
IMS user identity, 181
IMSI, 14
Instant Messaging, 197
Inter Cell Interference Cancellation, 73
Inter-RAT handover, 26
Interference, 132
Inverse Multiplexing over ATM, 140
IP Multimedia Subsystem, 169
IP tunnel, 46
IPTV, 153

IRAT Handover, 233
ISM Band, 77

Java, 281
Java Script, 319
Jitter buffer, 165

Long-tail economics, 305
LTE , 43
LTE RRC States, 62
LTE-Advanced, 68

Malicious programs, 291
Mashups, 304
Media Gateway, 152
Media Gateway Control Protocol, 195
Media Resource Function Controller, 180
Media Resource Function Processor, 180
MEGACO, 195
Memory management, 288
Microwave, 138
MIMO, 35, 59, 81
MME, 45
Mobile payment, 268
Mobility management, 13, 23, 26
Mobility Management Entity, 45
Modulation, 30
Motion sensors, 265
MSC, 13
Multitasking, 288
MVNO, 149

Near Field Communication, 268, 329
Network Address Translation, 76, 168
Network Allocation Vector, 83
Network capacity, 117
Network capacity per km^2, 118
Network refresh, 126
NFC, 268, 329
NodeB, 10

Object orientation, 344
OC, 17
OFDM, 53
OFDMA, 50
OLED, 260

OMA DM, 215
One-tunnel, 42
Open source, 307, 320
Open-GL ES, 254
Operational Expenditure, 96
Operator monopoly, 156
Optical Carrier, 17
Optical Ethernet, 143
Orthogonal Frequency Division
      Multiplexing, 79
Over-the-Top Voice, 236

P-CSCF, 173
Packet Core, 16
Packet Mobility Management, 26
Packet switching, 153
PAPR, 51
PCM, 165, 170
PDN-GW, 47
PDP, 16, 154
Peak to Average Power Ratio, 51
Per-User throughput, 114
Podcasting, 300
Point of Sales terminal, 270
Power consumption, 349
Power leakage, 259
Prepaid, 98
Privacy, 338
Process Shrinking, 259
Protocol Stack, 17
Pseudowire, 143
PSTN, 165
Pulse Code Modulation, 165
Push to Talk, 197

QAM, 30
QoS, 154, 175
QPSK, 30
Quality of Service, 154

Radio Network Controller, 13
Radio Resource Control, 62
RCS-e, 213
Real Simple Syndication, 314
Real Time Control Protocol, 165
Real Time Transfer Protocol, 162, 164

Reduced Instruction Set, 247
Reference Symbols, 55
Remote Radio Heads, 72
RFID, 268
RISC, 247
ROHC, 216
RRC, 234
RRC Connected, 62
RRC Idle, 63
RSS, 314
RTCP, 165
RTP, 162

S-CSCF, 175
SC-FDMA, 52
Scheduling grants, 58
SCP, 15
SDP, 162
Secondary PDP, 189
Serving GPRS Support Node, 17
Session Description Protocol, 162
Session Initiation Protocol, 157
Session tracking, 337
SGs Interface, 233
SGSN, 17, 154
Shannon-Hartley capacity equation,
     103
Short Message Service Center, 236
Signaling compression, 175
Signaling Gateway, 166
SIM, 182
Simple Traversal of UDP through NAT,
     168
Single Radio Voice Call Continuity, 202
SIP, 157
SIP Proxy, 160
SIP Registrar, 158
Sleep mode, 86
Slot, 54
Small Screen Flatrates, 97
SMS, 13
SMSC, 15, 236

SNR, 103
SoC, 250
Social networking, 303
Soft handover, 25
Softer handover, 25
Spectral efficiency, 101
Spreading code, 20, 28
SR-VCC, 202, 235
STUN, 168
Subscriber Identity Module, 4, 14
Synchronous Digital Hierarchy, 142
System on a Chip, 250

T-1, 11
Temporal Key Integrity Protocol, 85
Throughput, 109
TISPAN, 171, 207
Traffic estimation, 116
Traffic Indication Map, 86
Transcoding and Rate Adaptation Unit,
     151
Transmission power, 23, 108
TV-receivers, 272

UDP, 158
UICC, 182
UMTS, 10
Unlicensed Mobile Access, 228
User Agent, 158
User Datagram Protocol, 158, 164
User Equipment, 10
User Proxy, 173
User tracking, 353
UTRAN, 10

Viewing frustum, 255
VLR, 14
Voice Call Continuity, 200
Voice over LTE, 215
Voice over LTE via GAN, 235
VoIP capacity, 127, 156
VoLGA, 235

VoLTE, 215, 231
Volume charging, 97

W-CDMA, 20
Walled gardens, 284, 286
Web 1.0, 298
Web 2.0, 299
Web applications, 304

Wi-Fi, 74
Wiki, 302
Wireless Protected Access, 87
WPA, 87
WPA-2, 87

x86 architecture, 244
XML, 312